Practical Data Analysis
for Designed Experiments

```
JOIN US ON THE INTERNET VIA WWW, GOPHER, FTP OR EMAIL:

WWW:     http://www.thomson.com
GOPHER:  gopher.thomson.com
FTP:     ftp.thomson.com
EMAIL:   findit@kiosk.thomson.com
```

A service of I(T)P®

CHAPMAN & HALL TEXTS IN STATISTICAL SCIENCE SERIES

Editors:
Dr Chris Chatfield
Reader in Statistics
School of Mathematical Sciences
University of Bath, UK

Professor Jim V. Zidek
Department of Statistics
University of British Columbia
Canada

The Analysis of Time Series – An introduction
Fifth edition
C. Chatfield

Applied Bayesian Forecasting and Time Series Analysis
A. Pole, M. West and J. Harrison

Applied Non-parametric Statistical Methods
Second edition
P. Sprent

Applied Statistics – A handbook of BMDP analyses
E. J. Snell

Applied Statistics – Principles and Examples
D.R. Cox and E.J. Snell

Bayesian Data Analysis
A. Gelman, J. Carlin, H. Stern and D. Rubin

Beyond ANOVA - Basics of applied probability
R.G. Miller, Jr.

Computer-Aided Multivariate Analysis
Third Edition
A.A. Afifi and V.A. Clark

A Course in Large Sample Theory
T.S. Ferguson

Decision Analysis – A Bayesian approach
J.Q. Smith

Elementary Applications of Probability Theory
Second edition
H.C. Tuckwell

Elements of Simulation
B.J.T. Morgan

Essential Statistics
Third edition
D.G. Rees

Interpreting Data – A first course in statistics
A. J. B. Anderson

An Introduction to Generalized Linear Models
Second edition
A. J. Dobson

Introduction to Multivariate Analysis
C. Chatfield and A. J. Collins

Introduction to Optimization Methods and their Applications in Statistics
B. S. Everitt

Large Sample Methods in Statistics
P.K. Sen and J. da Motta Singer

Modeling and Analysis of Stochastic Systems
V. Kulkarni

Modelling Binary Data
D. Collett

Modelling Survival Data in Medical Research
D. Collett

Multivariate Analysis of Variance and Repeated Measures – A practical approach for behavioural scientists
D. J. Hand and C. C. Taylor

Multivariate Statistics - A practical approach
B. Flury and H. Riedwyl

Practical Data Analysis for Designed Experiments
B.S. Yandell

Practical Longitudinal Data Analysis
D.J. Hand and M. Crowder

Practical Statistics for Medical Research
D. G. Altman

Probability – Methods and measurement
A. O'Hagan

Problem Solving - A statistician's guide
Second edition
C. Chatfield

Randomization, Bootstrap and Monte Carlo Methods in Biology
Second edition
B. F. J. Manly

Readings in Decision Analysis
S. French

Statistical Analysis of Reliability Data
M. J. Crowder, A. C. Kimber, T. J. Sweeting and R. L. Smith

Statistical Methods for SPC and TQM
D. Bissell

Statistical Methods in Agriculture and Experimental Biology
Second edition
R. Mead, R. N. Curnow and A. M. Hasted

Statistical Process control – Theory and practice
Third edition
G. B. Wetherill and D. W. Brown

Statistical Theory
Fourth edition
B.W. Lindgren

Statistics for Accountants
S. Letchford

Statistics for Technology – A course in applied statistics
Third edition
C. Chatfield

Statistics in Engineering – A practical approach
A.V. Metcalfe

Statistics in Research and Development
Second edition
R. Caulcutt

The Theory of Linear Models
B. Jorgensen

Full information on the complete range of Chapman & Hall statistics books is available from the publishers.

Practical Data Analysis for Designed Experiments

Brian S. Yandell

University of Wisconsin
Madison
USA

CHAPMAN & HALL
London · Weinheim · New York · Tokyo · Melbourne · Madras

Published by Chapman & Hall, 2-6 Boundary Row, London SE1 8HN, UK

Chapman & Hall, 2-6 Boundary Row, London SE1 8HN, UK

Chapman & Hall GmbH, Pappelallee 3, 69469 Weinheim, Germany

Chapman & Hall USA, 115 Fifth Avenue, New York, NY 10003, USA

Chapman & Hall Japan, ITP-Japan, Kyowa Building, 3F, 2-2-1 Hirakawacho, Chiyoda-ku, Tokyo 102, Japan

Chapman & Hall Australia, 102 Dodds Street, South Melbourne, Victoria 3205, Australia

Chapman & Hall India, R. Seshadri, 32 Second Main Road, CIT East, Madras 600 035, India

First edition 1997

© 1997 Chapman & Hall

Printed in Great Britain by T.J. Press (Padstow) Ltd, Padstow, Cornwall

ISBN 0 412 06341 7

Apart from any fair dealing for the purposes of research or private study, or criticism or review, as permitted under the UK Copyright Designs and Patents Act, 1988, this publication may not be reproduced, stored, or transmitted, in any form or by any means, without the prior permission in writing of the publishers, or in the case of reprographic reproduction only in accordance with the terms of the licences issued by the Copyright Licensing Agency in the UK, or in accordance with the terms of licences issued by the appropriate Reproduction Rights Organization outside the UK. Enquiries concerning reproduction outside the terms stated here should be sent to the publishers at the London address printed on this page.

The publisher makes no representation, express or implied, with regard to the accuracy of the information contained in this book and cannot accept any legal responsibility or liability for any errors or omissions that may be made.

A Catalogue record for this book is available from the British Library

∞ Printed on permanent acid-free text paper, manufactured in accordance with ANSI/NISO Z39.48 - 1984 (Permanence of Paper).

Contents

Preface	**xiii**
Part A: Placing Data in Context	**1**
1 Practical Data Analysis	**3**
1.1 Effect of factors	4
1.2 Nature of data	6
1.3 Summary tables	7
1.4 Plots for statistics	9
1.5 Computing	13
1.6 Interpretation	16
1.7 Problems	18
2 Collaboration in Science	**21**
2.1 Asking questions	22
2.2 Learning from plots	23
2.3 Mechanics of a consulting session	24
2.4 Philosophy and ethics	26
2.5 Intelligence, culture and learning	28
2.6 Writing	29
2.7 Problems	32
3 Experimental Design	**35**
3.1 Types of studies	35
3.2 Designed experiments	37
3.3 Design structure	38
3.4 Treatment structure	42
3.5 Designs in this book	43
3.6 Problems	44

Part B: Working with Groups of Data — 47

4 Group Summaries — 49
- 4.1 Graphical summaries — 49
- 4.2 Estimates of means and variance — 53
- 4.3 Assumptions and pivot statistics — 58
- 4.4 Interval estimates of means — 60
- 4.5 Testing hypotheses about means — 63
- 4.6 Formal inference on the variance — 65
- 4.7 Problems — 67

5 Comparing Several Means — 71
- 5.1 Linear contrasts of means — 73
- 5.2 An overall test of difference — 77
- 5.3 Partitioning sums of squares — 78
- 5.4 Expected mean squares — 81
- 5.5 Power and sample size — 82
- 5.6 Problems — 86

6 Multiple Comparisons of Means — 89
- 6.1 Experiment- and comparison-wise error rates — 89
- 6.2 Comparisons based on F tests — 91
- 6.3 Comparisons based on range of means — 97
- 6.4 Comparison of comparisons — 100
- 6.5 Problems — 103

Part C: Sorting out Effects with Data — 105

7 Factorial Designs — 107
- 7.1 Cell means models — 107
- 7.2 Effects models — 108
- 7.3 Estimable functions — 110
- 7.4 Linear constraints — 114
- 7.5 General form of estimable functions — 117
- 7.6 Problems — 124

8 Balanced Experiments — 125
- 8.1 Additive models — 125
- 8.2 Full models with two factors — 129
- 8.3 Interaction plots — 133
- 8.4 Higher-order models — 139
- 8.5 Problems — 142

9 Model Selection — 145
9.1 Pooling interactions — 145
9.2 Selecting the 'best' model — 147
9.3 Model selection criteria — 150
9.4 One observation per cell — 151
9.5 Tukey's test for interaction — 155
9.6 Problems — 157

Part D: Dealing with Imbalance — 159

10 Unbalanced Experiments — 161
10.1 Unequal samples — 161
10.2 Additive model — 164
10.3 Types I, II, III and IV — 167
10.4 Problems — 175

11 Missing Cells — 177
11.1 What are missing cells? — 177
11.2 Connected cells and incomplete designs — 180
11.3 Type IV comparisons — 184
11.4 Latin square designs — 187
11.5 Fractional factorial designs — 189
11.6 Problems — 193

12 Linear Models Inference — 195
12.1 Matrix preliminaries — 195
12.2 Ordinary least squares — 196
12.3 Weighted least squares — 197
12.4 Maximum likelihood — 198
12.5 Restricted maximum likelihood — 199
12.6 Inference for fixed effect models — 200
12.7 Anova and regression models — 202
12.8 Problems — 205

Part E: Questioning Assumptions — 207

13 Residual Plots — 209
13.1 Departures from assumptions — 210
13.2 Incorrect model — 213
13.3 Correlated responses — 215
13.4 Unequal variance — 216
13.5 Non-normal data — 216
13.6 Problems — 219

14 Comparisons with Unequal Variance — 221
14.1 Comparing means when variances are unequal — 221
14.2 Weighted analysis of variance — 222
14.3 Satterthwaite approximation — 224
14.4 Generalized inference — 225
14.5 Testing for unequal variances — 226
14.6 Problems — 227

15 Getting Free from Assumptions — 229
15.1 Transforming data — 229
15.2 Comparisons using ranks — 234
15.3 Randomization — 235
15.4 Monte Carlo methods — 236
15.5 Problems — 237

Part F: Regressing with Factors — 239

16 Ordered Groups — 241
16.1 Groups in a line — 241
16.2 Testing for linearity — 243
16.3 Path analysis diagrams — 245
16.4 Regression calibration — 248
16.5 Classical error in variables — 250
16.6 Problems — 252

17 Parallel Lines — 255
17.1 Parallel lines model — 256
17.2 Adjusted estimates — 258
17.3 Plots with symbols — 261
17.4 Sequential tests with multiple responses — 263
17.5 Sequential tests with driving covariate — 266
17.6 Adjusted (Type III) tests of hypotheses — 268
17.7 Different slopes for different groups — 269
17.8 Problems — 274

18 Multiple Responses — 275
18.1 Overall tests for group differences — 276
18.2 Matrix analog to F test — 283
18.3 How do groups differ? — 284
18.4 Causal models — 289
18.5 Problems — 293

Part G: Deciding on Fixed or Random Effects — 295

19 Models with Random Effects — 297
- 19.1 Single-factor random model — 298
- 19.2 Test for class variation — 302
- 19.3 Distribution of sums of squares — 304
- 19.4 Variance components — 306
- 19.5 Grand mean — 309
- 19.6 Problems — 311

20 General Random Models — 313
- 20.1 Two-factor random models — 313
- 20.2 Unbalanced two-factor random model — 318
- 20.3 General random model — 322
- 20.4 Quadratic forms in random effects — 323
- 20.5 Application to two-factor random model — 324
- 20.6 Problems — 326

21 Mixed Effects Models — 327
- 21.1 Two-factor mixed models — 327
- 21.2 General mixed models — 330
- 21.3 Problems — 332

Part H: Nesting Experimental Units — 335

22 Nested Designs — 337
- 22.1 Sub-sampling — 338
- 22.2 Blocking — 341
- 22.3 Nested and crossed factors — 346
- 22.4 Nesting of fixed effects — 348
- 22.5 Nesting of random effects — 350
- 22.6 Problems — 354

23 Split Plot Design — 357
- 23.1 Several views of split plot — 357
- 23.2 Split plot model — 361
- 23.3 Contrasts in a split plot — 365
- 23.4 Problems — 367

24 General Nested Designs — 369
24.1 Extensions of split plot — 369
24.2 Strip plot — 372
24.3 Imbalance in nested designs — 373
24.4 Covariates in nested designs — 374
24.5 Explained variation in nested designs — 377
24.6 Problems — 378

Part I: Repeating Measures on Subjects — 381

25 Repeated Measures as Split Plot — 383
25.1 Repeated measures designs — 383
25.2 Repeated measures model — 386
25.3 Split plot more or less — 388
25.4 Expected mean squares under sphericity — 390
25.5 Contrasts under sphericity — 394
25.6 Problems — 396

26 Adjustments for Correlation — 399
26.1 Adjustments to split plot — 399
26.2 Contrasts over time — 402
26.3 Multivariate repeated measures — 405
26.4 Problems — 408

27 Cross-over Design — 411
27.1 Cross-over model — 412
27.2 Confounding in cross-over designs — 413
27.3 Partition of sum of squares — 415
27.4 Replicated Latin square design — 416
27.5 General cross-over designs — 418
27.6 Problems — 419

References — 423

Index — 430

Preface

This book is aimed at statisticians and scientists who want practical experience with the analysis of designed experiments. My intent is to provide enough theory to understand the analysis of standard and non-standard experimental designs. Concepts are motivated with data from real experiments gathered during over a dozen years of statistical consulting with scientists in the College of Agriculture and Life Sciences, augmented by teaching statistics courses on the 'Theory and Practice of Linear Models' (Stat 850) and 'Statistical Consulting' (Stat 998) at the University of Wisconsin–Madison. Students and colleagues have taught me much about what I tend to assume and about how to blend theory and practice in the classroom.

I had hoped to find a textbook geared to this subject. I began by using Scheffé's *Analysis of Variance* to establish the theoretical framework, and Milliken and Johnson's *Analysis of Messy Data* to provide the practical guidelines. What I wanted was half-way in-between. Searle's *Linear Models for Unbalanced Data* has much the flavor I desired, but seems too detailed in some aspects for the classroom setting. Several other texts have noteworthy strengths, in particular Neter, Wasserman and Kutner's *Applied Linear Statistical Models* (3rd edn., Irwin, Boston, 1990), but do not cover the material with my preferred emphasis.

This book can be used as a first or second semester text on linear models. Alternatively, it can serve as a reference for analyzing experiments which do not fit neatly into packages. The primary emphasis is upon analysis of factorial experiments rather than regression. Data and source code for all examples and problems are available via the Internet. The philosophy on problem sets is to give detailed computer code and focus on understanding of modeling and interpretation of results.

The first part examines broader issues of communication between statistician and scientist about the context of experiments. This is particularly useful in a consulting course. Parts B, C and D concern analysis of variance for a completely randomized design. These lay the groundwork for multiple comparisons of means and for analyzing factorial arrangements, important aspects of any designed experiment. Part E builds connections between regression and analysis of variance. Assumptions about linear models are questioned in Part F using graphical diagnostics and formal tests.

The final three parts of the text concern experiments with some structure to the design. Part G allows for two or more random effects, in which the important questions concern variability rather than mean values. This structure of mixed models allows a formal examination of blocking, subsampling and other aspects of nested designs in Part H. Part I investigates the important class of nested designs involving repeated measures on the same subject, in which observations may be correlated.

The material should be accessible to graduate students in statistics (MS and PhD) and to advanced students from other disciplines with some familiarity with regression. Word models are used to introduce concepts, which are later elaborated using a modest amount of mathematical notation. Readers should be familiar with summation notation and subscripts, and have some knowledge of the classical statistical distributions used in inference (normal, t, F and chi-square). The text uses matrix algebra only for more advanced topics, such as linear model inference (Chapter 12, Part D), multiple responses (Chapter 15, Part E), general random and mixed models (Chapters 20 and 21, Part G) and repeated measures (Chapter 26, Part I).

I wish to thank the many people who have encouraged me in this effort. Bland Ewing initiated me into the world of mathematical thinking about science. Douglas Bates supported the initial concepts of the course material that developed into this book. Biometry colleagues Murray Clayton and Rick Nordheim and fellow author Bob Wardrop applauded my efforts to keep the material relevant to practical consulting needs. John Kimmel's guidance and friendship was a constant reminder that I had something to say. Harold Henderson, Ron Regal and Walt Stroup provided many helpful comments during the review process. And of course, thanks are due to the many scientists who have collaborated with me and generously shared their data.

Yasuhiro Omori, Hong-Shik Ahn, Shih-Chieh Chang, Yonghong Yang, Chin-Fu Hsiao and Wendy Seiferheld were all helpful and enthusiastic teaching assistants. The many students who have taken my classes have kept me honest and have offered useful suggestions to make the material more accessible. Kari Veblen offered support and inspiration while finishing her own book at about the same time.

This book is dedicated to my parents, Wilson and Peggy.

Brian S. Yandell
yandell@stat.wisc.edu
www.stat.wisc.edu/~yandell
University of Wisconsin–Madison
June 1996

PART A

Placing Data in Context

Many scientific experiments compare measurements among groups. This part explores the interplay of design and analysis in the statistical interpretation of such data. An understanding of experimental design motivates data analysis within a logical framework.

Chapter 1 considers the framework of practical data analysis, identifying some of the pragmatic issues that concern scientists and statisticians. The scientific process involves crystallizing the key questions, determining the feasibility of an experiment, collecting and managing data, and developing an appropriate and easily explained strategy for data analysis. These aspects of experimentation are second nature to practicing scientists and statisticians, but are seldom addressed in a textbook.

Good consulting practice involves finding common ground to share ideas and uncover details of an experiment. The scientist should gain a clear understanding of the strengths and limitations of the experiment, and a cogent way to convey findings and defend results in his/her discipline. Chapter 2 explores how the knowledge discovery process engages scientist and statistician. Pragmatic issues include using plots to crystallize questions and concepts, writing clearly, adapting to different modes of communication, and understanding the ethical context of collaboration.

Chapter 3 outlines the main components of an experimental design, concepts that are developed in depth throughout the remainder of this book. An experiment contains factors which may be under the control of the scientist – the treatment structure – and others which are involved in the way the experiment is conducted – the design structure. The treatment structure is an arrangement of factors of interest, the simplest being a treatment and a control. The design structure highlights replication, randomization, and any blocking and/or sub-sampling in the experiment. Some factors may be considered part of the treatment or the design structure, depending on the questions being asked. Finally, experimental designs considered in this book are reviewed.

CHAPTER 1

Practical Data Analysis

How was the experiment conducted? What are the key questions? How are treatments applied? What are the experimental units and how were they assigned to factors? These are important practical questions that must be asked before embarking on data analysis. The discovery process of scientific experimentation hinges on asking key questions. It is easy to arrive at the right 'answer' to the wrong question. Careful study of the motivation for and design of an experiment is necessary in order to determine how to address appropriate questions.

Practical data analysis (PDA) in its broadest sense encompasses the pragmatic process of gaining insight about a scientific problem through data. PDA begins by placing data in the context of the scientific experiment, the statistical methods and the human interchange necessary to conduct research in an intelligent fashion. Central to early stages of an investigation is 'initial data analysis' (Chatfield 1995) using descriptive data-analytic methods to highlight important data features with tables and graphs. The inferential stage of PDA involves a healthy blend of exploratory data analysis (Tukey 1977) and confirmatory data analysis, recognizing that human judgement is an important part of the interpretation of evidence gleaned from indirect experimentation. Finally, PDA emphasizes the need for interpretation of mathematical results back in terms of the original scientific problem.

Often experiments are designed for one purpose but used for quite another. Some experiments may beg for a certain analysis but the scientist may be interested in rather different questions than those apparent to the statistician. It is incumbent upon both parties to come to consensus on the **key questions** which drive the subsequent data analysis. Most of this book assumes that key questions have been determined and concentrates on understanding the nature of the experiment and its impact on data analysis. However, some approaches which may help uncover key questions can be found in the next chapter.

Statisticians and scientists must find a common language for discussion, which requires some give and take from both sides. Often it is easier for the statistician to stretch, as statistical methods tend to translate readily from one discipline to another. Thus the statistician has a special responsibility to maintain communication and ensure that the exchange is equal. The scientist interested in really understanding his/her problem will have to stretch as well, learning a bit of notation and some of the pragmatic aspects of managing data using computers.

Section 1.1 introduces the idea of factor main effects and interactions, discussing the value of a compact system of notation. Section 1.2 examines the nature of data, identifying some of the mechanics of handling data. Tables of numbers can provide succinct summaries but Section 1.3 cautions against overuse. Rather, try to visualize data as much as possible using plots, as suggested in Section 1.4. Modern computing tools for data analysis, and the general organization of information, offer great opportunities for creativity as will as ignorant misuse, as noted in Section 1.5. Finally, Section 1.6 stresses the need for interpretation of results in the context of the original experiment and the assumptions which shape the analysis.

1.1 Effect of factors

The primary aim of designed experiments is to understand the sources of variation in one or more responses and to assess the effects of factors on those responses relative to other unexplained variation. A clear understanding of the various factors and the way they are organized in the experimental process is vital to proper data analysis.

Factors include aspects of the experiment which may be under the control of the scientist, such as drug or dietary treatments, and features which occur naturally and may or may not be of direct interest, such as gender, species and location. Factors typically have several **levels** – treatment and control for drugs, male and female for gender, the 50 states of the USA, etc. Sometimes it is helpful to consider a factor combination, or **cell**, examining one level of each factor.

Analysis of variance determines the variability in the measured response which can be attributed to the effects of factor levels. The remaining unexplained, or residual, variation is used to assess the strength of the effects. Significant evidence of differences in mean responses across the levels of a factor can support further detailed comparisons among those levels.

Many experiments focus on the interplay of several factors. Sometimes it is possible to isolate the **main effect** of each factor, allowing direct interpretation of the effect of changing levels of that factor on response without regard to any other factor. However, there can be **interaction** (synergy or antagonism) among factors, which requires careful interpretation of the effects of one factor on response. For instance, the drug is effective for younger males but harmful for older males and seems to have no effect on females of any age. Separation of main effects and interactions is straightforward when the experimental design is balanced, with equal numbers of observations per cell. However, analysis is more complicated with unbalanced designs. Experiments which have some cells empty, either by design or by accident, require further special care. Some experiments involving several factors do not readily lend themselves to interpretation in terms of main effects and interactions.

Some design factors may be of no direct interest but may account for important variation in response. For instance, a scientist may block on location, or subdivide plots assigned to varieties for application of different levels of fertilizer. These are important aspects of experimental design which must be understood in order to conduct proper data analysis.

Compact model notation allows ready interpretation and focus on the key concepts of design and analysis. Mathematics provides powerful tools for this, compressing complicated ideas into a compact language. Unfortunately, math notation can be daunting to the scientist whose own math experiences are a distant memory (or nightmare!).

This book encourages the translation of the basic ideas of an experimental design into compact **word models**,

yield = mean + fert + variety + fert*variety + error,

which can then be translated into **math notation**,

$$y_{ijk} = \mu + \alpha_i + \beta_j + \gamma_{ij} + e_{ijk},$$

or a **computing package language**,

```
proc glm;
    class fert variety;
    model yield = fert variety fert*variety;
```

for analysis. While the methods are explored in detail, key points are highlighted to provide an intuitive grasp of strengths and weaknesses of the approach. Models do not need to be filled with Greek symbols. Word models can be very effective and compact and are very useful to ground statistical concepts within the particular experiment. Model notation – whether based on words or symbols – helps to organize the key features of an experiment. Sometimes this highlights special aspects or problems not otherwise noticed. A little practice leads to easy movement from a model to an analysis of variance table to a statistical package such as SAS. This is actually easier with word models than with mathematical notation!

But **math symbols** are compact and have their place. At times, it is important to explore the meaning behind the symbols, especially when considering sources of variation in unbalanced experiments. Manipulation of symbols, particularly for sums of squares, provides useful insight into what test statistics are appropriate and what test statistics may be misleading. These results are sometimes not obvious and are at times counter-intuitive.

The statistician should be ready to translate math into words, formulas into concepts. The scientist who wants to reach a deeper understanding of how to analyze planned experiments may eventually appreciate notation, particularly if it is placed in proper context. On the other hand, the scientist uncomfortable with notation can gain by learning better ways to organize information and ask questions with the aid of someone trained in statistical methods.

1.2 Nature of data

Data analysis must be motivated by an understanding of design. But what is the nature of the data? How were they obtained and how are they maintained? Data are 'factual information, especially information organized for analysis or used to reason or make decisions' (American Heritage Dictionary 1992). While this information is primarily in the form of numbers, the processes surrounding the gathering and organizing of data are crucial to sound and practical data analysis.

Data quality greatly affects the chance of successfully addressing key questions. We must endeavor to understand how the data are gathered and what are the possible sources of error which arise during the experimental process. While some errors may arise from mistakes in reporting (e.g. omission, commission, transcription), others may represent more fundamental problems. Outliers are data which lie outside the ordinary range of response – do these represent fundamentally different experimental conditions (excessive moisture in one corner of a field, measured by a different person, etc.) or just unusually large or small responses? What about missing data – are they missing for some reason associated with the experiment which could introduce bias? Are there enough missing data to upset standard approaches to analysis?

The experimental design implies a certain **data structure** which identifies the experimental units, treatment groups and blocking. Documentation of the experimental methods should be considered an integral part of the data. Extended dialog between statistician and scientist may be needed to clarify these matters. In some situations there may be several responses of interest and these may by interrelated.

The **data mechanics** involves the process of manipulating the data. This may include transferring between media (paper to electronic, floppy disk to CD to Internet), verifying data quality and documenting the data structure. Special cases and handling of missing data need to be clarified. Complications can arise when dealing with very large data sets. Is it best to examine a subset of the data? Are there so many data that minute differences will be deemed significant? For some problems, manipulating the data can be more time consuming than the actual data analysis.

The mechanics of **data analysis** can proceed in a variety of ways. Usually there is some form of preliminary descriptive investigation involving graphs, tables and various summaries. The definitive inferential analysis focuses on the key questions. Follow-up may involve diagnostic checks of assumptions and investigation of side questions. Conclusions and interpretation are rendered in the language of the original questions in a manner accessible to the scientist.

Data are useless unless they can be conveyed effectively to a wider audience. Great care must be taken in the development of **data displays**. The next two sections introduce some ideas concerning tabular and graphical displays of information. Displays can be used initially for **description**, providing an uncritical summary of information, and later refined and/or annotated for **inference**, addressing the strength of the evidence for particular comparisons. Both description and inference have their place in analysis. Inference enhances results by showing their precision and/or the chance of differences as large as those observed, provided the assumptions needed to make these claims are believable. Descriptive statistics may suggest possible trends or relationships which, while not significant, could be further pursued in subsequent experiments.

Data analysis is distinct from **data mining** in which a large set of data is mined for every last bit of information. Data mining is essentially descriptive, with no clearly defined research questions. There is great danger of uncovering spurious relationships that look very strong because of the large quantity of data, but are not supported by later studies involving other, independent, data. However, if the process of data mining is viewed as a way to generate potentially interesting research questions, it can be rewarding at times.

Most of the methods discussed in this book consider responses to be roughly normally distributed around a group mean, with that mean depending in a linear fashion on the factor levels and other design considerations. The related problem of regression is nicely covered in Draper and Smith (1981) and Seber (1977). Other methods appropriate for nonlinear models (Bates and Watts 1988), robust models and exploratory data analysis (Mosteller and Tukey 1977; Tukey 1977), categorical data (Everitt 1977; Fienberg 1980; Agresti 1996) and nonparametrics (Conover 1980; Lehmann 1975) may be of further interest. Cox and Snell (1981) has many nice case studies of data analysis.

1.3 Summary tables

Numerical **tables** can provide succinct descriptive or inferential summaries. However, they can be dry and soon begin to look alike if they are too large. Helpful tables should account for the experimental design in an unambiguous way and address concerns about underlying assumptions for data analysis.

Large tables of numbers are frightening! Whenever possible, plot the

		fert		
variety	X	Y	Z	mean
A	3.18d	5.21bc	6.34a	4.91
B	4.89c	6.57a	5.99ab	5.82
mean	4.03	5.89	6.17	5.36

Table 1.1. *Table of means with letters for no significant difference*

Source	df	SS	MS	F	p
variety	1	4.9	4.91	16.0	0.0008
fert	2	21.5	10.8	35.2	0.0000
variety*fert	2	4.8	2.42	7.9	0.0034
error	18	5.5	0.306	---	---
total	23	36.8	---	---	---

Table 1.2. *Anova table for completely randomized design with four replications*

data. At the very least, use every trick to reduce the complexity of tables. This includes ordering by mean response rather than by the alphabet; showing only significant digits to reflect precision in data; removing repetitive phrases; organizing two-factor and three-factor information in cross-tabular form rather than as lists.

Many research papers in the scientific literature contain long **tables of means**, followed by some estimate of precision such as a standard deviation (SD), standard error (SE) or Fisher's LSD (all defined later in the text). Such tables are difficult to read and even worse when seen during an oral presentation. At the very least, consider ordering factor levels by mean response (see Ehrenberg 1982). Table 1.1 of means is organized by the two factors, fert and variety, with increasing marginal mean response across the columns and down the rows. Only significant digits are displayed. Cell means with the same small letter appended are not significantly different, as measured by the LSD, which is 0.82. This table is succinct and quite readable.

Table 1.2 summarizes the testing information in an analysis of variance (anova) table. Again, only significant digits are displayed. While it is fairly neat, is it necessary to publish it? The main information here is that the interaction variety*fert is very significant ($p = 0.0034$), which rejects simply reporting marginal means. Instead examine the combination of variety and fert when making recommendations about yield.

Tables are most useful when there are a modest number of levels; otherwise consider plots to convey information. Further, ask whether the audience actually needs the numbers themselves, or whether their qualitative and quantitative relationship to one another is more important. Even further, can the results be stated in the text rather than wasting page space with a table?

1.4 Plots for statistics

Graphical presentation of information is central to practical data analysis. Graphics can aid in initial discussions about key questions and later in uncovering further details of the experimental design. It can be very rewarding to view many plots and tables during data exploration and analysis. Final presentation of results, however, requires a careful selection of a very few graphics tailored to the appropriate audience. Developing and selecting useful plots and figures is more art than science, requiring repeated feedback between statistician and scientist, with an eye toward the ultimate use of graphics in conveying knowledge to others.

Factors affecting a response can be examined graphically using ideas dictated by formal analysis. In fact, it is often very helpful first to plot relationships and then to use them to guide analysis. The graphics can crystallize questions and point up lingering issues about experimental design. Scientists tend to be more comfortable with pictures than with equations or word models. Use sketches of potential graphs to explore possible relationships and key questions. These can be followed by plots using data, further substantiated by formal analysis of variance.

The scientist and the statistician have a responsibility to convey information in an even, unbiased way, without excessive emotional charge. The statistician, in particular, should encourage graphs which convey information about the sources of variation as well as about the central tendency. Every effort should be made to avoid graphics that might mislead interpretation (Huff 1954; Tufte 1983; 1990).

People have been visualizing information since the dawn of time. Our main knowledge of early culture comes from artifacts and cave drawings which can portray a surprising depth of material. Stonehenge, Newgrange, the pyramids of Greece and Central America and other 'pre-historic' human constructs showed sophisticated ways to organize knowledge about the world using very simple visual ideas. Historical approaches may use simple ideas in ways that seem novel even today.

Edward Tufte (1983) presents a vast display of ways humans have envisioned information. One overriding theme is the desire to cram as much information as possible into one display without confusing the viewer. Some of the most striking and enjoyable examples are quite dated: the progress of Napoleon Bonaparte's army across Russia and Japanese train schedules. In addition there are numerous diverse approaches to geographic information layering. Tufte's second book (1990) focuses on several aspects of graphics: escaping 'flatland', micro/macro readings, layering and separation, small multiples, color and information and narratives of space and time. The practicing statistical consultant often must organize a great amount of information, whether or not it is 'statistical'. Knowing useful ways to visualize information can expand possibilities for collaboration. Finally, valuable ideas for statistical displays arise from unlikely places. The key ingredient is being open to new ways to organize information. For example, many young people are now learning about statistical concepts through the humor of Larry Gonick's cartoons (Gonick and Smith 1993).

Graphics and pictures offer a refreshing visual break in a report or technical paper. Further, they can show the magnitude of differences in ways not possible in a table. Circles and arrows and other forms of annotation such as color and symbol size can enhance the value of graphics, provided this is done in a neat, uncluttered manner. Graphics improve with practice and with feedback from colleagues about what is effective. The remainder of this section provides a brief sketch of the use of several types of plots.

Many disciplines like **bar graphs** and **pie charts**, and many graphics packages have them. They provide quick graphic summaries but they can be very biased, depending on the ordering of levels as well as the 'creativity' of presentation. Huff (1954) and Tufte (1983) contain several humorous examples. Pie charts seem the worst offenders and are not considered further here. Bar graphs only present mean values without direct information about variation, and can bias interpretation by the ordering of groups along the horizontal axis. Simple improvements include ordering groups by mean value and adding some measure of precision. However, this is not pursued in this text.

It can be quite helpful to examine the distribution of responses, and eventually residuals, to gain some feel for the data beyond summary tables. **Stem-and-leaf plots** can present all the data in simple character plots (Example 4.1, page 50). However, when there are more than a hundred or so values, **histograms** (Figure 4.1, page 51) are more practical. Distributions for several groups can be compared by aligning axes in adjacent plots. An idealized comparison of distributions has several 'bell-shaped' hills which overlap slightly, but this is seldom the case in practice. Sometimes one histogram can be used for a few groups by employing different shadings or colors, although there is the danger that the choice of shading or color will influence subjective interpretation. Further, for groups sharing the same bin, should shaded regions be stacked upon each other or placed side by side? This idle question is actually important, as the ordering of groups and choice of shading can bias visual interpretation in subtle ways. What is the effect of shading, brightness and other types of color value on the viewer's perception? Further, there may be small sample sizes for some or all groups, making interpretation of shape tentative at best.

The biomedical community has become rather comfortable with **survival curves**, which are forms of cumulative histograms. These curves readily reveal changes of center, spread and shape, and are not subject to the problems of deciding on bin size. However, many in biology, engineering and business are not yet comfortable with cumulative distributions. These type of graphics are not used in this book.

Comparing several histograms can be quite effective, but it is cumbersome to examine more than three or four. Adjacent **box-plots** (Figure 4.2, page 52) convey the key features of distribution shape in a compact form, allowing quick comparisons of several groups within one figure. Sometimes it may be enough just to present the center and spread of the data, allowing even more groups, but with further loss of detail.

While it is possible to plot several factor combinations as if they were levels of a single factor, deeper understanding and simpler interpretation can often emerge from the use of **interaction plots** (Figure 8.2, page 135). Plotting the group means for one factor against the means for all two-factor combinations can highlight certain forms of interaction and suggest strategies for inference (Figure 10.1, page 166). The effects of three or more factors on a response can be examined with complementary sets of interaction plots. There is an art to selecting which factor to use on the horizontal axis and how to order its levels. Sometimes there is a natural ordering for one factor. Alternatively, examine both arrangements. For instance, consider the two interaction plots of Y means against A using symbols for B and of Y means against B with symbols for A, and subjectively decide which is more revealing. It may be useful to share both with other scientists and decide together.

Scatter plots seem generally acceptable across disciplines, requiring only a brief explanation for the uninitiated. Several authors (see Chambers and Hastie 1992) have suggested plotting the group mean against the individual responses, using different symbols for each group or identifiers along the horizontal axis to distinguish group membership. These displays provide an ordering of means, the relationship of variation to mean and some indication of distribution and shape (Figure 4.3, page 54). A similar effect is possible using box-plots with factor levels ordered by mean response, which may be preferable for large data sets and factors with few levels. When there are several responses of interest, or repeated measures, scatter plots of one response against another using plot symbols can reveal important relationships easily overlooked by summary tables. Scatter plots involving subsets, such as certain levels of a factor, or a random sample of the data can be very effective for large data sets by reducing the number of points plotted and tighten the understanding of key relationships. Various ideas have been proposed in recent statistical literature for the display of large quantities of data but that is beyond the scope of this text (see Cleveland 1993).

Residual plots are scatter plots of predicted values against residuals (Figure 13.1, page 211) which were originally developed as diagnostics for regression. The regression diagnostic ideas can be readily transferred to linear models. The predicted value is simply the expected mean, with residuals being the deviations. Using **plot symbols** for factor combinations can highlight problems with model assumptions. Two or more factors can be combined with appropriate symbols or can be presented separately in side-by-side plots with corresponding symbols for each factor. A factor which was not considered in a model can be examined by using plot symbols for its levels on the residual plot.

Nested designs present interesting problems for plotting, as there are usually two or more sources of variation. These components of variation of the experiment should play an important role in plotting. Learn how to remove variation that is not of direct interest for particular plots. For instance, if experimental units are arranged in blocks, with factor levels assigned at random within blocks, then it might be reasonable to remove any block differences before conducting diagnostic checks with a residual plot (Figure 22.2, page 343). The **split plot** design has two different sized experimental units and should be examined with two residual plots. That is, suppose factor A is assigned to whole plot units while factor B is assigned to split plot units. Estimate the mean response by level of A, determine residuals for each whole plot experimental unit, and plot these against one another using plot symbols for levels of A. Now remove the 'block' effect of the whole plot and examine the split plot effects in a similar fashion, using symbols for B levels or for combinations of A and B. Note that the effect of factor A assigned to the whole plot has been removed but the interaction $A * B$ has not. Plots to illustrate this can be found in Chapter 23, Part H. Thus ideas on interaction plots presented earlier can be adapted to good effect.

Repeated measures designs can involve taking repeated measurement on the same subject over time. Plots of the response or of residuals over time can reveal patterns which merit further study. Plots of measurements at one time against the next time can help identify correlation. More ideas from time series can lead to further diagnostic checks. Many repeated measures experiments do not have smoothly changing responses over time which could be handled by standard methods. In this case, graphical methods can provide an important focus for scientist and statistician to identify the nature of response and to customize analysis to the implicit questions. For instance, there may be an initial unstable period, followed by what appears to be treatment effect, concluding with degradation of treatment and washing out of differences. Formal analysis may focus only on those central periods, recognizing that in some senses there is only one degree of freedom (or perhaps a few) per subject anyway. Chapter 25, Part I, contains several plots highlighting repeated measures.

Care must be taken when examining summaries for designs that are very **unbalanced**. Unfortunately, the 'easy plots' using marginal means may be severely biased. Extra care is required to develop plots and tables which use appropriate estimators and test statistics. However, corrections for modestly unbalanced designs are often slight and may not be necessary for quick preliminary graphical summaries. For instance, in a split plot design with only a few missing data, it may be easier to use marginal means across subplot values, which are weighted means depending on the pattern of missing data, rather than getting the unweighted means. The inaccuracy may be small and well worth it to communicate the main point of separating sources of variation in a nested design.

1.5 Computing

Computational tools are now a part of everyday life. It is not important *per se* what package is used for data analysis, provided it has been well tested by the statistical community (see for instance reviews in the *American Statistician* and *Statistical Science*). A practical statistical package should be easy to use, flexible and extendible, easy to interpret by statistician and scientist and full of useful, elegant graphics. Some packages are easier to use than others for certain tasks. The best provide direct translation for most of the designs discussed in this book. Others may require several runs and a calculator or marking pen to reach the appropriate analysis and accompanying graphics. No package provides everything – some problems are so messy that they are best done by hand! Here are some quick (and biased) impressions of several packages.

This text uses SAS (SAS Institute 1992) as a basis for analysis of most examples because it is well suited to the types of designs considered here and is widely used in academia, industry and government. While SAS Institute produces vast volumes of manuals, there are now a few books of particular interest in this area. Di Iorio and Hardy (1996) offers a gentle introduction to the SAS system. Littell, Freund and Spector (1991) provides a comprehensive overview of linear models. SAS has excellent tools for analysis of variance but is not well designed for interactive work and regression. SAS takes a bit more work to learn initially than Minitab, but it has many features which are ideal for the purposes of this book. There is an interactive graphics module called SAS/INSIGHT which can be used for publication-quality plots.

The graphics in this book were developed with S-Plus. The S language was originally developed at Bell Laboratories (Becker *et al.* 1988; Chambers and Hastie 1992) and has since grown in popularity as a flexible tool for data analysis and graphical display. That growth has led to a name change (S-Plus) and commercial support (MathSoft). S-Plus contains tools, and the ability to extend and customize them easily, for most of the analyses discussed in this book. Their approach exploits the structure of the experimental design automatically into data storage, display and analysis. While S-Plus was originally designed for exploratory data analysis, it has evolved into an important system for practical data analysis, with contributed extensions for classical inference and Monte Carlo simulations. Venables and Ripley (1994) presents a useful introduction to S-Plus while having sufficient detail to serve as a primary reference. Many graphical ideas for interactive data exploration and analysis using S-Plus can be found in Cleveland (1985; 1993) and in Chambers *et al.* (1983). Public domain extensions to S-Plus are maintained by StatLib (http://lib.stat.cmu.edu).

Minitab (Ryan, Joiner and Ryan 1985) is quite handy for regression, including multiple regression, but until recently has been somewhat cumbersome for analysis of variance. One-factor and two-factor analyses of variance are readily obtained. Recent versions include general linear models along the lines discussed in this book.

The Statistical Package for Social Sciences (SPSS) is favored in some areas, particularly for research in the social sciences. Some biological departments also rely on this package. While SPSS has excellent data management capability, its documentation employs language that has developed independently from the classical approach espoused in the present text.

Systat is very popular among PC users and seems at first glance to have the required analysis of variance tools. However, this author has not had much success with more complicated problems, such as those presented in this book. Further, the inferential tools and graphical tools of Systat are located in separate programs, requiring the user to bounce back and forth in an interactive session.

All of the above packages have large audiences of statisticians and scientists, and are subject to close scrutiny with regard to the quality of numerical calculations. Many other data analysis packages are available, as well as several devoted to experimental design and analysis. More packages are being developed all the time. Unfortunately, fancy graphs and easy-to-use interfaces do not guarantee correct calculations or appropriate analyses. The reader should keep in mind that ultimately any practical data analysis will be presented to a community of scientific peers. It may be necessary to justify the choice of computing tools – for their proven accuracy as well as their accessibility – in addition to the choice of analysis. This book emphasizes ways to develop appropriate analyses and to understand the strengths and weaknesses of using statistical packages.

Dynamic graphics have recently enlarged the scope of interactive statistical graphics. Several groups have designed graphical tools which allow linking of different plots and tables (Lisp-Stat by Tierney 1990; Regard by Haslett *et al.* 1991; Data Desk by Velleman and Velleman 1988; JMP by SAS Institute 1994; Hurley 1993). Point-and-click identification of points on one plot highlights the same points on other graphs, tables or even a geographical map. This linking feature is currently available in SAS/Insight and in some contributed programs for S-Plus (check the StatLib collection described below).

Ideally, dynamic graphics tools will soon be more generally available, as they can be extremely powerful for exploration of data and development and execution of strategies of data analysis. The recent package JMP from SAS Institute (1994) shows great promise as a graphical system, with linked plots, data-driven choices of graphics and analysis and a module for designing experiments.

Recently there has been a new dimension added to visual display of information, in the form of 'hyper-media'. Newer workstation platforms allow great flexibility in visual display and manipulation in real time. In addition, recent public domain access tools such as the World Wide Web (WWW) have opened up the **Internet**, allowing searches of databases around the world in seconds or minutes. Krol (1992) gives a nice overview of current and projected resources for the Internet, carefully teaching the basics rather than listing thousands of items as some books are now doing. Higham (1993) highlights resources of particular interest for mathematicians. Interactive tutorials are readily available on line. Internet news groups are quite handy for quick inquiries about non-standard problems, yielding responses from around the globe. The main drawback is finding efficient ways to sort through all the information!

There are several math and stat related Internet libraries. The most important is the statistical library **StatLib**, which can be found on the World Wide Web at the address http://lib.stat.cmu.edu. StatLib has a wide collection of public domain software, data sets and publications which can be retrieved very quickly. In addition, it maintains addresses of statistical and mathematical resources around the world. A copy of its index is available by e-mail with just a subject entry:

    To: statlib@lib.stat.cmu.edu
    Subject: send index

Many statistics departments maintain their own Internet resources. In particular, the data sets for this book can be found under my home page at the University of Wisconsin–Madison Statistics Department

    http://www.stat.wisc.edu/~yandell/pda/

The data and statistical computing code (SAS and S-Plus for the examples found in this book can be viewed and copied from this site (and eventually via StatLib), giving proper credit as usual.

Databases are no longer confined to columns of numbers and pages of text or computer code. It is possible to grab photographs or recent (15 minutes old) satellite images and programs to display them or manipulate them locally. The advent of virtual standards for images (e.g. GIF and JPEG) allows confidence that visual tools will evolve at a rapid rate in the near future. Many WWW browsers support video and audio animation as well as static images and text.

Several interactive information systems are devoted to specific tasks which have, or will soon have, visual access tools for interactive research. Library systems around the world are already on-line, as well as some government systems (e.g. Congressional Record). However, many of these are largely keyword access to text, a very linear and tedious approach. Imaginative ideas for organizing information in object-oriented ways is evolving quickly in some fields such as molecular biology and geographic information systems.

Geographic information systems (GIS) are being used to layer data from geographic surveys, land-use maps, satellite images, census data, transportation networks, etc. Current systems are modest compared to the projected wishes and needs of private and governmental scientists but are still quite impressive in terms of detail and creativity. Each system has strengths and drawbacks and the meshing of information across different distance scales – from county to state to nation to world – is not ideal. However, ARC/INFO, TIGER and other systems provide direct visual access to data and are heavily used. It is interesting from a statistical viewpoint that these systems do not yet incorporate measures of the precision of measurements!

This information explosion begs organization. We need visual means to organize all this information and provide read access to those aspects which individuals most want at any given time. To some degree, this lies in the arena of computer science, but more and more it calls for careful thinking about the design. In particular, neither of the two interactive information systems highlighted above has adequate methods to date to store or use information about sources of variation and bias in measurements. Inferential tools for data analysis are likewise limited and in need of development with peer review within the field of statistics.

1.6 Interpretation

Interpretation of results must take account of the structure of the experiment, the purpose of analysis and the limitations of the methods employed. Inference is dictated by design. Data analysis requires certain assumptions, modelling the relationship between responses and experimental factors. Interpretation of results is limited by the scope of inference. That is, the data are a sample from a larger population of interest; hopefully inference from the sample to the population is relevant! The way in which the sample is drawn has a profound impact. For instance, selecting only fields owned by growers who are cooperative and have efficient management practices implies that these types of growers' fields constitute the **sampled population** even though all growers' fields may be the **target population** which is of interest for broader interpretation of results (Figure 1.1). Any restric-

INTERPRETATION

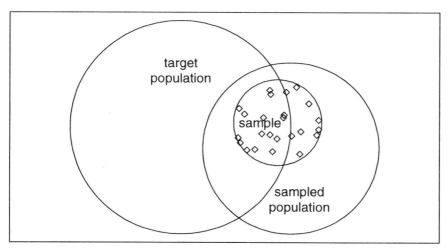

Figure 1.1. *Pragmatics of sampling from a population. It may not be possible to sample from the target population. Instead, a sample from a slightly different population may address some characteristics of interest, but may introduce some bias since the focus differs.*

tion in the process of random selection of items from a population can affect the scope. Some restrictions, such as blocking or stratifying, can be incorporated into a statistical model for analysis. Others, such as lack of randomization or bias in selection, cannot. These latter difficulties lead to caution in interpretation of results beyond the subjects of study.

Any treatment differences could systematically alter the distribution of measured responses in a variety of ways, as shown in Figure 1.2. For simplicity, introductory statistics courses consider simple shifts of mean response. This is a powerful method since other types of effects can be transformed approximately into mean response (e.g. relative differences become additive with a log transform). It is fairly 'robust', or insensitive to minor violations of model assumptions. Further, the math leads naturally to considering means and variation about the mean. While this may be the wrong approach for some problems, analysis based on means can yield useful insight. For instance, a nonparametric approach might be preferable due to major problems with the distribution assumptions, but the design may be too complicated for readily available tools. Replacing the observations by their ranks and using methods in this book offers a rough-and-ready alternative. Counting data violate several assumptions, but log counts or square root of counts may be adequate for a first pass; later analysis could use categorical data methods (e.g. generalized linear models), building on what is learned here.

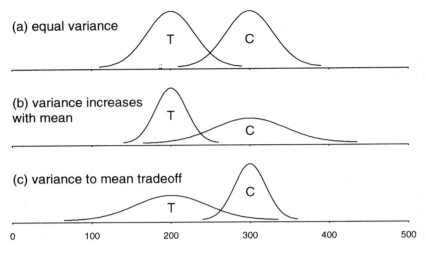

Figure 1.2. *Idealized histograms for group differences. Differences may in some cases (a) be characterized by a shift of center. However, there may be a change in spread as well (b), and these may be in opposite directions (c).*

Models may arise from a deep understanding of the problem, or may simply be convenient for explanation of factor effects. Linear models are often associated with this latter pragmatic, curve-fitting approach. However, it is much more satisfying from a scientific perspective to develop an explanatory 'mechanistic' model. This requires a deep understanding of the processes under study. Usually, there is only a partial understanding of underlying mechanisms. In practice, we must rely on a compromise. Hopefully we learn new insights about nature from the current experiment and can use this knowledge to design further experiments to uncover more structure. It is important that both scientist and statistician keep in mind this evolving understanding of the 'big picture'. Rarely, if ever, is an experiment performed in isolation of other studies past and future.

1.7 Problems

1.1 Design: Describe briefly in your own words what data analysis entails.

1.2 Interact: Become familiar with your computer system, and in particular the statistical packages you will use for the material covered in this book. Skim relevant introductory material such as the first three chapters of Venables and Ripley (1994) for S-Plus or the first seven chapters of di Iorio and Hardy (1996) for SAS. It is usually a good idea to create a course directory in which to place your work.

1.3 Interact: Locate StatLib on the Internet. Report on the primary resources available through this site.

1.4 Interact: The examples and problems throughout this book begin with a capitalized key word. All of these (with the exception of **Interact**) have associated directories on the Internet resource **pda**, which can be found through my home page. Each directory has a home page which describes the contents, with hot links to the files of data, source code and possible printouts or other explanatory documents.

Locate the **pda** resource at the University of Wisconsin–Madison (see Section 1.5). Examine the outline of this book found there. Copy the **Tomato** data set and accompanying source code that is used in Part B.

CHAPTER 2

Collaboration in Science

Practical data analysis usually involves collaboration between one scientist knowledgeable in a substantive field and another trained in statistics. Collaboration between such individuals, with different formal languages and probably different ways of thinking about science, is an acquired skill. It takes practice. This chapter provides some guidelines for improving collaboration. While it is aimed at an applied statistician, with a career of collaborative consulting, these ideas should prove useful to scientists trying to organize their thinking before and during a research investigation.

Healthy collaboration demands a comfortable environment for the interchange of ideas during the knowledge discovery process and an ability to convey concepts in simple, accessible language. Become an avid listener. The approach to any research problem should be neutral, in hopes of avoiding undue bias or emotional charge to discussions.

The statistical consultant encounters many scientists from a wide array of disciplines. This requires developing skills to learn quickly the framework and language of other world views. On the other hand, many scientists seldom talk about research with anyone outside their discipline. In this respect, the statistician has a singular responsibility to establish a healthy atmosphere for the exchange of information.

In a way, a practicing statistician might look at sessions with a scientist as a series of interviews. Initial discussions outline the science behind a project and provide an initial idea of the experiment and key questions. Later dialog delves deeper into specific aspects of design and analysis, with mutual education leading to common language for key concepts. During the process, scientist and statistician get to know one another, learning how each other organizes knowledge and reasons about the unknown.

Section 2.1 presents some general guidelines for analyzing data and asking questions. The mechanics of a consulting session are examined in Section 2.2. The 'big picture' of the ethical role of statistics and consulting is the subject of Section 2.3. Finally, Section 2.4 presents some guidelines for writing to provide a framework for communicating ideas in a lasting medium, so that other scientists can learn directly from the experience of analyzing data from the current experiments.

2.1 Asking questions

A statistician can provide valuable organizational skills and insights into ways of asking and addressing questions that transcend any particular scientific discipline. The scientist may grasp concepts of experimental design without being comfortable with its mathematical formulation. Alternatively, there may set of well-formulated research goals but some uncertainty about how to design and conduct a series of experiments. Thus it is incumbent upon the statistician to find other ways to share design concepts. How can this be done? Asking questions is a good start. Plotting raw data and plausible relationships among factors is simple to do, but so easy to overlook. The guidelines in this section rely largely on common sense and careful organization of time and materials.

One of the trickiest aspects of consulting is asking questions and learning what **key questions** drive the scientist's research. Asking after the general background of an experiment can uncover much more than blunt questions such as 'what is the design?' or 'what is your hypothesis?' Instead, consider asking:

- What results were (will be) expected from the data?
- How did (will) you gather the data?
- How did (will) treatments get assigned?

How can the statistician learn whether the experiment was designed or was simply an observational study without using those words? Further, what kind of questions will elicit information about whether and how randomization was employed, and about the nature of the experimental units for different factors. Begin with general questions, gradually becoming more specific, but returning to the general periodically to review the context of the experiment. It is important constantly to check that the line of questioning is pertinent to the scientific investigation. Avoid leading the scientist toward a particular statistical methodology too early. Instead try to:

- Ask neutral questions.
- Rephrase key questions.

This lets the scientist state key questions in appropriate language and encourages a dialog which can lead toward a common statistical framework for data analysis. At the apparent end of a session, ask:

- Is there anything else which may be relevant?
- Are there plans for subsequent experiments?

These questions can open up new and important lines of discussion. Following any session while ideas are fresh, both parties should write down a list of questions and any other notes on experimental design or proposed analysis, including plot sketches, and use these to begin dialog in future meetings.

2.2 Learning from plots

Response to questions can be greatly enhanced by using a variety of graphics during an interview. Initially these might be quick sketches. Later they can be refined with the aid of computers. Some graphics can be used for both description and inference, while others provide clues about the validity of assumptions or possible unusual features in the experiment. This is an opportunity for statistician and scientist to focus together on visualizing the experiment and evolving features of data analysis.

- Draw a diagram of the physical experiment.

Initial graphics should attempt to capture the experiment and key questions in broad strokes. The statistician must become familiar with the experiment, and the scientist needs an opportunity to clarify key questions. Placing the focus on pictures of the experiment and sketches of real or potential data grounds discussion.

- Plot responses against time or each other.
- Sketch possible interaction plots.
- Organize potential tables of means.

Enhance plots to highlight relationships of potential importance. This can be done by circling unusual points, adding titles and explanations in margins. Consider systematically highlighting other factors by the following methods:

- Use plot symbols to identify factor levels.
- Subdivide data by factor level and plot separately.
- Order factor levels by mean response in plots and tables.

These plots may raise further issues about the experiment, such as whether measurements were made in quick succession or under restricted conditions. These initial looks at raw data can illustrate the key questions which motivated the experiment. They may point out suspicious patterns which question assumptions.

Formal analysis can get quite complicated, even when the key questions were carefully framed ahead of time. It is easy to get mired in details of analysis and overlook important features of experimental design. To avoid this, gradually build up a mutual understanding of and experience with the statistical methods. Use plots to explain assumptions and demonstrate results. Build inference directly into plots by adding standard error yardsticks for multiple comparison of means.

- Start with simple models, using a subset of the data.
- Overlay model fit on data plots.
- Include precision estimates such as standard error bars.

The assumptions of formal statistical inference should always be checked using diagnostic plots. Plots of residuals (response minus fit) against fitted values, time order or other measurements can quickly highlight unusual cases or problems with the models. While there are formal tests of some model assumptions, they are not very useful for small data sets and can be very misleading even with larger data sets.

- Check model assumptions with residual plots.
- Identify outliers and double check raw data.
- Examine magnitude and sources of variation.

Suspicious patterns in residuals, such as trends or unequal variance, may suggest a transformation, weighted analysis, or the addition of new factors. This is an opportunity for statistician and scientist to review assumptions to determine if they are reasonable based on how the experiment was conducted.

Analysis must finish with interpretation of results. Written and oral presentations, whether to one person or to a large audience, demand clean graphics. Be selective, and keep it simple. Be sure to include cautions about design, assumptions and analysis.

- Present a few self-contained plots and tables.
- Annotate to highlight key features.

Plots, tables and other graphics can help build a mutual understanding of an experiment, from the phrasing of key questions through design features to description of data, formal analysis and checking of assumptions. Graphics can save time and avoid confusion, particularly when statistician and scientist may have different levels of comfort with mathematics and the substantive science under study. Keep graphics neat and well documented to ensure their lasting value.

2.3 Mechanics of a consulting session

Ideally, the statistician can provide new insights and improve the quality of science. The task is not merely to show how to use statistical techniques but to join in the scientific process of investigating an unknown phenomenon. During this process, the consultant and scientist can explore what sort of statistical methods have been employed and question the assumptions and requirements in a spirit of mutual education. Many things go on during a consulting session. Some involve the overall organization of time and responsibilities, others concern the science of the problem, still others involve the interpersonal climate of the meeting.

It is important that the statistician's contribution be given proper value. This may simply involve showing mutual respect but could include formal acknowledgement, co-authorship on publications or financial reimbursement. These issues should be raised early in a consulting relationship to avoid misunderstandings later.

Emotions and other aspects of interpersonal dynamics can profoundly affect any consulting session. Ideally, scientist and statistician meet as equals and conduct a balanced dialog based on **mutual respect**. The statistician should bypass tangential mathematical details which tend to confuse, striving for a simple exposition of subtle points. The scientist should be willing to explain seemingly irrelevant aspects of an experiment, respecting the statistician's judgement about the complexity of analysis, but should demand that the concepts be clearly explained. Both parties should encourage questions and interruptions, avoiding 'lecture mode'. Reduce the chance of interpersonal difficulties by listening carefully, speaking clearly and simply, dressing well and attending to personal appearance. Good personal hygiene and a strong self-awareness are instantly reassuring to one's colleagues.

Every session, and every project, has a beginning, a middle and an end. The beginning is the time to establish the reason for meeting and to get comfortable with each other. Review of past work may be in order. The middle entails the main body of work, clarifying goals, establishing the scientific framework, and developing a workable statistical approach. The end recognizes limited time to meet and the need to acknowledge what has been accomplished and what is left to do.

The climate at the **opening** of a session is very important to the success of any collaboration, as in any relationship. Make the beginning efficient by making sure the scientist is comfortable. Establish levels of expertise as the first step toward developing a common language. Set up a clear agenda. If this is a repeat session, both statistician and scientist should report on progress or lack thereof.

The **middle** of the session is usually devoted to the goals of the project, the science of the problem and the statistical approach. The central goal should be crystal clear, set in proper context. Be diplomatic, but do not hesitate to probe deeply to clarify issues. Secondary goals should be identified and prioritized. Examine how data were quantified and whether there were any constraints on the data, all in the context of the experimental procedures.

The **statistical approach** should be appropriate and understandable. Sometimes the statistician must dissuade a scientist from poor statistical practice or argue for a particular approach that may involve more work than the scientist would care to undertake. In such a situation, it may be better to find a simpler compromise while clearly identifying what assumptions are likely to be violated. Often, it is wise to write down concerns so that the scientist can refer to these later. Ideally, both parties can return to the issue at future meetings to point out strengths and weaknesses when there is tangible evidence such as residual plots, tables of means and standard deviations, etc. The statistician must value the expertise and intuition of the scientist while developing a strategy for analysis. Each session can be an effective learning experience for both parties if the scientist gains a better understanding of appropriate statistical methodology. The technical level should be appropriate to the task, the scientist's abilities and the time budget. Often, a simpler approach involving graphics and exploratory analysis can serve for 90% of the problem.

The **ending** of a consulting session is as important as the beginning, and should be just as clean. Review accomplishments and outline future responsibilities of both statistician and scientist. Keep in mind the ultimate goal and the time frame, adjusting as needed. Make sure plans are realistic and deadlines are reasonable. Are both parties satisfied?

Scientist and statistician may choose not to continue their dialog through to the completion of data analysis and reporting of results, especially if the statistician is viewed as an occasional advisor and not a collaborator or co-author. Temporary compromises suggested casually during a session can become permanent if there is no follow-through. Therefore it is very important that the statistician recognize each meeting is potentially the last chance to influence the scientist. If there are loose ends that require further contact, then explicit plans for meeting should be made and recorded on a calendar. Both parties need to take responsibility to follow through as needed.

2.4 Philosophy and ethics

Scientists with statistical training have tremendous opportunities for collaboration, with all its rewards and dangers. Great care is needed to assure that the consulting relationship is among peers rather than a power play. Some aspects of this are concerned with the different ways people learn ideas and express creativity. Others revolve around important ethical issues of honesty in science. Many of the ideas in this chapter arose from the work of Joiner (1982), augmented by the books of Boen and Zahn (1982) and Chatfield (1995).

Many articles examine philosophical issues inherent in the process of consulting, including some case histories. These help clarify the relative importance of different aspects of statistical consulting. Bross (1974) poses the question whether a consultant should be perceived as a technician or as a collaborator. Other writings about the art and practice of consulting from a statistician's view include Daniel (1969), Healy (1973), Hunter (1981), Kirk (1991), Marquardt (1979) and Sprent (1970). Deming (1965) is still relevant today. Lurie (1958) wrote about statisticians from a scientist's perspective. Two bibliographies of articles on statistical consulting (Baskerville 1981; Woodward and Schucany 1977) point to further work.

Healy (1984) and others ask serious questions about the training of statisticians for consulting. Several course-related articles focus on engineering and quality (Bisgaard 1991; Schilling 1987), government (Eldridge, Wallman and Wulfsberg 1981) and biostatistics (Zelen 1983). Consider browsing recent issues of *The American Statistician* and the *Proceedings of ASA* Sections on *Statistical Education* and on *Statistical Consulting* for new material as they appear.

Books on the history of statistics (Stigler 1986) and on prominent statisticians such as Neyman (Reid 1982) and Fisher (Box 1978; see also Savage 1976) provide interesting reading about how statisticians and scientists have worked together in the past. The journal *Statistical Science* regularly features interviews and other articles on the history of statistics.

Science does not proceed linearly from one success to the next. It moves in fits and starts, and can regress. These broader issues of the scientific method have been chronicled and debated in the philosophy and history of science. Kuhn's (1962) *Structure of Scientific Revolutions* and Feyerabend's (1988) *Against Method* question the fabric of the scientific method, arguing on the one hand that science has not always proceeded forward in knowledge discovery, and on the other hand that there is no guaranteed way to enumerate all the possible paths of scientific inquiry (see also Hoyningen-Huene 1993). Bronowski (1965) places science in society, drawing interesting parallels between art and science. Hofstadter (1979) cleverly weaves together unifying ideas of our current world view. Lerner (1993) reviews the tremendous unsung contributions of women to science and society, showing the cultural loss through reinvention of ideas due to the suppression of women's works over many centuries.

The statistician attempts to guide analysis and point out difficulties and cautions, while maintaining a professional distance from the results of the experiment. However, sometimes emotional, psychological or political issues interfere with this collaboration. These issues may be related to stress in the research laboratory, differing goals of the scientist and a superior, pressure for funding, or heated public debate on the topic of research. The statistician must keep in mind that ultimately the work belongs to the scientist. In some cases, the best advice may be to adopt a simple approach as a compromise. In extreme situations, the statistician may ask to be completely removed from the research program.

A statistician who believes a scientist is misusing information has the responsibility of disassociating from the work and documenting the abuse. This should be first shared, in writing, with the scientist, with the understanding that it will be communicated to superiors as appropriate. This is one example of an ethical dilemma that statistical consultants may encounter during a career. The Sigma Xi Society (1991) has developed a pamphlet entitled *Honor in Science* which gives an excellent overview of ethical issues, from control of intellectual property to fraud. Every scientist

	Cerebral Brain Component	
Left Brain	**Learning**	**Right Brain**
logical/factual	process	holistic/intuitive
quantify/analyze	style	explore/discover
lecture/read	method	video/experiment
	Limbic Brain Component	
Left Brain	**Learning**	**Right Brain**
plan/detail	process	emote/feel
organize/sequence	style	share/internalize
review/summarize	method	experience/discuss

Table 2.1. *Herrmann's complementary thinking processes*

should be aware of ethical guidelines in his/her discipline, such as those recently published by the American Statistical Association (1995). Examples of ethical dilemmas can be found in the books of Broad and Wade (1982) and Penslar (1995).

2.5 Intelligence, culture and learning

It can be quite helpful to examine the **learning process** and the concept of intelligence in order to understand how communication can be improved (Snee 1993). Recent concepts of multiple types of intelligence and creativity and of different modes of perception point out the need to recognize the variety of ways in which people think, learn and communicate. Skills for sharing information improve through practice and attention to patterns in oneself and others.

Snee (1993) used the work of Herrmann (1989) as a metaphor for examining the process of thought (Table 2.1). In his system, the notion that left and right brains function differently is augmented by a dichotomy of upper (cerebral) and lower (limbic) components, yielding four complementary thinking processes, with corresponding preferred learning styles and methods.

The classical approach to intelligence focuses on mathematical logic and defines the quotient of intelligence with an IQ test. Education in the sciences has stressed the value of math skills for the logical development of models of nature. As a result, a scientist who shows only modest mathematical intelligence is considered 'dumb'. This damaging stigma ignores the wealth of intelligences that humans can display. Gardner (1985; 1993) has proposed seven types of intelligences, which are briefly summarized in Table 2.2.

Markova (1991) built on diverse research in educational psychology and learning theory to notice that people differed in their use of perceptual

Intelligence	Characteristic Strengths
linguistic	memory/explain/rhetoric
musical	rhythm/melody/tone
logical/mathematical	deduce/abstract/analogy
spatial	orient/transform/imagery
bodily/kinesthetic	dance/objectify/manipulate
intra-personal	discriminate/understand/self-guide
inter-personal	grasp mood/motive/intent of others

Table 2.2. *Gardner's seven types of intelligences*

channels – visual, auditory or kinesthetic – for the conscious process of organizing, the subconscious process of sorting and the unconscious creative process. That is, individuals seem to use distinct perceptual channels for these three states of mind. The conscious channel focuses outwardly on the world, while the unconscious channel gathers information from a broad spectrum. The challenge is to engage all channels during the consulting process.

Hall (1966; 1981) and other anthropologists have explored the cultural dimensions which underpin the way people behave and organize ideas. Individuals from different cultures, especially those based on distinct language systems, may approach science in profoundly disparate ways. The 'culture' of statistics values logic and inference, while scientists in agriculture and life sciences value a comprehensive knowledge base. Mathematical statisticians can conduct research in isolation, while scientists in many disciplines need to work together due to the nature of the material being studied. These differences contribute to different 'world views', which can either confound or enhance understanding.

It is incumbent upon any statistical consultant to recognize that there are many forms of intelligence and to respect the way another's mind works. Statistician and scientist have an opportunity to complement each other and learn new ways of perceiving the world of science.

2.6 Writing

Writing is as much art as science. As any artist knows, improvement comes with practice, practice and more practice. Writing is especially valuable to statistical consultants, as better writing leads to improved communication. Writing helps organize thoughts, identifying the strengths and weaknesses in an argument. It provides a tangible product, something that can be handled and pondered.

Scientific writing should be unambiguous. It should state clearly what was done, how it was done and what was found. It should be set out in such a way that another scientist coming upon the report at a later time could repeat the work and reproduce the results. Good writing is concise and self-contained.

What are the components of good scientific writing? Proper spelling and grammar, along with the use of simple words rather than polysyllabic jargon, are central to easy reading. The written word should convey the big picture – why was this work important? – but at the same time should include specific details – what are the key findings? Good statistical writing communicates ideas by building trust and eliminating fear of black box techniques.

A scientist can improve research simply by being organized. Writing out protocols is vital. However, it is even more important carefully to articulate key questions in words. Planning ahead involves approaching a statistician well before an experiment with a written objective. Think hard about how to analyze data up front, and even whether the proposed data will address the key questions. Try to visualize the data, including a few sketches of anticipated results. Review the material with a statistician, using it as the basis for further discussion. During the meeting, make sure advice makes sense, restating ideas and tasks repeatedly. Keep good notes. Ultimately, it is the scientist's responsibility to understand what is done and to communicate key results to peers through scientific meetings and journals.

A well-written document speaks directly to the appropriate audience. Write for the scientist, as a report rather than as a personal letter or log of adventures with data. Imagine that the scientist may want to extract phrases, sentences or paragraphs to share with others, possibly with some rewriting for his/her own audience. Point out difficulties with experiments without finding fault, keeping in mind the real limitations of time, money and resources.

Several good guides for general writing include some inexpensive paperbacks. Strunk and White's (1979) classic *The Elements of Style* should be required reading. Gowers's (1988) reprint of another classic has further ideas. Goldberg (1986) has some delightful, elegant suggestions on how to write creatively. Turabian (1973) provides a writing style manual which many students find useful. Higham's (1993) handbook of writing, geared toward the mathematical sciences, includes chapters on English usage and all stages of writing a paper through publishing and presenting talks.

Virtually any project can be summarized in a half-page paragraph. This summary should identify the main ideas of the experiment and include specific information about results. A **summary** provides a quick condensation, conveying maximum information in minimum space. Comments about results should be as specific as possible while avoiding technical verbiage. Statements that 'significant differences among groups were detected' are meaningless. Instead, try something like 'mean yields of 37 and 39 bushels

Section	Contents
Title Page	descriptive title, name, date
Abstract	summary of key findings
Introduction	describe problem, key questions
Experimental Design	lay out experiment, identify design
Materials & Methods	describe method of analysis
Results	report findings from data analysis
Conclusions	interpret results, address key questions
References	include any cited literature
Appendix	optional (brief!) support material

Table 2.3. *Sample report sections*

per hectare for varieties A and B were significantly higher than that of the standard variety C (32 bushels)'. This conveys detail that the scientist can absorb immediately without having to open the report. Treat a summary as though it were a résumé, to be consumed in 15–30 seconds. Important information must be prominent or it is missed. Leave complicating details to the body of the report. Consider being asked to explain key findings from analysis in five minutes. It may have taken hours, days or months to perform analysis, but the critical reporting may be those five minutes. A well-edited summary organizes thoughts and gives the audience a concise but tangible product.

A report should have several sections, describing the key questions, the experimental design, methods of analysis, results and interpretation. Each section should begin with a paragraph highlighting what is to be found – the reader should not have to hunt for material. Tables and figures should be incorporated into the text where possible. Table 2.3 includes sections that are important for almost any report. It may be helpful to alter the titles to reflect the specific experiment.

A good report has several important components. It is neat and concise, and written for the scientist. The statistical problem is couched in the language of the field of study. Self-contained plots and tables are tastefully spread through the report, integrated into the text appropriately. The design is described precisely. Steps of analysis are laid out in a logical fashion.

A good report is developed in stages. Introduce the key ideas behind the science of the problem early and carefully document the experimental design. Share the written description with the scientist, changing as needed to get it right. Continue to modify the analysis section as understanding emerges. Update interpretation as well. Write the conclusion, interpreting main findings in the context of the goal of the experiment. Finally, compose a summary which presents the substance of the report in condensed form right up front.

2.7 Problems

2.1 Interact: Good writing comes with practice. Ideas flow when fingers connect with keyboard or pen touches paper. Use every opportunity to enhance your writing skills. The process of writing is messy. Rather than labor over details in the first draft, practice getting the ideas down on paper quickly.
(a) Take exactly five minutes (no more, no less) and write down the main ideas for an experiment. If you are involved with one now, report on that. If not, invent an experiment, but keep it simple. What are the key questions? How was the experiment done? How might one approach data analysis? Write down notes quickly, including a few plots.
(b) Go back and review this material for another five minutes.
(c) After a full review, modify as needed for another five minutes.
This 15 minute writing exercise, adapted from Goldberg (1986), is not that different from the dynamic of a consulting session. Consultants must learn to think on their feet, quickly grasping the whole idea and then digging for the details that make all the difference to design and later analysis. Writing engages three perceptual channels – visual through viewing, auditory through words and kinesthetic through the mechanics of typing – waking up many levels at once. The ideas may not appear at the beginning but the efforts put into organizing material on paper will be invaluable later on.

2.2 Interact: Repeat the above exercise on another experiment. This time use only plots and pictures, with a minimum of annotation so that everything is properly identified.

2.3 Interact: Seek out a peer who is unfamiliar with your work. This might be a person at a party, a parent or a housemate. Tell that person that you will try to explain what you do (an experiment, a course content) in plain language and in just five minutes.
(a) Do it. Use paper and pencil to assist with figures as needed. Mathematical and scientific jargon must be kept to a minimum.
(b) Repeat the process with another individual. This time, pay close attention to this person's reactions while you talk. Notice what keeps his/her interest and when boredom sets in. Try to anticipate this and modify your approach as you go.

2.4 Interact: Seek out a scientist and learn about his/her current research.
(a) Bring a pad of paper and take notes.
(b) Sketch pictures of laboratory or field setup. Share these with the scientist to verify accuracy.
(c) Ask for details about a current experiment. Write them down. Then relay them back to the scientist in your own words.

2.5 Interact: The previous two problems may be enhanced by recording the conversation. If you do this, play the tape back later while scanning your notes. What do you pick up this time, about science, statististics or human interactions, that you missed during the 'live' session?

CHAPTER 3

Experimental Design

This chapter outlines several types of scientific studies in Section 3.1, focusing in particular on designed experiments in Section 3.2. In order to understand an experiment, it may be helpful to consider separately those aspects which affect the way the experiment is conducted (Section 3.3) and those factors which are of direct interest to the experimenter (Section 3.4).

3.1 Types of studies

There are many ways to conduct a scientific study. Some involve largely gathering data on undisturbed natural processes, while others are completely enclosed in artificial environments with strictly controlled conditions. In a sense, the more the scientist controls the study, the more likely the results can be interpreted cleanly. However, the more control, the farther the experimental conditions stray from the true state of nature. This section briefly touches on several types of studies, noting their connections with designed experiments.

Pure observational studies involve observing and measuring items without interfering with natural processes. These arise, for instance, in anthropology, ethnography, natural history and environmental impact reporting. They can provide a baseline for comparison with subsequent changes. In some situations pure observation is all that is possible. However, it is often difficult to ascribe observed differences between groups, since they may be confounded with historical events which may not be well known or understood. Therefore, conclusions from such observational studies are of necessity tentative.

The pulse of the nation is regularly monitored by **sample surveys**, in which a small sample of a population is asked questions on a variety of topics. There are many ways to conduct surveys, and the manner of sampling greatly influences the proper form of analysis and the way in which results may infer back to the general population. Hidden biases and potential sources of variation must be addressed. Scientific surveys are regularly conducted on non-human populations: plant and animal communities, marine life, soil and leaf microbe micro-cultures, the stars in the sky and the earth below our feet. Sample surveys can suffer some of the same problems encountered with pure observational studies.

The primary focus of this text is on **designed experiments**. Here the scientist controls important aspects of the study, either by (1) randomly assigning subjects to groups (e.g. high or low fiber diet, hot or cold food); or (2) selecting subjects at random from different populations (e.g. male or female, urban or rural). These two different but complementary designs can lead to the same type of comparison. The first is a controlled experiment, with random assignment of experimental units to groups. The scientist controls the assignment to groups of experimental units drawn from a homogeneous population. The second approach entails a random sample from a natural experiment, in which experimental units are drawn at random from distinct natural populations. Here, the scientist controls the sampling from separate groups but not the group assignment.

Designed experiments formally randomize over extraneous factors which might influence the response, while controlling assignment to groups. This insures that, on average, any 'large' differences in response between groups can be attributed to groups rather than to other factors which were not controlled.

Designed experiments require careful attention to a **protocol** established before the experiment is run. This should declare how randomization is performed, including how samples are drawn from larger population(s). It should ensure that decisions and measurements are made by an objective, repeatable procedure.

An observational study or sample survey may be analyzed using methods in this book. However, interpret results with caution, especially if attempting to extrapolate beyond the data. When the scientist does not have control of key features of the study, discussion of the limitations of statistical inference may be much more important than the method of analysis.

Many health science studies arising in **biostatistics** fall naturally under experimental designs considered in this book. However, some of the same cautions about spurious association due to historical accident are warranted when working with long-lived organisms such as humans. For instance, prospective (or retrospective) studies follow selected individuals forward (or backward) in time to examine health differences between groups. Usually, people are not randomly assigned to groups, such as smokers or non-smokers, but instead have self-selected based on a confusing variety of genetic, behavioral and environmental factors unknown to the researchers. Further details of biostatistical methods in the health sciences can be found in Fisher and van Belle (1993).

3.2 Designed experiments

The purpose of a designed experiment is to provide a logical framework for comparing groups in terms of some response of interest. This section lays out the main elements of a designed experiment, distinguishing between treatment structures of interest to the scientist and design structures necessary for carrying out the experiment in practice.

In the simplest experiments, the **groups** correspond to different **levels** of a **factor** of interest, such as varieties of corn. Groups in more complicated experiments can be described in terms of combinations of factor levels. The term factor was borrowed from 'Mendelian factor' in genetics by Sir Ronald Fisher around 1926 (Fisher 1935, ch. 6; Box 1978, ch. 6). The units in these groups are (random) samples from real or potential populations with certain characteristics associated with the factor levels. Put another way, an **experimental unit**, or **EU**, is defined as that item to which a factor level is assigned.

Factors under study in an experiment may act in combination, or interact, in their effect on a response of interest. Analysis of variance, developed through this text, allows separation of main effects of each factor from any synergy or antagonism (interaction) among factors in combination. Factor combinations may be viewed as larger factors for some portion of analysis. It is sometimes helpful to arrange two or more factors in a grid to summarize information on factor combinations. For this reason, combinations of level from two or more factors are sometimes referred to as **cells**.

Example 3.1 Design: Consider an experiment to compare the yield of two different growing methods applied to three varieties.

```
              variety
    method  1  2  3
       A   ┌──┬──┬──┐
           │  │  │  │
       B   └──┴──┴──┘
```

There are two factors, **variety** with three levels and **method** with two levels. Together they determine six factor combinations, or cells. There is not enough information here to know how the experiment was performed. Were there ten plants of each variety for each method, or were there ten plants, with both growing methods applied to each plant? ◇

It is possible that different factors may have different sizes of experimental units to which their levels are assigned. It is vital to think deeply about the basic unit to which each factor is assigned. Examples of experimental units include a potted plant of a variety of interest, a field plot treated with a fertilizer under study, the day's production of an assembly line under specified conditions, etc. Learn as much as possible about how the experiment was performed and in particular how and to what object each factor level was applied. Was each crop treatment assigned (at random) separately to each plant in a field or to a portion of the field? Was the oven temperature changed between each run or left at one setting for several runs while other factors were changed? This concept is so important that it permeates the last three parts of this book.

The scientist is usually most interested in the **treatment structure**, that is the set of factors that the experimenter has selected to study and/or compare. Factors in the treatment structure may be analogous to a chemical application or mechanical procedure or may correspond to sets of preexisting populations such as species, gender, colors, locations or time intervals. Some designed experiments do not have 'treatments' in the usual sense. For instance several geographic locations may be under study, but they are not randomly applied to field plots. However, experimental units (e.g. seeds) may be assigned at random to plots in different locations much as plots are assigned to treatment groups. Gender is not assigned at random. Rather, there are typically two gender populations, with samples of individuals selected from each.

The process of assigning factors to experimental units (or selecting EUs from populations) and subsequently measuring one or more responses embodies the **design structure**. The design can control factors which are not of interest (e.g. water content in a sloping field) by grouping experimental units into homogeneous groups or blocks. Further, well-planned design structures include replication and randomized assignment of treatments to EUs to cancel out (on average) the effect of anything else which might modify the treatment response.

A **designed experiment** consists of the following steps: determine the key questions which drive the experiment; choose a treatment structure; select a design structure; conduct the experiment; and gather measurements (data) for each experimental unit. The assumptions implicit in these steps must be understood in order properly to conduct designed experiments and interpret subsequent data analysis.

3.3 Design structure

Design structure is the most important part of the experiment to understand because it is so often overlooked in the rush to examine the treatment structure. The key elements involve the way experimental units are assigned to treatment groups. Sometimes this is very simple, analogous to pulling slips of paper out of a hat. At other times, the experimenter may choose to block EUs into groups in some fashion – for convenience, because of limited resources, or to protect against some trend or factor which is not of direct interest. While a design may be very complicated, it can almost always be reduced to more manageable pieces which can be understood separately. These smaller pieces may suggest simpler analyses to elicit important results and build communication between statistician and scientist. As confidence comes with comprehension of the experimental design, data analysis can become more sophisticated.

Replication involves the independent assignment of several EUs to each group, or factor combination, leading to independent observations. Quite often the concept of what constitutes replication can be confusing. Multiple measurements of the same EU, known as sub-sampling or pseudo-replication (Hurlbert 1984), improve the precision of the measurement for that EU but do not increase the number of independent applications of the treatment. Time can also be problematic, particularly when there are repeated measurements on the same subject. However, this may be quite cost effective and may contain valuable information not otherwise available. Designs involving sub-sampling or conducted over time are discussed in detail in the latter parts of this book.

The manner and extent of **randomization** is a critical part of the design structure. Randomization involves selection of EUs from some population and assigning them to groups so that all possible arrangements are (usually) equally likely. This ensures that EUs have an equal chance of being assigned to each level of factor combinations. If group characteristics have no effect on response, then any observed differences among groups in the distribution of responses must be due to other aspects of the experimental units beyond the control of the experimenter. The particular random assignment of EUs to groups can be compared with all other possible assignments to assess the significance of observed differences. This is in fact the basis of the randomization test (see Chapter 15, Part F) and was the original justification for inference in the analysis of variance (Fisher 1935).

What does the word 'random' mean to the scientist? Unfortunately, it may indicate selecting the first subjects which are available, or some 'arbitrary' assignment of EUs to groups. It is much better to develop a very careful random procedure before conducting an experiment, perhaps using random number tables, a computer's pseudo-random number generator or a physical mechanism (slips of paper in a hat, coin toss). The assignment may involve some restrictions to randomization, such as blocking or stratifying EUs according to some features believed to influence response but of secondary importance. Understanding the allotment of EUs to groups and how they are drawn from those groups is central to understanding the experimental design, which in turn is crucial to appropriate data analysis.

Example 3.2 Design: Suppose there are three varieties of wheat to be grown in a field which can be subdivided for measurement of yield into 12 plots, say in a grid of four rows and three columns. Three ways to lay out the plots are illustrated in Table 3.1 and described below. Each of these three design structures requires a different analysis even though the treatment structure does not change.

B	A	C
B	A	C
B	A	C
B	A	C

(a) Sub-sample

B	A	C
C	B	C
A	C	B
B	A	A

(b) CRD

B	A	C
A	B	C
C	A	B
B	A	C

(c) RCBD

Table 3.1. *One-factor designs*

(a) One approach would be to assign each variety randomly to a column (making planting easy) and later take the 12 measurements. But what are the units to which varieties were assigned? How many separate units are there? The plots are in fact subsamples of the columns and do not represent true randomization. The EUs are therefore columns and are not replicated.

(b) Suppose instead that each variety is written on four separate slips of paper; and that the plots are each assigned a variety by drawing the slips from a hat. That is, each variety is replicated four times. Now what are the experimental units for variety? Would you want to analyze these two different experiments in the same fashion or not? This simple design structure is known as a **completely randomized design** (CRD) in which one assigns factor levels completely at random. This is appropriate if the experimental units are (fairly) homogeneous, yielding similar responses if given the same treatment.

(c) Another approach would be just to have three slips of paper in the hat, one for each variety, and determine a random ordering separately for each of the four rows down the field. This can help protect against a known or suspected variation among the EUs available for the experiment. That is, divide the EUs into equal sized groups, or blocks, and assign treatments at random within each block. This is known as a **randomized complete block design** (RCBD). ◊

In an RCBD, it is important to randomize factor level assignment within each block, independent of all other blocks. Subsequent analysis must account for the possible block-to-block differences when assessing the strength of treatment response. Note that complete randomization is a special case of this design, with only one block containing all the EUs. Usually each block contains exactly one EU for each treatment, but some designs have two or more replicates within each block. The important feature for later analysis is to have the design as **balanced** as possible, with the same number of replicates of each factor level within each block.

There are many other types of design structures which build on the CRD

DESIGN STRUCTURE   41

B1	A1	C1
B2	A2	C2
B2	A2	C2
B1	A1	C1

(a) Strip plot

B1	A2	C2
C2	B2	C1
A1	C1	B2
B1	A1	A2

(b) CRD

B1	A1	C1
A2	B2	C2
C2	A2	B2
B1	A1	C1

(c) Split plot

Table 3.2. *Two-factor designs*

and the RCBD. However, a clear understanding of CRD and RCBD alone can greatly facilitate the construction of almost any design structure. Most designs can be decomposed into smaller components which are CRD or RCBD.

Example 3.3 Design: Suppose there are three varieties of wheat under study assigned to 12 plots as before but, in addition, there are two types of fertilizer to compare. See Table 3.2.

(a) One could randomly assign each variety to a column and each fertilizer type to two rows (making planting easy) and later take the 12 measurements. What are the units to which varieties were assigned? What units for fertilizer? This is an example of a **strip plot design**, requiring some care in analysis. There is in fact no true replication of varieties in this **nested design**.

(b) Suppose instead one wrote each of the possible six variety and fertilizer combinations on two separate slips of paper, placed all 12 slips in a hat, and drew them out one at a time, assigning them to the 12 plots. Now what are the experimental units for variety? For fertilizer? Would you want to analyze these two different experiments in the same fashion or not? This design structure has 12 experimental units, with the EU for variety and fertilizer being a plot. This is in fact another example of a CRD, now with a crossed two-factor treatment structure.

(c) Another approach would be to just have three slips of paper in the hat, one for each variety, and determine a random ordering separately for each of the four rows down the field. Then one could randomly assign the two fertilizer types to the four rows. This presents a third design structure for the same treatment structure, and each design requires a different analysis. This is an example of a **split plot** design which arises often in experimental science.   ◇

Design structures involving different types of experimental units for different key questions are the subject of the latter parts of this book. Other designs involving Latin squares and balanced incomplete blocks are briefly examined there. For more complete coverage, see texts on experimental design such as Cochran and Cox (1957) or Box, Hunter and Hunter (1978). The principal focus here is on the implications of design for analysis, and in particular developing a facility with tools to handle designs that are not nicely balanced.

3.4 Treatment structure

The simplest treatment structure consists of comparing several groups as a **single factor**. Key questions may initially ask if groups are all the same. If they differ, the goal may be to order them, compare subsets, or pick the 'best'. Most treatment structures can be viewed as single-factor arrangements for some portion of data analysis by ignoring relationships among treatments.

The **two-factor model** consists of factor combinations in which the experimenter controls two types of factors (e.g. light and temperature levels, or varieties and fertilizers). For some purposes, this can be viewed as a single factor – are all treatments the same? which is best? – but other questions can now be asked. For instance, is there strong evidence that one variety always yields more than the others, regardless of fertilizer used, or is it necessary to qualify statements for certain fertilizers (A is better with fertilizer 1, but B is better with fertilizer 2)? The randomized complete block design introduced in the previous section could be viewed as a two-factor layout with factors being the treatment and blocking. For instance, block on location when comparing varieties. However, interest may focus on how location affects the comparison of varieties.

More complicated treatment structures build up higher-order **factorial arrangements**. However, the nature of problems – statistical and conceptual – are already present in the design with two factors. In some situations, logistical constraints of resources may force inclusion of only some of the possible treatment combinations in a (balanced) **fractional factorial arrangement**. The choice of fraction and design of such experiments is the subject of other texts such as Box, Hunter and Hunter (1978).

Some experiments look like factorial arrangements with a slight twist. For instance, there may be added controls. Consider examining several levels of two drugs, A and B, which readily dissolve in saline. The scientist may include saline as one control (0 level of both A and B) but may want another control (e.g. EDTA) for comparison with earlier experiments. Such an experiment may seem obvious to the scientist but can be confusing to a statistician who has only learned the textbook factorial arrangements. Again, it is useful for some questions to view this as a single factor, while other questions may be best addressed by setting aside the extra control and examining the two factors in a balanced experiment.

It is often instructive to examine subsets of the treatment structure to address certain questions of interest. Thus, an initial 'simple' approach may help build understanding in a modular fashion about the scope of the whole experiment.

3.5 Designs in this book

The completely randomized design with a single factor is the prototype upon which all other designs are built. Part B considers comparison of groups as if they represented levels of a single factor. Issues raised here about comparing group means carry over to more complicated designs in a natural fashion.

Treatment structures become more interesting in Parts C and D. The two-factor treatment structure allows the possibility of separating main effects from interaction among factors. The completely randomized design structure is coupled with progressively more complicated factorial arrangements. These designs correspond to simple random samples, with random assignment of groups, or levels of combinations of factors, to experimental units. Incomplete designs including such as fractional factorials and Latin squares are introduced.

Part E reflects on assumptions tacitly accepted to this point. Several graphical approaches to checking assumptions are examined, along with inference when assumptions are not reasonable. While these concepts can be carried over to the later parts of this book, they are easier to consider in completely randomized designs.

Part F considers experiments with measurements on more than one characteristic of each experimental unit. Some covariates may reflect preexisting conditions or be otherwise unaffected by group assignment. Other measurements may be multiple candidates for response and/or covariates affecting another response. This part concerns ways to draw inference about treatment groups adjusted by the covariate and ways to infer the association of covariate and response adjusted for treatment and design structures.

Up until this point, interest has focused on the fixed effects, centering around comparing means. Part G begins with an examination of random effects, considering groups as random elements from a larger pool rather than as the items of direct interest. Here, the attention is on the variance among elements. Fixed and random effects are combined in mixed models. An example is the randomized complete block design, considered briefly in Part B without interpretation of the blocks themselves. Many designs in practice involve mixed effects, and some of these comprise the remaining parts.

Part H examines the way nesting of factors enforces a design structure on the random assignment of experimental units to factor levels. Sometimes similar units are randomly assigned to treatment within blocks. Other experiments may have sub-sampling from experimental units, with possible further assignment of levels of another factor to those subunits. Blocking and sub-sampling are the key features of nested designs. The split plot design illustrates the ways nested designs can affect data analysis. This provides a natural bridge back to random and mixed models, and suggests how to address general nested designs.

Many experiments involve repeated measures on subjects, often over time. These designs in Part I can be viewed as more general forms of nested designs, allowing correlation among measurements. Experiments which change some treatment assignment over time are known as crossover designs. These include certain types of replicated Latin squares.

All of these concepts can be blended in a modular fashion. The challenge is to view any experiment from a variety of (simple) perspectives and build an understanding of the whole in stages. There are a number of excellent texts on experimental design, beginning with the classics, Cochran and Cox (1957) and Cochran (1977). Many factorial designs are covered in Box, Hunter and Hunter (1978), with further development of response surface methods in Box and Draper (1987).

3.6 Problems

3.1 Design: Briefly, neatly and concisely (no more than one legible page) describe a particular designed experiment that you might conduct. Be specific about what questions you are asking and how you would conduct the experiment, with attention to components listed below. Keep it simple; you may consider an experiment with only one factor.
(a) Identify in your own words:
  key questions
  experimental units
  design structure
  treatment structure
  method of randomization
  method of replication.
(b) State assumptions important for analysis.
(c) Examine potential problems with assumptions that may arise during the course of your experiment.

3.2 Design: Below are some descriptions of actual (or possible) experiments. Your job is to piece them apart, identify the important components of each experimental design and describe that design succinctly. If it helps, part of your answer could examine subsets of the design which have structures you readily grasp. Your 'answer' may include a word model or math model with all terms identified. You may assume designs are balanced unless otherwise indicated. It is not important *per se* to use the 'right' buzz words for a particular design. Better to identify the design structure and treatment structure in your own words.

(a) Suppose a company wants to measure the 'quality' of a product by using the ratio of the maximum to minimum diameter of a particular hole punched in plastic (the ratio should ideally be 1.000 ± 0.001). A number of products have this hole, although they are made at different stations depending on the product. The statistician selects five worker stations and three different products at each station. The same product is measured every Monday in February, and then the results are analyzed.

(b) A dairy scientist wants to investigate the effect of warm vs. cold water fed to cows on their milk productivity. He is concerned about the large cow-to-cow variation and adjusts for that by giving each cow both treatments, in random order, with sufficient time between treatments to remove any residual effect. To complicate matters, some cows were studied in the summer and some in the fall.

(c) A plant scientist is measuring the amount of soluble sugars in cranberries in a bog in northern Wisconsin. The rectangular bog is divided along its length into six 'samples' (her term) and a randomly selected branch from each sample is measured on six dates throughout the reproductive season (July–October). Soluble sugar is measured in the vegetative and flowering parts of each branch. The researcher wants to examine the time course of sugar production to document differences between vegetative and flowering parts. [Ignore possible correlation between plant parts or over time.]

(d) Research on putting plants in space to feed astronauts is conducted at the Biotron, a controlled environment facility. In one such experiment, 4 chambers are employed to compare the effects of light, relative humidity (RH) and variety of potato on total biomass. Five plants of each variety are placed in each chamber (ten plants per chamber). Two levels of light and two of RH are selected for study. The chambers are randomly assigned to the environmental treatment combinations. Three weeks later the plants are measured for total biomass.

PART B

Working with Groups of Data

This part develops initial ideas about mathematical models for comparing groups of data. The concepts developed in this part arise again and again through the remainder of this book.

Chapter 4 develops summaries for groups within the context of the cell means model. Inference questions focus primarily on estimates and tests for cell means. Degrees of freedom, precision, strength of evidence (p-value) and power are briefly introduced. Some attention is devoted to inference for the variance as well.

The central theme of this book revolves around comparing groups, and in particular group means. Chapter 5 quickly moves from comparing two means to linear contrasts. The overall test of significance leads to partitioning the total variation in response, or sum of squares, into explained and unexplained variation. Many questions about comparisons of means can be phrased readily in terms of sums of squares, and in particular expected mean squares. Finally, the relationship between power and sample size is developed for the comparison of two groups for practical purposes of planning subsequent experiments.

Selected methods of multiple comparison of means are discussed in Chapter 6, in particular those motivated by classical test statistics and those motivated by examining the spread among several means. While both statistician and scientist might like to have an 'answer' for all situations, there is in fact no best approach to comparing several means which works for all situations.

CHAPTER 4

Group Summaries

Comparison of groups of measured responses can be greatly facilitated by good summaries. The means and variances of groups capture key features of central tendency and spread, and have nice properties under suitable conditions. This chapter briefly reviews this material as a prelude to formal comparison of groups.

Numerical summaries are inadequate on their own, however. Graphical summaries in the form of histograms, stem-and-leaf plots and box-plots provide useful information about the center, spread and other features of each group. Nevertheless, histograms can be cumbersome for quick comparison of many groups, or of a few groups over many responses.

This chapter begins with a graphical comparison of groups. The comparison of groups leads naturally to considering the cell means model in Section 4.2. Estimates of means and variance arise in Section 4.3. Section 4.4 develops interval estimates for means, with some attention to precision of reported values. Formal testing of hypotheses for means comprises Section 4.5. Finally, inference for a common variance across groups is examined in Section 4.6.

4.1 Graphical summaries

Sometimes it is valuable to look at all of the data, or the distribution of the responses. The stem-and-leaf plot offers a way to organize data to reveal the shape of the distribution while maintaining a few digits of each measurement. With larger samples, the histogram works well to approximate the shape of the distribution. Comparison of shapes of several distributions is often easier with box-plots, which capture the shape in broad strokes and can be stacked side to side. These and other graphical ideas can be found in Cleveland (1985).

The **stem-and-leaf plot** is constructed by using the most significant digit of each response as a 'stem' and accumulating the next digit as a 'leaf' on the stem. Thus all responses with the same stem appear on the same line. Adjustments can be made in practice if there are too many or too few leaves per stem. Typically this involves splitting a leaf into two halves (for leaves

0-4 and 5-9) or into five parts (0-1, 2-3, etc.). Extreme values which would require several empty stems in between are sometimes reported separately. These diagrams can be stacked alongside one another, as in the example below, although they are generally used to present concisely one set of data.

Example 4.1 Tomato: A tomato breeder is interested in finding out the location of genes which control various plant growth attributes (Goldman, Paran and Zamir 1995). He is using molecular markers to do this. He thinks he has located a major gene for fruit weight (fw) close to a marker designated as tg430. The marker is a DNA fragment which is cut to different lengths by an enzyme according to parent type (A or B). The parents were crossed to form an F1 individual, which was selfed to create F2 offspring, called entries here, and coded as A for one parent type, B for the other parent, or H for a hybrid with both types.

```
         A allele        Hybrid    B allele
      ---------------    ------    --------
   -3 :                       :    : 0
   -2 :                       :    :
   -1 :                       :    :
   -0 :                       :    :
    0 :                       :    :
    1 : 15                    :    : 9
    2 : 4                     :    :
    3 :                       :    : 455
    4 : 0024                  :    : 46
    5 : 2456677          : 06      : 03
    6 : 1124568          : 125     : 8
    7 : 011334777        : 0036    :
    8 : 023458           : 018     : 0
    9 : 0678                  :    :
   10 : 00247888         : 28      :
   11 : 01               : 377     : 6
   12 : 6                : 26      :
   13 : 12               : 6       :
   14 : 69               : 2       :
   15 :                       :    :
   16 :                       :    :
   17 : 4                     :    :
```

The original experiment covered two years (yr), with 93 unique plant entries (entry). There were several replications each year. Unfortunately, the data now available to the scientist consist of the mean fruit weight (mfw), averaged across the replicates for each year. Several offspring had missing data for fruit weight, leaving 89 for year 1 and 76 for year 2. [The raw data are in notebooks halfway around the world!] The scientist believes a $\log_{10}$ transformation (mfwlog) is reasonable, and has presented data in

GRAPHICAL SUMMARIES

that way. The tomato breeder actually wanted to see all the data. The stem-and-leaf diagram above provides a few digits of each observation in addition to the general shape of the distribution. The decimal place is one place to the left of the colon. That is, the smallest value is −0.3, for a plant with the B allele, while the largest value is 1.74 for an A allele plant. ◇

The **histogram** approximates the true distribution of the response. It can be constructed of rectangles of equal area, one per response, stacked over a horizontal axis spanning the range of the response values. Since the total area for a histogram is 1.0, each rectangle has area one over the number of responses. Alternatively, divide the range of responses into adjacent bins (usually of the same width) and count the number of responses per bin. The height of the bin is set as that count over the total number of observations. Sometimes the rectangles or bins are sliced off to produce a smooth polygon rather than rectangular steps.

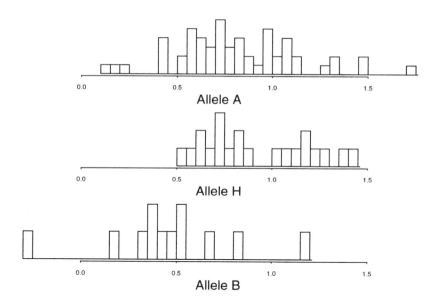

Figure 4.1. *Tomato histograms by group: flowering times (on a log scale) are longer for plants with A or H (hybrid) genotype than for B. This suggests that the A allele is dominant in a genetic sense.*

Example 4.2 Tomato: The histograms of the tomato data by genotype in Figure 4.1 show the basic character of the roughness in the data. With larger sample sizes, the stem-and-leaf diagrams would be cumbersome, while the histogram would tend to follow the distribution shape. ◇

There is a fair degree of art in selecting the shape of rectangles, or the width of bins. A common guideline suggests that the number of bins be about the square root of the number of observations, modifying this accordingly if there are large gaps as in allele B of Figure 4.1. Using more bins (tall, skinny rectangles) leads to rough histograms which may be too 'noisy' to show any real pattern. However, using fewer bins (short, fat rectangles) produces smoother histograms which may not show enough features of the data. Fancier approaches involving smooth 'density estimates' are possible, but require data sets of at least several hundred observations. These are beyond the scope of this text.

Some researchers prefer the simpler **dot-plots** to histograms. Dot-plots consist of vertical strings of dots above the response scale, with one dot for each response with the corresponding value. One advantage of dot-plots is that they can be constructed with standard characters. That is, for a horizontal dot-plot, use a period (.) for one value or a colon (:) for two on each line. Several dot-plots can be easily aligned on a page in this way. Dot-plots can also be presented vertically, side by side, as was done earlier for the stem-and-leaf plot.

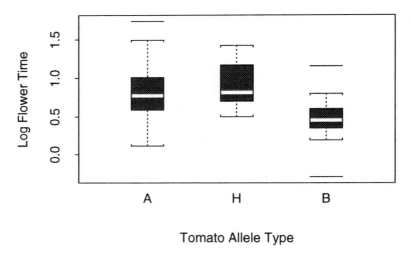

Figure 4.2. *Tomato box-plots by group: note similar spreads in data across the allele types. There is some evidence of skew in the center of the distributions, but there is no consistent pattern.*

Box-plots reduce the distribution of data to a few key features. However, different statistical packages may use slightly different features – it never hurts to read the fine print. Typically, box-plots include the median flanked by the lower and upper quartiles (25th and 75th percentiles, respectively). These form the box. 'Whiskers' extend out some distance, one to two box widths depending on the package, until they include all the data in that region. Any points beyond the whiskers are flagged as outliers, sometimes as extreme outliers, and plotted separately. Thus, box-plots capture the central shape, including skewness, as well as some features of the 'tails' of the distribution. Several box-plots can be stacked together in a single plot. Further, they can be used in conjunction with other plotting methods to represent concisely subsets of data that are too large to present individually.

Example 4.3 Tomato: The first year's tomato data are presented as box-plots in Figure 4.2. The box-plots suggest that the A allele for this gene is dominant for higher fruit weight. That is, the hybrid (H) plants have similar values of mfwlog, both being larger than that of the B allele type. Other features of the data, such as the extreme values, are readily apparent. The general shape of the distribution, as indicated by the spread in the central 50% of the data and the approximate symmetry between the tails, is quickly grasped. However, the fine structure of the distribution is lost. For example, the Figure 4.1 histograms suggest the H allele data are either skewed or are a mixture of two groups, but the box-plot of Figure 4.2 only suggests skewness, with the median close to the lower quartile. ◇

4.2 Estimates of means and variance

The simplest treatment structure has one factor at several levels, while the simplest design structure is a completely randomized design. Combining these yields the cell means model. This section discusses design considerations, assumptions and the basic framework of the cell means model.

An experiment may involve either assigning units to factor levels or drawing units at random from already existing levels. In either case, the number of units per group may be fixed in advance or may depend on the random process of assignment. Suppose there are a groups of interest – that is, a levels of the factor of interest. The scientist might randomly assign experimental units to a sets, with n_i units in the ith set, for a **total sample size** of $n. = \sum_{i=1}^{a} n_i$ units, then randomly assign the a group labels to the sets of units. Alternatively, the scientist might draw n_i units at random from the ith population, again for a total of $n. = \sum_{i=1}^{a} n_i$ units. Randomization in either situation should lead (roughly) to a groups of homogeneous experimental units. The only difference between groups would be the effect of the group, the factor level difference. The first approach has the advantage that differences among groups could be directly attributed to group assignment. In the second case, it is possible that groups differ for a variety of reasons associated with group membership. The statistical model and data analysis are the same for both situations, but the interpretation of results by the scientist might be rather different.

The easiest experiments to analyze have a **balanced design structure**, with **equal sample sizes** in each group ($n_i = n$ and $n. = na$). In some cases, an experiment may begin with equal numbers of units per group but some units may be lost during the course of study. Sometimes the process of losing units can affect interpretation of results. If the chance of loss depends on group assignment, then results can be seriously biased.

Theory suggests that the best chance to detect differences among treatment groups occurs when the sample sizes are equal or nearly equal. Further, the math gets more arcane when the sizes are unequal. For more complicated experiments, imbalance leads to complications in the choice of procedures to address key questions. In general, the more balanced an experiment is at the outset, the easier the data analysis. However, this is difficult to achieve in practice in many disciplines. If the imbalance is somehow related to the treatment, this can introduce unwanted bias making interpretation of later results tricky.

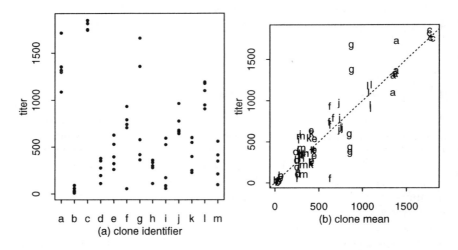

Figure 4.3. *Cloning data by group. The first figure (a) is ordered by the clone identifier, which has more to do with lab procedures than data analysis. The second, square figure (b) plots means against values using plot symbols for clone identifier. Here it is possible to examine center and spread, and to consider the position of clones relative to each other.*

Example 4.4 Cloning: A plant scientist measured the concentration of a particular virus in plant sap using enzyme-linked immunosorbent assay (ELISA) (Novy 1992). The study included 13 potato `clones` (plants reproduced asexually), with two commercial `cultivars`, five somatic hybrids, five progeny of the somatic hybrids, and one clone of *Solanum etuberosum* (a species related to potato). Of the five progeny of the somatic hybrids, two were classified as susceptible and three as resistant to the virus. The scientist wants to understand the resistance to the virus among these 13 clones. Plant sap was taken from five inoculated plants of each clone, for a total of 65 measurements of titer. Unfortunately, one measurement was lost during processing of the samples. Figure 4.3 shows the raw data by clone. Figure 4.3(a), ordered by clone identifier, indicates considerable range in center and spread but is otherwise not very informative. Figure 4.3(b) orders by clone mean, jittered to mitigate overlap, using a separate plot symbol for each clone. The dashed identity line helps track the spread around the mean. This latter plot suggests there is less spread in titer for clones at the extremes than for those in the middle of the range. ◇

During the experiment, the scientist measures a response, denoted in this book by the letter y. The response for the jth experimental unit in group i is denoted by y_{ij}. That is, the n_1 responses $y_{11}, \cdots, y_{1n_1}$, are a random sample of possible responses from a population with treatment 1. The distribution of responses in this population has some true **mean** μ_1 and **variance** σ_1^2. The responses for the ith group come from a distribution with mean μ_i and variance σ_i^2. The mean and variance characterize, in a mathematical way, the center and spread of a population. Typically, the **standard deviations (SDs)** $\hat{\sigma}_i$ are reported instead of variances, as SDs are in the same units as the original data. However, the variances have nice properties, including natural estimates based on squared deviations from the mean.

In mathematical terms, the means and variances can be expressed as

$$\mu_i = E(y_{ij})$$

and

$$\sigma_i^2 = V(y_{ij}) = E[(y_{ij} - \mu_i)^2] ,$$

with E and V being the expectation (mean or average) and variance (mean squared deviation from mean), respectively, over the population of all possible responses. In many situations, these summaries adequately characterize the group behavior of response.

The basic objective of a designed experiment is to address the key questions. This may involve estimating parameters such as means $\mu_1, \cdots, \mu_a$, and variances $\sigma_1^2, \cdots, \sigma_a^2$, and drawing inference about these parameters. Formal inference usually involves hypothesis testing, confidence intervals and/or multiple comparisons.

Generally, it is desirable to obtain the best estimates possible of model parameters by using the data at hand. Statisticians acknowledge that the definition of 'best' depends on the context and the manner of comparison of estimation procedures. Consider the following estimates:

$$\hat{\mu}_i = \bar{y}_{i\cdot} = \sum_{j=1}^{n_i} y_{ij}/n_i$$
$$\hat{\sigma}_i^2 = \sum_{j=1}^{n_i} (y_{ij} - \bar{y}_{i\cdot})^2/(n_i - 1)$$

with the convention in notation that a dot at a subscript signifies summation over that subscript ($y_{i\cdot} = \sum_j y_{ij}$) while a bar indicates a mean ($\bar{y}_{i\cdot} = y_{i\cdot}/n_i$). The cell sample means $\hat{\mu}_i$ are the **least squares estimators**, minimizing the sum of squared deviations within group i, $\sum_j (y_{ij} - \mu_i)^2$, and hence minimizing the sum of squares

$$SS = \sum_{i=1}^{a} \sum_{j=1}^{n_i} (y_{ij} - \mu_i)^2 \ .$$

This can be verified by setting the partial derivatives of SS with respect to μ_i to 0,

$$0 = \frac{\partial SS}{\partial \mu_i} = -2 \sum_{j=1}^{n_i} (y_{ij} - \mu_i) \ ,$$

which yields the **normal equations**,

$$n_i \mu_i = \sum_{j=1}^{n_i} y_{ij} \ , \ i = 1, \cdots, a \ ,$$

whose unique solutions are give by $\hat{\mu}_i = \bar{y}_{i\cdot}$, provided $n_i > 0$. These least squares estimators are unbiased since $E(\hat{\mu}_i) = \sum_j E(y_{ij})/n_i = \mu_i$.

Example 4.5 Cloning: Summaries for the 13 cloning experiment in terms of means $\hat{\mu}_i$ and SDs $\hat{\sigma}_i$ appear in Table 4.1. Notice that the susceptible progeny and the cultivars have higher titer than the resistant, with the parents being intermediate. The clones with smallest SDs apparently have the highest (cultivar c) and lowest (etb b) mean, while the clone with largest spread (cultivar g) has an intermediate mean. This pattern is evident in Figure 4.4(a) as well as in the earlier plot of raw data (Figure 4.3(b)). While the SDs are quite variable, there is no discernible relationship between mean and SD. Note, however, that there are patterns of center and spread associated with type in Figure 4.4(b). It is entirely possible that the apparent fluctuation of variances is due to the small sample sizes per clone. This could be examined using methods considered in Part E. ◊

The a mean estimates each contain one piece of information, or **degree of freedom** (df), about that group, its center of mass. The remaining ($n_i - 1$) degrees of freedom for the ith group are used to measure the mean squared deviation from the mean, the variance. These may be referred to as the model and error degrees of freedom, $df_{\text{MODEL}} = a - 1$ and $df_{\text{ERROR}} = n_\cdot - a = \sum_i (n_i - 1)$, respectively. Note that there must be at least one

ESTIMATES OF MEANS AND VARIANCE   57

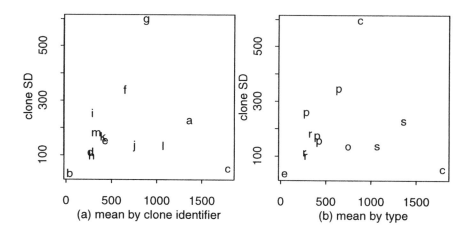

Figure 4.4. *Cloning means and SDs by group. The suggestion of increasing spread with increasing mean in (a) is far from perfect. However, the two cultivars identified in (b) have the highest and lowest SD.*

Clone	n	Mean	SD	type
c	4	1789	49	cultivar
a	5	1358	226	susceptible
l	5	1068	134	susceptible
g	5	876	594	cultivar
j	5	746	133	odd
f	5	645	340	parent
e	5	428	150	parent
k	5	406	169	parent
m	5	329	180	resistant
i	5	288	253	parent
h	5	279	98	resistant
d	5	265	111	resistant
b	5	45	32	etb

Table 4.1. *Cloning summary statistics*

observation in a group to estimate the mean and at least two to estimate the variance.

Under the assumption of equal variance across groups, the best estimate

tg430	n	mean	SD
A	56	0.806	0.324
H	21	0.912	0.283
B	12	0.458	0.350

Table 4.2. *Tomato means and SDs of log flowering times*

of the common variance is

$$\hat{\sigma}^2 = \sum_{i=1}^{a}(n_i-1)\hat{\sigma}_i^2 / \sum_{i=1}^{a}(n_i-1)$$
$$= \sum_{i=1}^{a}\sum_{j=1}^{n_i}(y_{ij}-\bar{y}_{i.})^2/(n.-a)$$

with $df_{\text{ERROR}} = n. - a$ degrees of freedom. Counting degrees of freedom is one of the tricks of the trade and will be exploited repeatedly in this book. It can save a lot of time in the planning stages and can uncover problems which arise during data analysis.

Example 4.6 Tomato: Estimates of allele type means and standard deviations in Table 4.2 provide a concise summary of the box-plots shown in Figure 4.2. Sometimes a small table can replace a series of plots for comparing groups. The combined estimate of standard deviation is $\hat{\sigma} = 0.319$ with degrees of freedom $df_{\text{ERROR}} = 56 + 21 + 12 - 3 = 86$. ◊

Example 4.7 Cloning: The estimate of the common SD is $\hat{\sigma} = 239$ with $df_{\text{MODEL}} = 64 - 13 = 51$. However, the cultivars and the anomalous odd and etb clones seem to have a range of variances. Dropping these should result in a more homogeneous variance, with common estimate of SD as $\hat{\sigma} = 198$ with $df_{\text{MODEL}} = 54 - 11 = 43$. ◊

4.3 Assumptions and pivot statistics

These estimates provide good guesses for the 'true' group means and variances. How good? That is, how close are they to the true parameters? How can they be used to compare combinations of true parameters? Inference formally addresses questions about larger populations based on data measured on random samples drawn from such populations. It is certainly possible to estimate sample means and sample variances without consideration of design. However, inference requires close attention to experimental design and the assumptions inherent in sampling and measuring processes.

Several **assumptions** are commonly employed but not always explicitly stated. First, and most important, is the belief that the **model is correct**, that the characterization of mean and variance correctly captures the key

ASSUMPTIONS AND PIVOT STATISTICS

features of the response. Second, the design structure supposes that the observations are **independent** of one another and that observations from different groups are independent. When considering several groups it is common (but not necessary) to assume **equal variances**, that is $\sigma_1^2 = \cdots = \sigma_a^2 = \sigma^2$. Finally, it is often assumed that the distribution of the response y_{ij} is **normal** (with a bell-shaped histogram).

With these assumptions, it is possible to show that

$$\hat{\mu}_i \sim N(\mu_i, \sigma^2/n_i), \ i = 1, \cdots, t,$$

$$\hat{\sigma}^2 \sim \sigma^2 \chi^2_{n.-a}/(n.-a).$$

If the y_{ij} are normally distributed, then the estimators $\hat{\mu}_i$ are normal and independent of $\hat{\sigma}^2$; further, $\hat{\sigma}^2$ has a distribution proportional to a χ^2. If the data are not normal, then $\hat{\mu}_i$ and $\hat{\sigma}^2$ may be dependent and $\hat{\sigma}^2$ may be far from χ^2. For large samples, the central limit theorem ensures that $\hat{\mu}_i$ have approximately normal distributions. However, normality is rather important for the distribution of the group means for small samples; for moderate samples it helps if the distribution of responses is roughly symmetrical.

Inference typically revolves about a **pivot statistic**, a specially chosen random variable which summarizes information about the data and model. The trick is to frame questions about model parameters using estimates by constructing a statistic with known distribution, under specified conditions, that does not depend on the unknown model parameters. Then either use properties of the statistic to guess reasonable values for unknown parameters (interval estimation) or determine how strong the evidence is in favor of or against a particular set of conditions (hypothesis testing).

Suppose interests focuses on estimating a particular group mean μ_i. The pivot statistic is a centered version of the sample group mean $\hat{\mu}_i$ around that mean divided by an estimate of its standard error

$$T = \frac{\hat{\mu}_i - \mu_i}{\sqrt{\hat{\sigma}^2/n_i}} = (\hat{\mu}_i - \mu_i)\sqrt{n_i}/\hat{\sigma} \sim t_{n.-a}$$

which has a t distribution with $n. - a$ degrees of freedom provided model assumptions hold.

Example 4.8 Power: Small degrees of freedom for error lead to uncertainty in estimating the variance σ^2. The pivot T is more likely to have extreme values (far from zero). This is reflected in the 'heavier tails' for t distributions with small sample sizes (Figure 4.5). If the sample size is large enough, the distribution is approximately normal, $T \approx N(0,1)$, whether or not the responses are assumed to be normal. This pivot uses data from all a groups to estimate σ^2. In some cases (when equal variance is in doubt), it may be prudent to use only some groups, or just the ith group. ◊

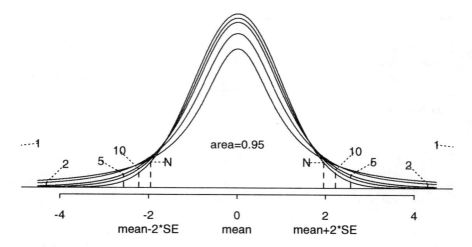

Figure 4.5. *Normal and t distributions. The t distribution is flatter in the middle with heavier tails for smaller degrees of freedom. As the degrees of freedom increase above 20 or so, the t and normal distributions differ markedly only in the extreme tails. Dashed lines indicate lower (2.5%) and upper (97.5%) critical values by degrees of freedom (N = normal), used for 5% two-sided test and 95% confidence intervals.*

The pivot statistic is very convenient since many texts and software packages contain tables of the **critical value**. For the t distribution with $n.-a$ degrees of freedom, the critical value $t_{\alpha/2;n.-a}$ is determined by

$$\text{Prob}\{|T| > t_{\alpha/2;n.-a}\} = \alpha \ .$$

That is, the chance of observing a pivot statistic T larger than the critical value $t_{\alpha/2;n.-a}$ is only $\alpha/2$. With probability $1-\alpha$ a realization from the $t_{n.-a}$ distribution is between the critical values $-t_{\alpha/2;n.-a}$ and $t_{\alpha/2;n.-a}$. In terms of the original units of the response $|\hat{\mu}_i - \mu_i|$ is less than $t_{\alpha/2;n.-a}\hat{\sigma}/\sqrt{n_i}$ with probability $1-\alpha$.

4.4 Interval estimates of means

The estimate of a group mean contains no information about the precision of that estimate. Interval estimates consist of an interval of values around the point estimate, with the width of the interval depending on the standard error of the estimate. Confidence intervals provide a formal level of confidence that an interval covers the unknown group mean.

The **standard error (SE)** of a statistic, such as the estimator $\hat{\mu}_i$ of a group mean, is the square root of the variance of that statistic. It provides a measure of precision in the original units of the response. The SE of $\hat{\mu}_i$ is

INTERVAL ESTIMATES OF MEANS

thus $\sigma/\sqrt{n_i}$. Published reports commonly include standard errors alongside means for ready assessment. The SEs should be reported if the focus is on the precision of a statistic or on estimates of model parameters, while the standard deviation (SD) is appropriate for discussion of the variability for a single experimental unit.

Simple interval estimates for group means consist of sample means plus or minus their standard errors, $\hat{\mu}_i \pm \sigma/\sqrt{n_i}$. A practical consequence of thinking about interval estimates leads to considering the number of **significant digits**, that can reasonably be reported for estimates. It is a good idea to report means up to the first or second significant digit of their standard errors.

Example 4.9 Infer: Suppose a mean is estimated as 10.69734 with standard error 1.29316. Suppose the SE is based on 120 degrees of freedom. It is appropriate to report 10.7 ± 1.29 rather than 10 ± 1.3 or 10.69734 ± 1.29316. While some may prefer 10.697 ± 1.293, reporting 10.7 reflects the precision of the mean better. ◊

The pivot statistic can be easily transformed into a formal confidence interval for an unknown parameter:

$$\alpha = \text{Prob}\{|T| > t_{\alpha/2;n.-a}\}$$
$$= \text{Prob}\{|\hat{\mu}_i - \mu_i| > t_{\alpha/2;n.-a}\hat{\sigma}/\sqrt{n_i}\}$$

with $t_{\alpha/2;n.-a}$ the critical value of the $t_{n.-a}$ distribution. Thus, the interval

$$\left(\hat{\mu}_i - t_{\alpha/2;n.-a}\hat{\sigma}/\sqrt{n_i},\ \hat{\mu}_i + t_{\alpha/2;n.-a}\hat{\sigma}/\sqrt{n_i}\right)$$

covers the true μ_i with probability $1 - \alpha$, providing a $100(1 - \alpha)\%$ **confidence interval** for the unknown group mean. Actually, any particular confidence interval resulting from an experiment either covers the true value or misses it. The probability statements refer to the chance process inherent in the experiment. On average, if the experiment were repeated 100 times, about $100(1 - \alpha)$ of the confidence intervals would cover the true means.

Example 4.10 Infer: The 2.5% critical value $t_{\alpha/2;n.-a}$ for a 95% confidence interval is roughly 2 for moderate to large sample sizes. For instance, $t_{0.025;20} = 2.09$, with the value decreasing to a limit of 1.96 as the degrees of freedom rise. This is shown in Figure 4.6(a), along with similar trends for other confidence levels. Turning this around, the confidence level for a critical value of 2 rises from 70% for 1 df to about 90% for 10 df, tending toward a limit of 95.45%. Figure 4.6(b) shows similar trends for several rounded values. A common approach for summary tables, or for quick calculation purposes, is to approximate a 95% confidence interval by the sample mean plus or minus two standard errors, $\hat{\mu}_i \pm 2\hat{\sigma}/\sqrt{n_i}$, or as an interval $(\hat{\mu}_i - 2\hat{\sigma}/\sqrt{n_i}, \hat{\mu}_i + 2\hat{\sigma}/\sqrt{n_i})$. ◊

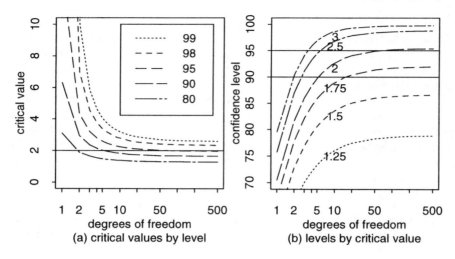

Figure 4.6. *Rough guide for confidence levels. The 95% confidence level (upper 2.5% critical value) based on the t distribution is about 2, give or take 0.2, when error df are greater than 10. This could be examined by plotting degrees of freedom against (a) critical values by confidence level or (b) confidence levels by critical value.*

```
tg430    mean     (SE)      rough 95% CI
A        0.806    (0.043)   0.720-0.892
H        0.912    (0.070)   0.772-1.052
B        0.458    (0.092)   0.274-0.642
```

Table 4.3. *Tomato summary statistics*

Example 4.11 Tomato: Rough 95% confidence intervals for the mean log fruit weights by tomato allele type can be quickly calculated from Table 4.3. The standard error estimates use the combined variance estimate $\hat{\sigma}^2 = 0.102$. It appears that types A and H do not differ while type B has lower fruit weights than either. However, it is not enough to check if (rough) confidence intervals overlap. Generally, if these intervals do not overlap, then means can usually be shown to be significantly different; but if they do overlap, as with A and H, they may still be significantly different. ◊

Example 4.12 Cloning: Figure 4.7 shows 95% confidence intervals for the 13 clones sorted by mean value. Figure 4.7(a) uses a common estimate of variance while Figure 4.7(b) has each interval constructed using only data from that group, allowing variances σ_i^2 to be different. Since there are only a few degrees of freedom (three or four per group) for each confidence

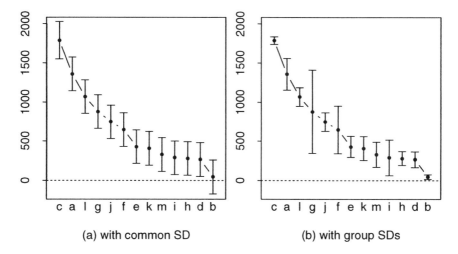

Figure 4.7. *Cloning confidence intervals by group. Groups are ordered by decreasing mean values. The first figure (a) uses a common SD while the second (b) uses separate SDs by group.*

interval, the intervals should be interpreted with caution. However, notice how Figure 4.7(b) highlights the relationship between center and spread seen earlier in Figure 4.4(a). ◇

4.5 Testing hypotheses about means

In some cases a guess μ_{i0} may be based on previous experience, such as research on similar systems, a pilot study, etc. The **test of hypothesis** $H_0 : \mu_i = \mu_{i0}$ uses the null mean μ_{i0} in the pivot,

$$T_0 = \frac{\hat{\mu}_i - \mu_{i0}}{\sqrt{\hat{\sigma}^2/n_i}} ,$$

and looks for sufficient evidence to reject the hypothesis H_0. Note that $T_0 \sim t_{n.-a}$ provided $\mu_i = \mu_{i0}$. However, if the null hypothesis H_0 is false, then T_0 has a distribution whose center is shifted away from 0. Thus, reject H_0 for 'large' $|T_0|$, determined by the probability under the null hypothesis,

$$\text{Prob}\{|T_0| > t_{\alpha/2;n.-a} : H_0\} = \alpha .$$

The interpretation of 'large' depends on what alternatives the scientist wishes to consider. 'Two-sided' tests are often used, rejecting for large $|T|$, when there is no clear alternative to the null (μ_{i0}). Sometimes, a 'one-sided' test is appropriate if there is prior information about the nature of alternatives, rejecting H_0 in favor of, say $H_1 : \mu_i > \mu_{i0}$ with probability

$$\text{Prob}\{T_0 > t_{\alpha;n.-a} : H_0\} = \alpha \;.$$

In practice, it usually makes sense to conduct a **'three-sided' test**, identifying either which side is large or that it is not possible distinguish from the null mean given the evidence in the data.

If the null hypothesis $H_0 : \mu_i = \mu_{i0}$ is true, it will be rejected with probability α. Failure to reject it leads to the tentative conclusion that the mean could be μ_{i0}. If the hypothesis H_0 is false, then the statistic T_0 would be large (far from 0) more often. The further the true value of μ_i is from μ_{i0}, the more often T_0 will be large. Thus there is more resolving **power** the further the true value is from the hypothesized one. Power is discussed in more detail toward the end of the next chapter.

The **significance level** α must be chosen in advance of the experiment. A value of $\alpha = 0.05$ (5%) is very popular, to the point of being the default or only choice for most statistical packages. This fixation on 5% seems to date back to Fisher (1935, sec. II.7):

> It is usual and convenient for experimenters to take 5 per cent. as a level of significance, in the sense that they are prepared to ignore all results which fail to reach this standard and, by this means, to eliminate from further discussion the greater part of the fluctuations which chance causes have introduced into their experiments.

The classical choice of significance level depends on the perceived **strength of evidence** needed, which can be determined before doing the experiment. This is a judgement call and in some disciplines the choice $\alpha = 0.05$ is written in stone. In most disciplines there is more of a sliding scale following:

$\alpha = 0.10$ 'mild evidence'
$\alpha = 0.05$ 'modest evidence'
$\alpha = 0.01$ 'strong evidence'
$\alpha = 0.001$ 'very strong evidence'.

While $\alpha = 0.05 = 5\%$ is very popular, the preferred level varies from discipline to discipline. A physical scientist or engineer, demanding near exactness, might consider $\alpha = 0.01$ to be modest evidence and may not be satisfied with any larger value. A biologist used to working with small samples and high variability may consider $\alpha = 0.05$ to be strong and may be quite content with $\alpha = 0.10$ for certain preliminary studies. A physician working with life and death may have differing criteria depending on the health prospects of the target population.

A different approach, in favor with many statisticians, is to determine the strength of evidence in the data to make a judgement call but display the evidence for other scientists reading the report or paper to interpret as they see fit. Under the hypothesis H_0, the statistic T_0 is centered about 0. In a realized experiment conducted by the scientist, a value of the statistic

FORMAL INFERENCE ON THE VARIANCE   65

T_{DATA} is computed,

$$T_{\text{DATA}} = (\bar{y}_{i\cdot} - \mu_{i0})\sqrt{n_i}/\hat{\sigma} \ .$$

The strength of the evidence can be assessed in an objective fashion by finding the p-value such that

$$\text{Prob}\{|T_0| > |T_{\text{DATA}}| : H_0, \text{ data }\} = p\text{-value}.$$

A *small* p-value is evidence against the null hypothesis H_0. In other words, consider a new experiment. The probability of observing as extreme a value as that found in the current experiment is the p-value. This is the strength of evidence against the null hypothesis.

Note that the p-value is a **random variable** which is uniformly distributed between 0 and 1 if H_0 is true. That is, about 5% of the time (if H_0 is always true), the p-value is below 0.05. More precisely,

$$\text{Prob}\{p < \alpha : H_0\} = \alpha \ .$$

Thus a p-value can be compared to classical testing levels. It is often sufficient for publication to report the sample mean and its standard error $\hat{\mu}_i \pm \hat{\sigma}/\sqrt{n_i}$, along with the p-value for a clearly specified test.

4.6 Formal inference on the variance

This section examines formal inference on the common variance, assuming that all groups have the same variance. Issues involving possible differences among group variances are deferred to a later part of this book. The estimated error variance $\hat{\sigma}^2$ is proportional to a χ^2 variate with $n. - a$ degrees of freedom provided the usual assumptions hold. This section uses a general result about sums of squares, which is shown by induction in Searle (1987, sec. 2.14a) and using matrix algebra in a variety of standard texts.

Example 4.13 Infer: The result, simply stated, is that the weighted sum of squares of normal variates, centered about their weighted mean, has a χ^2 distribution, provided the weights are the inverses of the variances of those normal variates. In mathematical terms, suppose that there are n independent normal variates with possibly different variances, $z_i \sim N(\mu, 1/w_i)$, $i = 1, \cdots, n$. Let $\bar{z}_n = \sum_{i=1}^{n} w_i z_i / \sum_i w_i$ be their weighted mean. Then

$$S_n = \sum_{i=1}^{n} w_i (z_i - \bar{z}_n)^2 \sim \chi^2_{n-1} \ .$$

This is easy to show for $n = 2$. Assume it is true up to $n - 1$. The induction step involves showing that

$$S_n = S_{n-1} + (z_n - \bar{z}_{n-1})^2 \times w_n(W_n - w_n)/W_n \ ,$$

with $W_n = \sum_{i=1}^{n} w_i$. The terms in this sum are both χ^2 variates and are independent since z_n and $\bar{z}_{n-1}$ are independent of the $(z_i - \bar{z}_{n-1})$, $i = 1, \cdots, n-1$. Therefore their sum has a χ^2 distribution with degrees of freedom $n - 1 = (n - 2) + 1$. ◊

Now consider samples from a groups. For the ith group, the sum of squared deviations about the group mean $\bar{y}_{i\cdot}$ is proportional to a χ^2 variate by the above result:

$$\sum_{j=1}^{n_i}(y_{ij} - \bar{y}_{i\cdot})^2 \sim \sigma^2 \chi^2_{n_i-1}.$$

Since these a terms are independent and the sum of independent χ^2 variables has a χ^2 distribution,

$$\sum_{i=1}^{a}\sum_{j=1}^{n_i}(y_{ij} - \bar{y}_{i\cdot})^2 \sim \sigma^2 \chi^2_{n.-a}.$$

In other words, the error variance has distribution

$$\hat{\sigma}^2 \sim \sigma^2 \chi^2_{n.-a}/(n. - a)$$

with mean $E(\hat{\sigma}^2) = \sigma^2$ and variance $V(\hat{\sigma}^2) = 2\sigma^4/(n. - a)$.

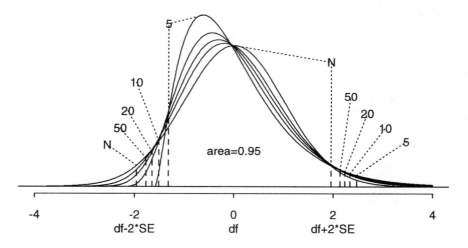

Figure 4.8. *Normal and standardized χ^2 distributions. Chi-squared distributions have been standardized to have mean 0 and variance 1. Note the skew to the right which diminishes for large degrees of freedom. This suggests potential problems using the normal approximation for inference on variance components if there are only a modest number of degrees of freedom available. Dashed lines indicate lower (2.5%) and upper (97.5%) critical values by degrees of freedom (N = normal), used for 5% two-sided test and 95% confidence intervals.*

Example 4.14 Power: A confidence interval for σ^2 can be derived using upper and lower percentiles of the χ^2 distribution as

$$\left(\hat{\sigma}^2(n.-a)/\chi^2_{\alpha/2;n.-a},\ \hat{\sigma}^2(n.-a)/\chi^2_{1-\alpha/2;n.-a}\right).$$

Notice that such intervals are asymmetric, reflecting the fact that variances must be positive (or zero). If the degrees of freedom are large, then the distribution of $\hat{\sigma}^2$ is roughly symmetric – in fact, it approaches a normal distribution. Figure 4.8 shows the normal against various chi-square distributions, all standardized to have mean 0 and variance 1. It appears that a normal approximation to the chi-square is inadequate unless there are at least 50 df or more. However, inference concerning the variance using the chi-square is still very sensitive to the assumption of normality, as discussed in Part E. ◇

In practice, interval estimates of the variance may not be a central concern. However, they can indicate the **precision**, or number of significant digits, that should be reported for estimates. The standard error of the variance is $\sigma^2\sqrt{2/(n.-a)}$. However, the standard deviation (SD = $\hat{\sigma}$) is usually reported instead, as it is in the units of observation. Using an approximation similar to that employed for variance stabilizing transformations (Section 15.1), the standard error of the SD is roughly $\sigma/\sqrt{2(n.-a)}$. In other words, σ is known within a fraction of $1/\sqrt{2(n.-a)}$. If $n.-a$ is at least 50, then the SD is known to within 10%, or one digit. In practice, it is a good idea to report one extra digit: one digit for 1–5 degrees of freedom, two digits for 6–50 degrees of freedom and three digits for 50–500 degrees of freedom for error.

4.7 Problems

4.1 Forage: The data for this problem are abstracted from a research project at the US Center for Dairy Forage (Dhiman and Satter 1995). The experiment involved comparing the effects of five different `diets` on dry matter intake (`dmi`), the amount of food eaten. The animals under study were heifers (cows with their first offspring) and mature cows during the lactation (milk-producing) cycle following birth of a calf. The five diets ranged from low (`diet=1`) to high (`diet=5`) alfalfa content. The experiment was actually fairly complicated. The proportion of alfalfa for the low group began at 45% and increased to 65% over course of the lactaction cycle, while the high began at 85% and increased to 95%. However, the order of the groups (low to high) remained the same. Interest here focuses on dry matter intake for cows. Your task is to analyze only the mature cows.

(a) Plot a histogram (or stem-and-leaf plot or box-plot) of all the `dmi` data for mature cows. Plot separate histograms by `diet`. Comment briefly on

the patterns.
(b) Find means and standard deviations (SDs) for each diet group.
(c) Estimate a common SD. Does this seem appropriate for this problem?
(d) Mark the means for the diet groups next to the common histogram of (a). Add a bar as long as the estimated common SD to indicate precision.
(e) Produce a scatter plot of group mean against individual values, identifying groups by symbols or with some annotation.

4.2 Diet: This experiment (Wattiaux, Combs and Shaver 1994) involved 60 cows initially, although some were lost during the study. The cows were randomly assigned to one of six diets and followed for a number of weeks. Diets were begun after the third week, allowing the animals some initial time to adjust to their new environment. Interest focuses on the effect of diet on the average dry matter intake (dmi), the amount of food eaten by each cow. The data you have are the average dmi over 7-14 weeks. For now, we treat this experiment as if it were a completely randomized design, ignoring the fact that cows were blocked.
(a) Plot a histogram (or stem-and-leaf diagram) of all the dmi data. Comment on the pattern.
(b) Find means and standard deviations for each diet group.
(c) Estimate a common SD. Does this seem appropriate for this problem?
(d) Along a line next to the histogram of (a), identify the means for the treatment groups. Add a bar for the estimate of common SD to indicate precision.
(e) Produce a scatter plot of group mean against individual values, identifying groups by symbols or with some annotation.

4.3 Infer: Confidence intervals are convenient graphical displays about the degree of uncertainty in an estimate of a group mean. However, caution is needed to use confidence intervals for comparisons among means. Tukey suggested that a lack of overlap of two confidence intervals implies that the means are not 'compatible'. However, compatible means, with overlapping confidence intervals, may not be significantly different. Show that with equal sample sizes a graphical approximation to a test of differences can be achieved by reducing the widths of the confidence intervals by 30% (multiplying by $1/\sqrt{2}$).

4.4 Infer: This problem demonstrates the first step for the induction Example 4.13, for weighted and centered sums of normal variates. You can do this problem by showing each result on its own or by showing the more general result (c) and demonstrating that (a) and (b) are special cases.
(a) Suppose $y_1 \sim N(0,1)$ and $y_2 \sim N(0,1)$ and let $\bar{y}. = (y_1 + y_2)/2$. Show that

$$(y_1 - \bar{y}.)^2 + (y_2 - \bar{y}.)^2 \sim \chi_1^2 .$$

PROBLEMS

This is used to show that the sample variance $\hat{\sigma} = \sum_{j=1}^{n}(y_i - \bar{y}.)^2/(n-1)$ has a chi-square distribution.

(b) Show that the properly weighted and centered sum of two sample group means has a chi-square distribution with 1 degree of freedom.

$$[n_1(\bar{y}_{1.} - \bar{y}_{..})^2 + n_2(\bar{y}_{2.} - \bar{y}_{..})^2]/\sigma^2 \sim \chi_1^2$$

with $\bar{y}_{i.} \sim N(\mu, \sigma^2/n_i)$ and $\bar{y}_{..} = (n_1\bar{y}_{1.} + n_2\bar{y}_{2.})/(n_1 + n_2)$ their weighted mean.

(c) Show that the properly weighted and centered sum of two normal variates has a chi-square distribution with 1 degree of freedom:

$$w_1(z_1 - \bar{z}_2)^2 + w_2(z_2 - \bar{z}_2)^2 \sim \chi_1^2$$

with $z_i \sim N(\mu, 1/w_i)$ and $\bar{z}_2 = (w_1 z_1 + w_2 z_2)/(w_1 + w_2)$ their weighted mean.

CHAPTER 5

Comparing Several Means

The scientist would usually like to obtain an overall assessment of the effectiveness of an experiment, followed by examination of specific questions about comparing groups. It may seem natural to compare means two at a time. However, this is impractical for experiments with many groups. Further, the risk of making claims that are not really supported by the data can increase with the number of comparisons. Pooling variance estimates across all groups increases the power for detecting differences in any particular comparison.

This chapter examines the problem of comparing several means simultaneously. Section 5.1 develops linear contrasts of means. An overal test of differences among means is developed in Section 5.2. Section 5.3 explores partitioning the total sums of squared deviation around the sample grand mean. Expected values sums of squares comprise Section 5.4, with power and sample size addressed briefly in Section 5.5.

In many experiments, the interest is more tightly focused on comparing several means rather than on statements about individual means. A pivot statistic can be employed for the hypothesis $H_0 : \mu_i = \mu_j$ by writing it as $H_0 : \mu_i - \mu_j = 0$ so that if H_0 is true

$$T = \frac{\hat{\mu}_i - \hat{\mu}_j}{\hat{\sigma}\sqrt{\frac{1}{n_i} + \frac{1}{n_j}}} \sim t_{n.-a} \; .$$

Hypothesis tests and confidence intervals arise in an analogous manner to the above. For instance, a confidence interval for the difference would be

$$\hat{\mu}_i - \hat{\mu}_j \pm t_{\alpha/2; n.-a} \; \hat{\sigma}\sqrt{\frac{1}{n_i} + \frac{1}{n_j}} \; .$$

Example 5.1 Tomato: Pairwise t tests of differences in log fruit weight for the tomato gene allele types find supporting evidence for the differences noted in the box-plots, as shown in Table 5.1. The standard error estimates use the combined variance estimate $\hat{\sigma}^2 = 0.102$. It appears that types A and H are not significantly different ($p = 0.20$) while type B has signif-

Grouping	Mean	N	TG430	SE
A	0.912	21	H	0.070
A	0.806	56	A	0.043
B	0.458	12	B	0.092

Table 5.1. *Tomato pairwise t tests*

icantly smaller fruit weight than either ($p < 0.001$). The grouping letters are commonly used to mark means that do not differ significantly at the 5% level.

Figure 5.1 shows (a) 95% confidence intervals contracted by $\sqrt{2}$ and (b) notched box-plots which can both serve as rough graphical tests for differences. That is, types A and H cannot be distinguished, while type B has significantly smaller fruit weight. The confidence intervals contraction reflects the fact that the standard error for a difference of means $\sigma\sqrt{1/n_1 + 1/n_2}$ is roughly $\sigma(\sqrt{1/n_1} + \sqrt{1/n_2})/\sqrt{2}$ provided that n_1 and n_2 are close. The notched box-plots were suggested by John Tukey (see Hoaglin, Mosteller and Tukey 1991) as an alternative. In the present case, the lower notches slightly exceed the box limits. ◇

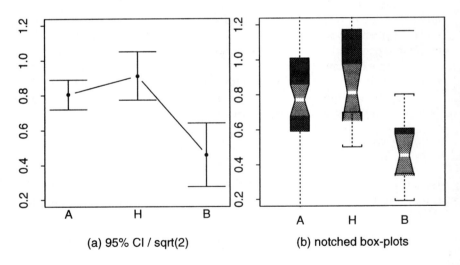

Figure 5.1. *Tomato graphical inference: (a) classical 95% confidence intervals are symmetric about the mean; (b) notched box-plots reflect features of the data more directly. Both suggest where group centers may overlap, and both should be interpreted with some skepticism.*

LINEAR CONTRASTS OF MEANS

When attention shifts to more than one comparison, several issues arise. An overall test of a compound hypothesis concerning all means is highly recommended. Subsequent questions about comparison of means can be addressed formally (if evidence supports significant differences) or on an ad-hoc basis (to suggest possible future studies). **Compound hypotheses**, involving several simultaneous comparisons, require a bit more machinery. Consider the hypothesis

$$H_0 : \mu_1 = \mu_2 = \ldots = \mu_a = \mu ,$$

with μ the common but unknown population mean. There is no longer an easy comparison. The general alternative to this hypothesis is that not all means are the same.

There are several possible alternatives, even for $a = 3$, including

$$H_1 : \mu_1 = \mu_2 \neq \mu_3 \quad H_1 : \mu_1 \neq \mu_2 = \mu_3$$
$$H_1 : \mu_1 = \mu_3 \neq \mu_2 \quad H_1 : \text{all } \mu_i \text{ are distinct.}$$

Three tests seem natural, comparing each pair of means, but there are only 2 degrees of freedom given the grand mean. Typical choices by statistical packages are:

$$H_0 : \mu_1 - \mu_3 = 0, \ \mu_2 - \mu_3 = 0 .$$

If these two hypotheses hold, then the compound hypothesis of all three being equal is true. However, knowledge concerning one may reveal something about the other. For instance, if $\hat{\mu}_3$ were large, then both $\hat{\mu}_1 - \hat{\mu}_3$ and $\hat{\mu}_2 - \hat{\mu}_3$ could be large and negative.

5.1 Linear contrasts of means

Any linear combination of means $\sum c_i \mu_i$ can be handled in a similar way. Setting $c_i = 1/a$ leads to questions about the population grand mean $\bar{\mu}. = \sum \mu_i/a$. The sample grand mean $\bar{y}.. = \sum_{ij} y_{ij}/n.$ can be examined by setting $c_i = n_i/n.$ and focusing on $E(\bar{y}..) = \tilde{\mu}. = \sum n_i \mu_i/n..$ Setting $c_2 = 1$ and all others to zero focuses on μ_2. Similarly, examining differences of two means is accomplished by setting $c_i = 1$ and $c_j = -1$, with all others being zero. Combinations with $\sum c_i = 0$ are often called **contrasts**, since they do not depend on the population grand mean $\bar{\mu}.$.

Under the compound null hypothesis $H_0 : \mu_i = \bar{\mu}.$ of all means equal, all contrasts ($\sum c_i = 0$) have expectation 0 ($\sum c_i \mu_i = \sum c_i \bar{\mu}. = 0$). Further, the population grand mean and the expectation of the sample grand mean coincide ($\bar{\mu}. = \tilde{\mu}.$). Tests of hypotheses such as $H_0 : \sum c_i \mu_i = m$ (usually $m = 0$) revolve around the pivot statistic

$$T = \frac{\sum c_i \hat{\mu}_i - m}{\hat{\sigma} \sqrt{\sum c_i^2/n_i}} \sim t_{n.-a} .$$

Parameter	Estimate	T	p-value	SE
additive	0.347	3.42	0.0009	0.101
dominance	0.280	3.25	0.0016	0.086

Table 5.2. *Tomato contrasts for genetic effects*

Parameter	Estimate	T	p-value	SE
susc vs cults	-0.120	-1.09	0.28	0.110
susc vs res	-0.922	-9.46	0.0001	0.098
res vs cults	1.042	10.30	0.0001	0.101
prog vs parents	0.218	3.04	0.0037	0.072

Table 5.3. *Cloning contrasts for disease resistance*

Note that if the $\hat{\mu}_i$ are uncorrelated, $V(\sum c_i \hat{\mu}_i) = \sum c_i^2 V(\hat{\mu}_i) = \sigma^2 \sum c_i^2/n_i$. Confidence intervals of level $(1-\alpha)$ follow in an analogous fashion:

$$\sum c_i \hat{\mu}_i \pm t_{\alpha/2; n.-a} \hat{\sigma} \sqrt{\sum c_i^2/n_i} \ .$$

Example 5.2 Tomato: The tomato breeders wanted explicitly to address the contrasts referring to additive and dominance effects. That is, (1) is there a genetic effect, i.e. difference between the parent types ($\mu_A = \mu_B$)? and (2) does the gene show evidence of dominance, or is the hybrid halfway between the parents ($\mu_H = (\mu_A + \mu_B)/2$)? The estimates for the corresponding contrasts $C_{\text{ADD}} = \mu_A - \mu_B$ and $C_{\text{DOM}} = \mu_H - (\mu_A + \mu_B)/2$ are shown in Table 5.2. ◇

Example 5.3 Cloning: Novy (1992) was interested in several contrasts among the clones. In particular, he had identified certain clones from progeny as being susceptible (**susc**) and others as resistant (**res**) to a particular plant disease. Further, some of the clones were from commercial 'cultivars' (**cult**). The clones from the parent lines (**parents**) were distinguished from their progeny (**prog**). Table 5.3 shows four of the contrasts of interest. All but the first were significant at the 0.005 level. ◇

A wide variety of linear contrasts of means is possible. However, there are only $a-1$ degrees of freedom among the means (removing one for the grand mean). Some would argue for only as many comparisons as degrees of freedom. Ideally the questions in an experiment might suggest $a-1$ contrasts which are orthogonal to one another to get full value from such an approach. This can be done for a balanced experiment, by considering

LINEAR CONTRASTS OF MEANS

orthogonal contrasts. That is, any two contrasts $\sum_i c_i \mu_i$ and $\sum_i d_i \mu_i$ are orthogonal if $\sum c_i \hat{\mu}_i$ and $\sum d_i \hat{\mu}_i$ are uncorrelated. If the responses y_{ij} are normally distributed, then the contrast estimators are independent.

For equal sample sizes ($n_i = n$) orthogonality implies that $\sum_i c_i d_i = 0$. Consider three means and the contrasts $\mu_1 - \mu_2 = 0$ and $\mu_1 + \mu_2 - 2\mu_3 = 0$. Any contrast among these three means can be generated from these two. For instance, $\mu_3 - \mu_1 = -(\sum c_i \mu_i + \sum d_i \mu_i)/2$.

Orthogonal contrasts have the advantage that evidence from one contrast provides no information about other contrasts. This can be very useful in an experiment amenable to predetermined contrasts that are both orthogonal and sensible in terms of the scientific questions. Sometimes, orthogonal contrasts are not possible or sensible for the scientist's questions. In this case, there are more generic methods of comparison, considered in the next chapter. For instance, contrasts of μ_1 with each of μ_2 and μ_3 are not orthogonal but may be scientifically meaningful.

Linear, quadratic and higher-order trends among group means arise naturally when the factor levels are ordered. The strength of linear trend among four means, assuming equally spaced factor levels, can be investigated using the contrast $c_1 = -3, c_2 = -1, c_3 = 1, c_4 = 3$. Quadratic, cubic and higher trends are usually examined using **orthogonal polynomial** contrasts. In practice it is difficult to interpret trends higher than quadratic or cubic.

Example 5.4 Infer: Consider an experiment in which the `factor` has five equally spaced levels, such as 0,10,20,30,40 units of `nitrogen` fertilizer. Here are two ways to use SAS to examine orthogonal linear and quadratic polynomial contrasts. If the data are balanced and factor levels are few and equally spaced, orthogonal polynomials can be readily derived. Here is an SAS example with five equally spaced levels:

```
proc anova;   /* explicit orthogonal contrasts */
   class nitrogen;
   model resp = nitrogen;
   contrast 'linear'     nitrogen -2 -1  0  1  2;
   contrast 'quadratic'  nitrogen  2 -1 -2 -1  2;
   contrast 'cubic'      nitrogen -1  2  0 -2  1;
   contrast 'quartic'    nitrogen  1 -4  6 -4  1;
```

These contrasts are not orthogonal if the data are unbalanced.   ◊

These polynomial contrasts can be constructed for factors with equally or unequally spaced ordered levels using standard methods (see Draper and Smith 1981, sec. 5.5; Seber 1977, sec. 8.2; Snedecor and Cochran 1989, sec. 19.6). However, if the design is unbalanced, such contrasts are not orthogonal even for equally spaced levels. Since the main issue is testing rather than estimation, it is possible to use the Gram–Schmidt approach from regression. That is, fit a linear trend, followed by a quadratic trend orthogonal to that, followed by higher-order polynomials if desired. This approach is an example of the Type I approach explained in general terms in Chapter 10, Part D. The following example shows how to accomplish this with very little work. No contrasts need be explicitly derived!

Example 5.5 Infer: Reconsider the nitrogen fertilizer experiment. This time, construct 'regressors' from the `nitrogen` levels. In a statistical package, this may involve defining new variables and not making them `class` variables, as in the SAS example below:

```
data b; set a; /* regressors for first two contrasts */
   linear = nitrogen;
   quad   = nitrogen * nitrogen;
proc glm;
   class nitrogen;
   model resp = linear quad nitrogen / ss1;
```

Higher-order effects, in this case cubic and quartic, are lumped together as two degrees of freedom for the `nitrogen` factor. This approach leads to orthogonal contrasts whether or not the experiment is balanced. However, the sums of squares for the contrasts developed in Example 5.4 will not agree with these unless sample sizes are equal. ◇

It may be desirable to select **orthogonal contrasts**, such as

$$H_0 : \mu_1 - \mu_2 = 0 \, , \, \mu_3 - (\mu_1 + \mu_2)/2 = 0 \, ,$$

which completely specify the compound hypothesis. For instance, if 1 and 2 are new but similarly formulated treatments and 3 is an established treatment, the above orthogonal contrasts would examine whether the new treatments differ from each other, and whether the average of the new treatments differs from old. It is always possible to select $a - 1$ orthogonal contrasts but they may not be meaningful. This issue arises in important ways when considering factorial models in unbalanced designs in Part C.

Orthogonal contrasts are mathematically appealing but may not agree with the scientist's aims. If the means are not all the same, what questions are particularly interesting to the scientist? While some questions may be formulated ahead of data analysis, others may be more general, such as which group is best or how groups can be distinguished. Thus it can be important to maintain some flexibility about how to investigate comparisons among means. This is the subject of the next chapter.

Example 5.6 Cloning: Contrasts in Example 5.3 are not orthogonal. With no missing data, the contrast `prog vs parents` would be orthogonal to `susc vs res`. The other five pairs of contrasts are not orthogonal regardless of missing values. Table 5.4 shows the contrast coefficients. ◇

contrast	a	b	c	d	e	f	g	h	i	j	k	l	m
susc vs cults	1	0	-1	0	0	0	-1	0	0	0	0	1	0
susc vs res	-3	0	0	2	0	0	0	2	0	0	0	-3	2
res vs cults	0	0	3	-2	0	0	3	-2	0	0	0	0	-2
prog vs parents	4	0	0	4	-5	-5	0	4	-5	0	-5	4	4

Table 5.4. *Cloning contrast coefficients*

5.2 An overall test of difference

Overall tests of difference among group means require a bit more machinery. Since there are several possible 'directions' in which group means may differ, as summarized for instance in orthogonal contrasts, it makes sense to think about the overall magnitude of differences among groups. A natural way to measure that when comparing a means is to consider squared deviations from an overall mean across groups. This is equivalent to examining Euclidean distance in the $(a-1)$-dimensional space of contrasts among means.

An overall test for the hypothesis of differences among a group means relies on the quadratic form

$$\sum_{i=1}^{a} n_i[(\bar{y}_{i\cdot} - \bar{y}_{\cdot\cdot}) - (\mu_i - \tilde{\mu}_{\cdot})]^2 \sim \sigma^2 \chi^2_{a-1} ,$$

with expected sample grand mean $\tilde{\mu}_{\cdot} = \sum_i n_i y_{ij}/n_{\cdot}$, is a pivot statistic which is statistically independent of $\hat{\sigma}^2$. Under the null hypothesis H_0 that all means are equal ($\mu_i = \tilde{\mu}_{\cdot} = \mu$) the mean square for the model is

$$MSH_0 = \sum_{i=1}^{a} n_i(\bar{y}_{i\cdot} - \bar{y}_{\cdot\cdot})^2/(a-1) \sim \sigma^2 \chi^2_{a-1}/(a-1).$$

This is approximate if the data are not normally distributed but quite good in practice if all n_i are 'large'. Normality further implies that MSH_0 and $\hat{\sigma}^2$ are independent, leading to the test statistic

$$F = MSH_0/\hat{\sigma}^2 = \left[\sum_{i=1}^{a} n_i(\bar{y}_{i\cdot} - \bar{y}_{\cdot\cdot})^2/(a-1)\right]/\hat{\sigma}^2 ,$$

with

$$F \sim F_{a-1, n_{\cdot}-a} \text{ under } H_0 : \mu_i = \mu .$$

Again, the distribution result is approximate for non-normal data but is increasingly reliable as sample sizes increase.

Typically, rejection of the null hypothesis of all means equal requires sufficient evidence of differences. That is, F has to be 'large enough'. The observed F value can be compared to tabled values of the upper tail of $F_{a-1,n.-a}$ (either using readily available tables printed in many texts or using built-in utilities in statistical packages). Rejection of H_0 suggests multiple comparisons among the means, which is the subject of the next chapter.

5.3 Partitioning sums of squares

The basic idea for developing F tests of model parameters is as follows. Use the estimator of error variance in the denominator and find an appropriate numerator sum of squares (SS) involving estimates for parameters in the hypothesis of interest. Under the hypothesis H_0, the statistic has an F distribution; under an alternative H_1, the statistic has a 'more extreme' distribution. Thus large values of F raise suspicion that the hypothesis H_0 could be false. However, in practice the 'truth' is never known. In fact, violation of model assumptions can lead to large (or very small) F values whether or not the groups actually differ! In practice, the full model is assumed correct for formal testing. Assumptions can be checked as best as possible by design (before) and diagnostic examination of residuals (after) and experiment.

The partition of the centered response into a group effect plus error,

$$y_{ij} - \tilde{\mu}. = (\mu_i - \tilde{\mu}.) + e_{ij},$$

suggests a natural partition based on sample estimates of group means and residuals as

$$(y_{ij} - \bar{y}..) = (\bar{y}_{i.} - \bar{y}..) + (y_{ij} - \bar{y}_{i.}).$$

Squaring both sides and summing over all $n.$ observations results in a partition of the total variation as

$$\sum_{i=1}^{a} \sum_{j=1}^{n_i} (y_{ij} - \bar{y}..)^2 = \sum_{i=1}^{a} n_i (\bar{y}_{i.} - \bar{y}..)^2 + \sum_{i=1}^{a} \sum_{j=1}^{n_i} (y_{ij} - \bar{y}_{i.})^2,$$

with the sum of cross-product terms cancelling out,

$$\sum_{ij} (\bar{y}_{i.} - \bar{y}..)(y_{ij} - \bar{y}_{i.}) = 0.$$

Assigning names to each sum of squares (SS) helps interpretation:

$$SS_{\text{TOTAL}} = SS_{\text{MODEL}} + SS_{\text{ERROR}},$$

PARTITIONING SUMS OF SQUARES

source	df	SS	MS = SS/df	F
model	$a-1$	SS_{MODEL}	MS_{MODEL}	$MS_{\text{MODEL}}/\hat{\sigma}^2$
error	$n.-a$	SS_{ERROR}	$MS_{\text{ERROR}} = \hat{\sigma}^2$	–
total	$n.-1$	SS_{TOTAL}	–	–

Table 5.5. *One-factor anova*

with

$$SS_{\text{MODEL}} = \sum_{i=1}^{a} n_i (\bar{y}_{i\cdot} - \bar{y}_{\cdot\cdot})^2 ,$$
$$SS_{\text{ERROR}} = \sum_{i=1}^{a} \sum_{j=1}^{n_i} (y_{ij} - \bar{y}_{i\cdot})^2 ,$$
$$SS_{\text{TOTAL}} = \sum_{i=1}^{a} \sum_{j=1}^{n_i} (y_{ij} - \bar{y}_{\cdot\cdot})^2 .$$

This partitions the total variation into that between groups, explained by the model or treatments, and that within groups, the error or residual which is unexplained.

Under the usual assumptions, SS_{MODEL} and SS_{ERROR} are statistically independent. The last section showed that these quantities are (asymptotically) distributed as χ^2 and that the ratio of mean squares ($MS = SS/df$) has an $F_{a-1, n.-a}$ distribution under the hypothesis H_0 of no group differences:

$$F = \frac{SS_{\text{MODEL}}/(a-1)}{SS_{\text{ERROR}}/(n.-a)} .$$

Organizing this information into an **anova table** as in Table 5.5 provides a convenient summary of the sources of variation, degrees of freedom and test statistics. Selected p-values (e.g. 0.01, 0.05, 0.10) can be found in published table of $F_{a-1, n.-a}$ (see Rohlf and Sokal 1995) or by using a statistical package. The expected value of the F statistic under the null hypothesis is close to one (see Searle, Casella, McCulloch 1992, app. S.4),

$$E(F) = df_{\text{ERROR}}/(df_{\text{ERROR}} - 2) .$$

The 5% critical value is roughly 4 unless the degrees of freedom for error (df_{ERROR}) are very small. Thus a quick assessment without handy tables would be that F values near or below 1 are not significant, while values above 4 probably are.

Example 5.7 Tomato: A formal test for the tomato fruit weight gene, presented in Table 5.6, shows significant differences among the three groups. If the sample sizes per allele type were identical, then the contrasts considered earlier would be orthogonal. Since they are not, the sums of squares

source	df	SS	MS	F	p-value
tg430	2	1.643	0.821	8.08	0.0006
error	86	8.743	0.102		
total	88	10.386			

Table 5.6. *Tomato analysis of variance*

contrast	df	SS	F	p-value
additive	1	1.193	11.73	0.0009
dominance	1	1.074	10.57	0.0016
tg430	2	1.643	8.08	0.0006

Table 5.7. *Tomato contrast sums of squares*

explained by these contrasts (1.193 + 1.074) do not add up to the explained sum of squares (1.643) shown in Table 5.7. ◇

Example 5.8 Cloning: The analysis of variance in Table 5.8 shows that the overall test is highly significant ($p < 0.0001$). Measurements were first rescaled by 1000. The number of digits for most entries has been substantially reduced to reflect the precision of values more closely. ◇

Another way to think about this process is to view the original model as the **full model** and the model subject to the null hypothesis H_0 as the **reduced model**. The test of the null hypothesis compares the error variation in the full model to the difference in unexplained variation between the reduced model ($SSE_{\text{REDUCED}} = SS_{\text{TOTAL}}$ and $dfE_{\text{REDUCED}} = n. - 1$ for the cell means model) and full model ($SSE_{\text{FULL}} = SS_{\text{ERROR}}$ and $dfE_{\text{FULL}} = n. - a$ here),

$$F = \frac{(SSE_{\text{REDUCED}} - SSE_{\text{FULL}})/(dfE_{\text{REDUCED}} - dfE_{\text{FULL}})}{SSE_{\text{FULL}}/dfE_{\text{FULL}}}.$$

source	df	SS	MS	F	p-value
clone	12	13.84	1.1537	20.22	0.0001
error	51	2.91	0.0570		

Table 5.8. *Cloning analysis of variance*

This so-called **general F test** is the same test as described above but written in a form which has general utility. Later, this idea of comparing reduced and full models, relative to unexplained variation in the full model, is used repeatedly. It is the principal inferential tool for analysis of variance.

5.4 Expected mean squares

Nearly all the approaches to statistical inference discussed in this book concern sums of squares. That is, formal investigation of many experiments involves characterizing sources of variation in terms of mean squares for factors and factor combinations. The distributions of these mean squares are usually associated with the family of χ^2 distributions. Under simplifying null hypotheses H_0, many (but not all!) mean squares considered later are distributed proportionally to central χ^2. However, under alternative hypotheses H_1, the distribution may be non-central χ^2.

The central χ^2 with r degrees of freedom is defined as the sum of r independent squared standard normal variates. In contrast, the non-central χ^2 is distributed as the sum of r independent normal variates with variance 1 and mean $\delta/\sqrt{r}$. The first two moments yield

$$E(\chi^2_{r;\delta^2}) = r + \delta^2 \text{ and } V(\chi^2_{r;\delta^2}) = 2r + 4\delta^2$$

which reduce to r and $2r$ for the central case ($\delta^2 = 0$). The distribution of $\chi^2_{r;\delta^2}$ is completely specified by the degrees of freedom r and the non-centrality parameter δ^2. Note that some authors use a 'prime' to distinguish the non-central variate, $\chi'^2_{r;\delta^2}$. Further, the literature is split between those who define the non-centrality parameter as the above value and those who use $\delta^2/2$; some authors refer to δ as the non-centrality parameter.

Consider the cell means model. Under an alternative hypothesis H_1, the distribution of the quadratic form is proportional to a non-central χ^2,

$$MSH_0 = \sum_{i=1}^{a} n_i(\bar{y}_{i.} - \bar{y}_{..})^2/(a-1) \sim \sigma^2 \chi^2_{a-1;\delta^2}/(a-1) \ ,$$

with the non-centrality parameter δ^2 determined by replacing the sample means by their expectations and dividing by the variance,

$$\delta^2 = \sum_{i=1}^{a} n_i(\mu_i - \tilde{\mu}_{.})^2/\sigma^2 \ .$$

The non-centrality parameter measures the spread among the cell means relative to the variance. The expected value of the model mean square is

$$E(MSH_0) = \sigma^2(1 + \delta^2/(a-1)) = \sigma^2 + \sum n_i(\mu_i - \tilde{\mu}_{.})^2/(a-1) \ ,$$

which reduces to σ^2 under the null hypothesis of all means equal ($\delta^2 = 0$).

As experiments get more complicated, it is more difficult to determine appropriate pivot statistics. Sometimes it is possible to ascertain hypotheses by examining expected mean squares. The non-centrality parameters summarize deviation from the null hypothesis, which can be used to examine the relative power of tests against classes of alternatives, as mentioned in the next section.

Later chapters consider some inferential situations where the sums of squares are not proportional to either central or non-central χ^2 variates. This can happen even under the null hypothesis for some questions involving unbalanced designs, which makes it very difficult to develop tests. Various approximations and more sophisticated approaches are considered at those points.

5.5 Power and sample size

A well-designed experiment should have sufficient resolving **power** to detect effects or differences which are important to the scientist. Even if the main interest is in showing no differences, there must be a decision about the smallest magnitude of effect which would lead to doubts concerning that belief. Greater power is achieved by increasing sample size, reducing variance or considering effects of larger magnitude. Usually there is a tradeoff between more power and higher cost, with little exact knowledge available before an experiment.

In practice, power is seldom computed to any precision. Rather, statistician and scientist try to agree in general terms about varied alternatives. This can be expressed in terms of differences or effects relative to the standard deviation. If computations are done, they usually concern power for comparing two treatment groups. This makes sense, since the primary result from a comparison of many groups often comes down to comparing two. An experiment which cannot tell the difference between them is a wasted effort.

A test should be powerful enough to reject H_0 in favor of an alternative. In practice, significant results about a new variety may only be interesting if it substantially increases yield (say by 2-3%). On the other hand, it is important to know ahead of time whether the experiment is large enough to detect that big a difference. This power can be investigated by considering the conditional probability of rejecting H_0 if a specified alternative H_1 is true. (Recall that the significance level α is the conditional probability of rejecting H_0 if the null H_0 is actually true.)

Under H_0, the F statistic is distributed as $F_{a-1, n.-a}$ while under an alternative H_1, $F \sim F_{a-1, n.-a; \delta^2}$ with δ^2 depending on the specific alternative. The significance level is

$$\alpha = \text{Prob}\{\text{reject } H_0 | H_0 \text{ true}\} = \text{Prob}\{F_{a-1,n.-a} > F_{\alpha; a-1, n.-a}\} \,.$$

POWER AND SAMPLE SIZE

Under the alternative H_1,

$$\text{power} = \text{Prob}\{\text{reject } H_0 | H_1 \text{ true}\} = \text{Prob}\{F_{a-1,n.-a;\delta^2} > F_{\alpha;a-1,n.-a}\} \ .$$

The power increases as the alternative is moved farther from the null hypothesis, as measured by δ^2. However, as seen in the previous section, the non-centrality parameter is a complicated term which may involve many group means. With several groups, there is seldom an obvious way to determine reasonable values of δ^2. Thus even though there are tables and figures to examine power for F tests, they seldom are of much practical use.

However, there may be particular interest in comparing certain groups, or in knowing if any pair of treatments are at least so far apart. This is more easily addressed by reverting to pairwise comparisons of interest. Thus, it is often enough to simplify power and sample size considerations to the 'worst case scenario', in which all means coincide except for two (say $\mu_1 \neq \mu_2$ and $\mu_k = \mu_3$ for $k > 2$). The non-centrality parameter can be reduced to

$$\delta^2 = [n_1 n_2 (\mu_1 - \mu_2)^2 / (n_1 + n_2) + n.(\mu_3 - \tilde{\mu}.)^2]/\sigma^2 \ .$$

If $\mu_3 = (n_1 \mu_1 + n_2 \mu_2)/(n_1 + n_2)$, and hence $\mu_3 = \tilde{\mu}.$, the other groups drop out. A little algebra reduces the non-centrality parameter to

$$\delta^2 = \left[\frac{\mu_1 - \mu_2}{\sigma \sqrt{\frac{1}{n_1} + \frac{1}{n_2}}} \right]^2 ,$$

which is just the square of the non-centrality parameter for the t test. Further, it may be appropriate to consider only the equal sample size case $n_1 = n_2 = n$, as this yields the greatest power (smallest δ^2).

Suppose a scientist wishes to have power to detect an absolute difference of $|\mu_1 - \mu_2| > d$, or a relative difference of d/σ using equal sample sizes. The test of size α for the null hypothesis $H_0 : \mu_1 = \mu_2$ against the two-sided alternative $H_1 : \mu_1 \neq \mu_2$ rejects H_0 if

$$|\hat{\mu}_1 - \hat{\mu}_2| > c = z_{\alpha/2} \sigma \sqrt{2/n} \ ,$$

with $z_{\alpha/2}$ the upper $100\alpha/2$ percentile of the standard normal. That is, $|\hat{\mu}_1 - \hat{\mu}_2|\sqrt{n/2}/\sigma$ has a standard normal distribution under H_0. Ideally, the critical difference (c) is much less than the desired difference for detection (d). In practice, this is accomplished by setting the sample size large enough so that c is small. Put another way, for a given c, the required samples size would be

$$n = 2(z_{\alpha/2}\sigma/c)^2 \ .$$

Example 5.9 Power: The power depends on how far the critical value c is from a desired detectable difference d. Formally, under the alternative

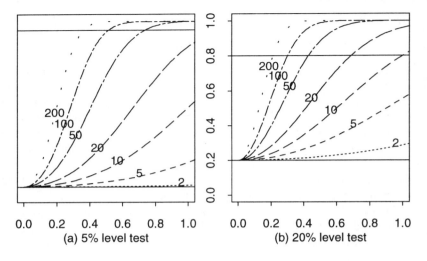

Figure 5.2. *Power curves for comparing two groups. Power increases as the detectable difference increases or the degrees of freedom increase for a given detectable difference. Power is always lower if the size of the test is smaller, as in (a) with a 5% chance of Type I error as opposed to 20% in (b).*

$H_1 : \mu_2 = \mu_1 + d$, the probability of rejecting H_0 is the power $1 - \beta$,

$$\begin{aligned}\text{power} &= \text{Prob}\{|\hat{\mu}_1 - \hat{\mu}_2| > c : H_1\} \\ &= 1 - \text{Prob}\{-(c+d)\sqrt{n/2}/\sigma < Z < (c-d)\sqrt{n/2}/\sigma\} \\ &= 1 - \beta \approx \text{Prob}\{Z > -z_\beta\} \ .\end{aligned}$$

with $z_\beta = z_{\alpha/2} - d\sqrt{n/2}/\sigma$. The approximation involves ignoring the small chance that the observed difference $(\hat{\mu}_2 - \hat{\mu}_1)$ is large and negative. Thus, taking power into consideration, the required sample size is

$$n = 2[(z_{\alpha/2} + z_\beta)\sigma/d]^2 \ .$$

The appropriate critical value c is typically found after deciding on size α, power $(1-\beta)$ and group sample size n. Power curves for $\alpha = 0.05$ and 0.20 and $n = 2, 5, 10, 20, 50, 100, 200$, are shown in Figure 5.2 for $\sigma^2 = 1$ and a range of detectable differences d.

If a scientist decides to set the critical value c at the desired detectable difference d, the test would have only a 50% power to detect the relative difference d/σ. That is, $z_\beta \approx 0$ if $c = d$. However, if the detectable difference is set at roughly twice the critical value $(d = 2c)$, the power is about

$1 - \alpha/2$, roughly symmetric in the chance of rejecting H_0 and H_1. That is,

$$\begin{aligned} \text{power at symmetry} &= \text{Prob}\{|\hat{\mu}_1 - \hat{\mu}_2| > d/2 : \mu_2 = \mu_1 + d\} \\ &= 1 - \text{Prob}\{-3z_{\alpha/2} < Z < -z_{\alpha/2}\} \\ &\approx 1 - \alpha/2 \ . \end{aligned}$$

Any larger real difference would probably be detected. Figure 5.3(a) shows the power against the detectable difference d in multiples of the critical value c for the same range of degrees of freedom used in Figure 5.2. Notice that the curves with few error degrees of freedom have low initial power, but all are about 50% when $c = d$ as indicated above. The distributions of the t statistic under various hypotheses are shown in Figure 5.3(b). ◇

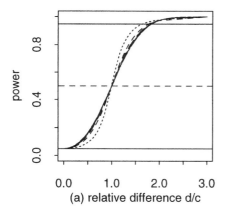 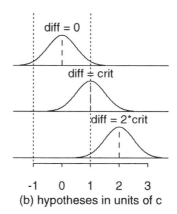

Figure 5.3. *Power for difference relative to critical value. (a) The relative power hardly depends on digrees of freedom, beyond some initial steepness for small degrees of freedom (dotted lines). When* `crit = diff` *the power is roughly 50%, and when* `2*crit = diff` *it is over 95%. (b) shows the shift in distribution for these two situations.*

Slight imbalance in sample sizes among groups has a modest effect on power calculations. Adjustments can be made for multiple comparisons, as presented in the next chapter, by suitable inflation of the critical value $z_{\alpha/2}$, leading to larger sample sizes in most situations. This provides a quick tool which could then be interpreted in light of other important factors, such as time and resources, when designing an experiment.

5.6 Problems

5.1 Infer: (a) Verify that $V(\sum_{i=1}^{t} c_i \hat{\mu}_i) = \sigma^2 \sum_{i=1}^{t} c_i^2/n_i$ under the usual conditions for the one-factor completely randomized design.
(b) For $t = 4$ verify that the following three sets of coefficients c_i for linear comparisons are orthogonal. In other words, show that these lead to orthogonal contrasts, which would be uncorrelated for equal sample sizes.

i	1	2	3	4
linear	-3	-1	1	3
quadratic	1	-1	-1	1
cubic	-1	3	-3	1

(c) Why are these comparisons labelled as polynomials? Justify your answer.

5.2 Infer: Generate 50 trial simulations from a one-factor layout in a completely randomized design with $n_1 = n_2 = n_3 = 4$, $\sigma = 1$ and $\mu_i = i, i = 1, 2, 3$. For each simulation record the SS_{MODEL} and SS_{ERROR}.
(a) Plot histograms (or stem-and-leaf diagrams) of SS_{MODEL} and SS_{ERROR}. For each, compare the 80%, 90% and 98% points with the critical points of the appropriate distribution.
(b) Plot SS_{MODEL} against SS_{ERROR} (possibly using some transformation). Examine whether or not they appear to be statistically independent.
(c) Plot a histogram of the F statistic for equal means. Compare with 80%, 90% and 98% points for the null (H_0) distribution and comment on your findings.

5.3 Tree: A student in plant pathology examined the growth of a soil pathogen in several different soils. He was interested in comparing the effect of tree root systems on the pathogen. Three of the soils (vermix, promix and grnhous) are in a certain sense controls, the first two being artificial and the third coming from a controlled environment. A fourth soil contains only grass. The other four are of particular interest, as the scientist wants to know whether soil from tree nm-6 has lower score than those of the other three trees.

For each type of soil, there were six soil samples (rep). These samples were inoculated with pathogen and then split in half. One half was autoclaved (that is, sterilized) and the other half was not. Here we only consider the latter, non-autoclaved half. Each of these six samples (per soil) was placed in storage and scored for pathogen (by some objective criterion not of concern here) on six different dates. The data in hand are the averages of those scores. [Other measurements were taken periodically, but are to be ignored here.]

Your task is to assess whether nm-6 has a significantly lower score averaged over the six dates than the other tree soils for the non-autoclaved

samples. In the process, it is important to characterize the difference between the four tree soils and the four controls.
(a) State assumptions and key questions clearly and concisely.
(b) Compare all eight soils for pathogen score. Conduct an overall test.
(c) Plot the data. Include the means and SDs or SEs. State clearly how these are derived so that someone else could reproduce your work.
(d) Select a particular type of multiple comparison. You must state why you chose your approach: what are its strengths and weaknesses (be brief!)? Follow the overall test by your chosen method of multiple comparisons and interpret the results.
(e) Contrast the four controls with the soils of interest. Follow this with a contrast of nm-6 with the other four soils.

5.4 Tree: The scientist wants another analysis with only the four tree soils.
(a) Reanalyze with only the four tree soils of interest.
(b) Argue why this might be appropriate. Alternatively, why do you think this is inappropriate? Consider Problem 5.3(a)–(c) in your argument.

5.5 Cloning: Consider the contrasts reported in Example 5.3 and made explicit in Example 5.6. Conduct formal analysis to verify these results.

5.6 Forage: Conduct a formal analysis for the mature cows.
(a) Clearly state assumptions, model and hypotheses.
(b) Indicate the form of the test statistic and its distribution under the null and alternative hypotheses.
(c) Plot predicted against residuals. Add horizontal lines for zero residual as well as 1 SD on either side for convenient viewing.
(d) Interpret results in light of graphical and/or tabular summaries as in Problem 4.1.

CHAPTER 6

Multiple Comparisons of Means

Consider an experiment comparing ten treatments and imagine that an F test rejects the hypothesis that all ten treatment means are the same. Is it 'valid' simply to compare any pair of means? What are the cautions and pitfalls?

There are many ways to answer these questions and ultimately the answers rest on compromises. There is no universally 'best' way to proceed to compare treatment means. However, some ways are preferable, depending on the goal. For instance, did the scientist decide only after finding significant differences that it was important to compare the best and worst treatment, or was this comparison an integral part of the experimental plan?

A key feature of compromise involves the tradeoff between experiment-wise and comparison-wise error rates (Section 6.1). The classical approaches are based on the standard t and F statistics (Section 6.2), while most others begin by considering the range of treatment means (Section 6.3). The last section examines the relationship among these methods of multiple comparisons.

Several other approaches have been developed but are not discussed in depth here. Miller (1981) presents an excellent detailed account of all the methods in this chapter (with citations to the original work) and is the standard reference for multiple comparisons. SAS (1992) presents many methods of multiple comparison in an accessible manner.

6.1 Experiment- and comparison-wise error rates

In some situations, it may be desirable to control the overall **experiment-wise error rate**, the probability of rejecting the hypothesis

$$H_0 : \mu_1 = \cdots = \mu_a$$

when it is in fact true. That is, if there really are no differences among the group means, it would be nice to have only a small chance of detecting *any* differences at all. Typically, it is inadvisable to compare group means in detail unless H_0 has been rejected, say with an F test.

Difficulties emerge, however, when comparing many groups. Suppose that two groups are different from one another but neither is that different from the other eight on test. An experiment conducted with just those two groups might uncover evidence for differences. But a test for all ten group means may not have sufficient power to detect this. Here, it is important to make the distinction between comparisons that are planned *before* conducting the experiment and those that appear interesting *after* examining the data.

Alternatively, the hypothesis that all means are the same may not be all that interesting. Pilot studies or related experiments may have already shown some evidence or indication of differences. The current experiment may have been designed precisely to sort those differences out. A formal test of H_0 may be necessary to satisfy critics but the real interest focuses on individual comparisons. Here, interest probably focuses on controlling the **comparison-wise error rate**, the probability of Type I error (concluding differences exist when they do not) for each comparison.

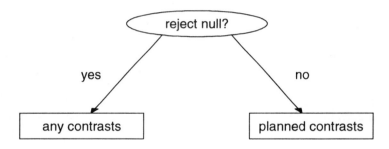

Figure 6.1. *Multiple comparisons general procedure: perform overall test followed by planned or unplanned contrasts.*

Example 6.1 Infer: Still another situation may best describe the scientist's dilemma. Suppose that the current experiment is a pilot study, or a small study which will later be followed by more in-depth studies with selected treatments. The scientist may want to find promising differences among groups for later study while being cautious about reading too much into the data. An overall test of the null hypothesis H_0 can control the experiment-wise error rate, protecting the scientist against unwarranted data mining. A **general procedure** which many follow in practice is to conduct any planned or unplanned multiple comparisons if H_0 is rejected, say at a 5% level. However, even if the null hypothesis is not rejected, certain pre-planned comparison of means may provide useful evidence to guide future studies. Thus it may be prudent to proceed with some individual comparisons whether or not an overall test is definitive, as illustrated in Figure 6.1. Significance for the overall test justifies further individual comparisons. Comparisons following a non-significant overall test should be interpreted with appropriate caution. ◇

Depending on a scientist's preference with respect to 'conservative' or 'liberal' comparisons, investigations may find few or many significant differences. Conservative methods tend to rely on controlling the experiment-wise error rate at the expense of power, while liberal methods focus on comparison-wise error rates. Some tests do an admirable job of balancing these two extremes but have the disadvantage of being more complicated and consequently harder to convey to colleagues and editors.

In practice, scientists often use Fisher's least significant difference (LSD) for planned contrasts, protected by an overall F test. Those interested in 'post-hoc' comparisons not supported by results of the F test often turn to the BONFERRONI method, or simply use the unprotected LSD with appropriate cautionary verbiage in reporting.

Most methods of mean comparison have been adapted to unbalanced designs by using the **harmonic mean**

$$\tilde{n} = a / \left[\sum 1/n_i \right] .$$

The statistical package SAS has an option `lines` for means comparisons which forces use of $\tilde{n}$. This works fairly well if sample sizes are nearly identical. However, it is possible to get very odd results if designs are very unbalanced. If there are only two experimental units in a group, can comparisons involving their mean be very reliable? This problem should be kept in mind whatever method is used for comparison of means.

While many problems may have particular comparisons in mind, quite a few experiments suggest focusing part of the analysis on comparing and ordering groups by their means. It may be somewhat dissatisfying to have only one critical value with which to compare all means. Surely comparisons among subsets of means would be better. A variety of **multiple-stage tests** have been developed with this in mind. The multiple-stage test idea is first to compare all a means and then to examine all subsets of size $a - 1$ only if a significant difference is found. Comparisons continue with progressively smaller subsets, down to sets of two means, until no more significant differences are found.

6.2 Comparisons based on F tests

Three different methods for comparison of means are adapted directly from the t or F statistic. Fisher's protected LSD is the most liberal, corresponding to an array of t tests. SCHEFFÉ's method is analogous to the F test, yielding significant comparisons only if the F test would reject H_0. However, despite its theoretical appeal, it is often the most conservative approach in practice. BONFERRONI's method is also conservative, corresponding to the LSD with reduced error rate. These methods are quite flexible, allowing for unequal sample sizes between groups. Further, arbitrary contrasts involving group means may be examined.

Fisher's least significant difference (protected LSD) involves first conducting an overall F test of H_0 and then using t tests with a common estimator of variance $\hat{\sigma}^2$ for comparisons of interest, provided the overall test was significant. The F test controls the experiment-wise error rate while the individual t tests control comparison-wise error rates. For pairwise comparisons, group means are considered different if

$$|\hat{\mu}_i - \hat{\mu}_j| > t_{\alpha/2; n.-a}\hat{\sigma}\sqrt{\frac{1}{n_i} + \frac{1}{n_j}}$$

with $t_{\alpha/2; n.-a}$ the critical value from the $t_{n.-a}$ distribution. LSD and other means comparisons are often summarized by ordering groups by means and marking non-significant differences with common letters as in Table 6.2.

If the design is balanced, with $n_i = n_j = n$, then the comparison of means is reduced to

$$|\hat{\mu}_i - \hat{\mu}_j| > \text{LSD with LSD} = t_{\alpha/2; n.-a}\hat{\sigma}\sqrt{2/n} \ .$$

A single value for LSD allows simple reporting of means and their comparison in a concise table. Approximate LSDs for slightly unbalanced designs can be quickly computed by replacing n by the harmonic mean $\tilde{n}$. However, it is highly recommended to double-check comparisons for unbalanced designs by using the actual samples sizes.

Example 6.2 Infer: The SAS general linear models procedure `glm` has two ways to perform LSD comparisons. The `means` phrase can either perform the quick calculation based on the harmonic mean of the sample sizes, or give all pairwise differences. The `lsmeans` phrase reports p-values for all pairs in a concise table. In practice, the `means` phrase is nice because it orders the groups by mean values.

```
proc glm;
   class group;
   model resp = group;
   /* LSD using harmonic mean sample size */
   means group / lsd lines;
   /* SEs and p-values for differences */
   lsmeans group / stderr pdiff;
```

The `lsmeans` gives more precise information on paired comparisons based on actual sample sizes, possibly alterating significance and/or order. ◊

It is possible to consider arbitrary contrasts involving means, $\sum c_i \mu_i$, in a similar fashion. Confidence intervals for significant differences and for

COMPARISONS BASED ON F TESTS

	nominal	number of orthogonal comparisons (c)				
		2	5	10	20	50
error	0.01	0.0199	0.0490	0.0956	0.182	0.395
rate	0.05	0.0975	0.226	0.401	0.642	0.923
α	0.1	0.190	0.410	0.651	0.878	0.995

Table 6.1. *Multiple comparisons selection bias*

contrasts can also be constructed, respectively, as

$$\hat{\mu}_i - \hat{\mu}_j \pm t_{\alpha/2; n.-a} \hat{\sigma} \sqrt{\frac{1}{n_i} + \frac{1}{n_j}},$$

$$\sum c_i \hat{\mu}_i \pm t_{\alpha/2; n.-a} \hat{\sigma} \sqrt{\sum c_i^2 / n_i}.$$

In practice, many comparisons might be made with the LSD approach. Therefore, it is possible to detect some differences which are not real. Note that if the F test rejects H_0, then *some* comparison(s) must be declared significant at level α. However, they may not be among those chosen for examination. For instance, c independent comparisons at rate α with none of them real would have a chance of detecting at least one significant differences as

$$\alpha_c = 1 - (1 - \alpha)^c.$$

Table 6.1 shows how large such probabilities can be. For $\alpha = 0.05$, there is a 40% chance of finding at least one significant difference in ten independent comparisons. With 20 comparisons there is a 64% chance of finding a significant difference even if none exist among the comparisons being considered.

Comparisons are usually not all independent. For instance, evidence that $\mu_A > \mu_B$ and $\mu_B > \mu_C$ suggests that the evidence would support $\mu_A > \mu_C$ (this is not always true!). These three comparisons are dependent, or more formally, the random variates $\hat{\mu}_A - \hat{\mu}_B$, $\hat{\mu}_B - \hat{\mu}_C$ and $\hat{\mu}_A - \hat{\mu}_C$ are correlated. In such situations, $\alpha_3 = 1 - (1 - \alpha)^3$ over-estimates the combined error rate for these three comparisons, each done at level α.

The problem of inflating error rate that arises with the LSD method can be quietly addressed for some problems. Note that for small α and c, $\alpha_c \approx \alpha \times c$. The **Bonferroni** method uses the LSD idea for examining c comparisons, but with the level α replaced by α/c, finding significant differences only if

$$|\hat{\mu}_i - \hat{\mu}_j| > \text{BONFERRONI} = t_{\alpha/2c; n.-a} \hat{\sigma} \sqrt{\frac{1}{n_i} + \frac{1}{n_j}}.$$

This approach can be very handy when the number of comparisons c is small, say 5 or fewer. It is sometimes used for planned comparisons when the F test is not significant. This method can be fairly conservative, but is quite useful in practice if there are a modest number of comparisons to be made.

The **Scheffé** method controls the experiment-wise error rate for all possible comparisons. It is very handy for conducting a large number of unplanned comparisons, or 'data snooping'. It is always more conservative than the LSD method and usually more conservative than the BONFERRONI method, unless interest focuses on many comparisons. The SCHEFFÉ method is mathematically equivalent to the F test in the sense that (i) no comparison of means is significant with the SCHEFFÉ method if the F test fails to reject H_0 and (ii) if the F test rejects the null hypothesis, then at least one comparison is significant using this method. Unfortunately, there are no clues in the theory to tell which comparisons may be significant and no guarantees that the scientist can quickly find significant comparisons if the F test rejects H_0. Considering an arbitrary contrast $\sum c_i \mu_i$, a $1 - \alpha$ SCHEFFÉ confidence interval can be constructed as

$$\sum c_i \hat{\mu}_i \pm \sqrt{(a-1)F_{\alpha;a-1,n.-a}} \; \hat{\sigma} \sqrt{\sum c_i^2/n_i}$$

which may or may not cover 0, corresponding to a hypothesis test. Pairwise contrasts for the balanced case have critical value

$$\text{SCHEFFÉ} = \sqrt{(a-1)F_{\alpha;a-1,n.-a}} \; \hat{\sigma} \sqrt{2/n} \; .$$

A multiple-stage test generalization of the SCHEFFÉ method (commonly known as REGWF after its authors – Ryan, Einot, Gabriel and Welsch) overcomes some of the SCHEFFÉ method's conservative features. Suppose the means are ordered, $\hat{\mu}_1 \leq \cdots \leq \hat{\mu}_a$, and consider testing whether the first k of these means are identical,

$$H_0 : \mu_1 = \cdots \mu_k = \mu \; .$$

The usual F test for this hypothesis uses the hypothesis sum of squares

$$SSH_0 = \sum_{i=1}^{k}(n_i\hat{\mu}_i - \hat{\mu}_{(k)})^2$$

having $k-1$ degrees of freedom, with

$$\hat{\mu}_{(k)} = \sum_{i=1}^{k} n_i \hat{\mu}_i / \sum_{i=1}^{k} n_i = \sum_{i=1}^{k}\sum_{j} y_{ij} / \sum_{i=1}^{k} n_i \; .$$

The corresponding test statistic would simultaneously compare the first k means. The significance level for comparisons is adjusted to control the experiment-wise error rate by using

$$\alpha_k = 1 - (1-\alpha)^{a/k} \; , \; k = 1, \cdots, a-2,$$

COMPARISONS BASED ON F TESTS

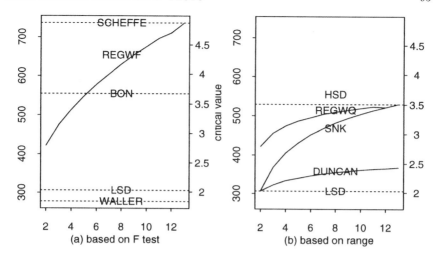

Figure 6.2. *Cloning unbalanced multiple comparisons: dashed lines indicate 5% critical values for multiple comparison methods not depending on the number of groups. Critical values increase for methods sensitive to the size of subset of groups being compared (solid lines). Methods are divided into (a) those based on the F test and (b) those based on the range statistic. The left vertical axes show cloning units; the right vertical axes reflect unitless critical values.*

and $\alpha_k = \alpha$ for $k = a-1, a$. The F statistic is compared with $F_{\alpha_k; k-1, n.-a}$. Thus, arguing as in the SCHEFFÉ method, confidence intervals for arbitrary contrasts among these k means would be of the form

$$\sum_{i=1}^{k} c_i \hat{\mu}_i \pm \hat{\sigma} \sqrt{(k-1)F_{\alpha_k; k-1, n.-a} \sum_{i=1}^{k} c_i^2 / n_i}\ .$$

The critical value for the REGWF method increases as the number of means to be compared increases and corresponds to the SCHEFFÉ method when $k = a$. The critical value for comparing two means from a set of $a > 2$ is larger than Fisher's LSD because it allows for the multiple comparisons but smaller than BONFERRONI's because it takes a less conservative approach to the selection of significance level.

The **Waller–Duncan** (WALLER) approach minimizes the Bayes risk under additive loss instead of controlling error rates. It has the same form as Fisher's LSD or the BONFERRONI method, but the tabled critical value depends on a user-supplied (or default) ratio of the relative importance of Type I and Type II errors. Unfortunately, it tends to be more liberal than Fisher's LSD (critical value is 278 as compared with 306) while being more complicated. In addition, it is only appropriate for equal sample sizes.

Clone	n	Mean	LSD	BONFERRONI	SCHEFFÉ	REGWF	WALLER
c	4	1789	A	A	A	A	A
a	5	1358	B	AB	AB	B	B
l	5	1068	BC	BC	ABC	BC	C
g	5	876	CD	BCD	BCD	CD	CD
j	5	746	D	CDE	BCDE	CDE	D
f	5	645	DE	CDE	BCDE	CDEF	DE
e	5	428	EF	DEF	CDE	DEFG	EF
k	5	406	EF	DEF	CDE	EFG	EF
m	5	329	FG	DEF	DE	EFG	F
i	5	288	FG	EF	DE	EFG	FG
h	5	279	FG	EF	DE	FG	FG
d	5	265	FG	EF	DE	FG	FG
b	5	45	G	F	E	G	G
Critical Value			306	555	737	*	278

Table 6.2. *Cloning multiple comparisons based on F*

Example 6.3 Cloning: The plant pathologist ran all 13 clones together and arrived at the following multiple comparisons using methods based on the F test. Clones with the same letter are not significantly different at the 5% level using the indicated method. Since there was a missing data value for clone c, the harmonic mean of sample sizes, $\tilde{n} = 4.91$, was used in calculations below.

The tabled critical value is $t_{0.025;53} = 2.01$. For the set of 13 clones, using the harmonic mean, the approximate LSD was 306. Results using exact sample sizes would be virtually the same in this case. That is, instead of using $\sqrt{2/\tilde{n}} = 0.638$ for all comparisons, those with clone c (having only four points) should use $\sqrt{1/4 + 1/5} = 0.671$, requiring a 5% increase in critical value (e.g. to 322 for LSD), while comparisons between all other clones (having five points) should use $\sqrt{2/5} = 0.632$, a 5% decrease (303 for LSD).

There are 78 paired comparisons among 13 treatment means. Thus the BONFERRONI method would use $.05/78 = 0.00064$ and critical value $t_{0.00064;51} = 3.64$, or 555 in concentration units, which is 80% larger than that used for the LSD method. The SCHEFFÉ method has critical value $\sqrt{12F_{0.05;12,51}} = \sqrt{12 \times 1.95} = 4.84$, leading to a comparison value of 737, which is more conservative than either the LSD or BONFERRONI method.

The last line of Table 6.2 contains the critical values for the methods based on F tests, except for REGWF. The critical values for this latter multiple-stage test are depicted in Figure 6.2(a) against the number of means. Notice the tradeoff between REGWF and BONFERRONI which is reflected in the table of multiple comparisons. The Waller–Duncan (WALLER) method is included for comparison but not recommended. ◇

6.3 Comparisons based on range of means

A rather different type of comparison was developed by examining the range of group means. Theoretical properties are only known for the balanced case of equal sample sizes, although simulation results indicate that the harmonic mean can be used for mildly unbalanced designs (Miller 1981). The basic idea is that the a group means have distribution

$$\bar{y}_{1\cdot}, \cdots, \bar{y}_{a\cdot} \sim N(\mu, \sigma^2/n)$$

under H_0. Instead of examining the squared deviations of $\bar{y}_{i\cdot}$ about $\bar{y}_{\cdot\cdot}$ as is done for the F test, consider the range

$$Q = \sqrt{n}[\max(\bar{y}_{i\cdot}) - \min(\bar{y}_{i\cdot})]/\hat{\sigma} \sim q(a, \nu)$$

with $\nu = n_{\cdot} - a$. The studentized range $q(a, \nu)$ has been tabled (see Scheffé 1959, Table 2.2, or Milliken and Johnson 1992, Table A.4).

This could be used as a test, rejecting H_0 and comparing means, if $Q > q(\alpha, a, \nu)$, with $q(\alpha, a, \nu)$ the upper α critical value of $q(a, \nu)$. However, recall that the F test is uniformly most powerful for H_0. Because of this, methods based on the range are usually considered separately from the test of H_0. Miller (1981) has a good discussion of this pivot statistic.

Six range methods are highlighted below. Tukey's HSD method is simple to use and quite effective for comparing pairs of means, although it can be rather conservative. Dunnett's many-to-one statistic is a slight variation for comparing many groups to one control. The Student–Newman–Keuls (SNK) method adapts Tukey's approach to control both experiment-wise and comparison-wise error rates. The REGWQ test is similar to REGWF except that it is based on the range statistic rather than the F test. Duncan's multiple range method is more liberal than SNK, gaining power and sensitivity while losing control of the experiment-wise error rate.

Tukey's honest significant difference (HSD) is better than the SCHEFFÉ method for comparisons involving only pairs of means, in the sense of yielding shorter confidence intervals. While it controls both the experiment-wise and comparison-wise error rates, it may be overly conservative. Compare mean differences to

$$\text{HSD} = q(\alpha, a, \nu)\hat{\sigma}/\sqrt{n} = [q(\alpha, a, \nu)/\sqrt{2}]\hat{\sigma}\sqrt{2/n},$$

and declare a significant difference if

$$|\hat{\mu}_i - \hat{\mu}_j| > \text{HSD}.$$

Approximate HSD comparisons for unequal sample sizes can be performed by replacing $2/n$ by $1/n_1 + 1/n_2$ or $2/\min(n_i, n_j)$. For several comparisons, it may be convenient to use the harmonic mean $\tilde{n}$. The experiment-wise error rate in such situations is unknown.

Dunnett's many-to-one method is similar to Tukey's, but geared toward comparing many groups to one control. It focuses on the distribution of the statistic

$$\sqrt{n} \max_i |y_{i\cdot}) - y_{0\cdot})|/\hat{\sigma}$$

with $y_{0\cdot}$ the mean response for the control level. Miller (1981, sec. 2.5) discusses this statistic in detail, showing that its distribution arises from a multivariate analog of the t distribution. SAS and many other statistical packages now make this readily available, but it will not be discussed further here.

The **Student–Newman–Keuls (SNK)** method is named after the three people who independently developed it between 1927 and 1952. [As the cognoscenti of statistics all know, 'Student' was the pseudonym for William Gosset, a statistician who worked for the Guinness brewery but was not allowed to use his own name for publications because his employers feared other breweries would catch on to their secret weapon!]

The SNK method controls the experiment-wise and comparison-wise error rates by a succession of HSD-like comparisons. It has been adapted for unequal sample sizes but the error rates are unknown in this case.

The method is very elegant and is available in several packages. Its major drawback is the difficulty in explaining it. First, consider all a ordered means and use $\text{HSD}_a = q(\alpha, a, \nu)\hat{\sigma}/\sqrt{n}$, with $\nu = n_\cdot - a$, to compare the smallest and the largest. If they are not significantly different (and sample sizes are equal), then no other pairwise means comparison can be significant, which concludes the SNK procedure.

If the largest and smallest are different, then consider the two sets of $a - 1$ consecutive ordered means. Again, compare the smallest and largest in each $(a-1)$-tuple using $\text{HSD}_{a-1} = q(\alpha, a-1, \nu)\hat{\sigma}/\sqrt{n}$ (but the common estimate $\hat{\sigma}^2$ with $\nu = n_\cdot - a$ degrees of freedom as before). If there is no significant difference for a particular tuple of consecutive means, they are not significantly different and no subset of non-significant group means can later be deemed significant. For significant tuples, consider the next smaller sub-tuples and so on. For k-tuples, use $\text{HSD}_k = q(\alpha, k, \nu)\hat{\sigma}/\sqrt{n}$.

The smallest possible tuple is a pair. Here the comparison reduces to Fisher's LSD, with $\text{HSD}_2 = \text{LSD}$. That is, $q(\alpha, 2, \nu) = t_{\alpha/2;\nu}\sqrt{2}$. Thus by construction, the SNK method is between HSD and LSD in terms of its power and conservative nature.

Duncan's multiple range method is like SNK in stepwise construction but it uses different significance levels for each size group. It tends to be more liberal than SNK for higher-order levels, giving it greater sensitivity and power. However, Duncan does not control the experiment-wise error rate, which is in fact unknown. It does provide greater protection (is more conservative) than the LSD (see Table 6.3).

The idea of Duncan's method is to use a significance level which is consistent with $k-1$ independent tests at level α, namely $\alpha_{k-1} = 1-(1-\alpha)^{k-1}$. Comparisons would be made as in SNK but using $q(\alpha_{k-1}, k-1, \nu)$.

	Experiment-wise error rate for k tests					
k	2	3	4	5	10	20
Duncan	0.05	0.0975	0.143	0.185	0.370	0.623
LSD	0.05	0.122	0.203	0.286	0.627	0.918

Miller (1981) points out that the independence of hypotheses plays the major role in properties of tests, rather than the independence of test statistics as argued by Duncan. This test and the related Waller–Duncan (WALLER) method, described in the previous section, are largely discouraged by many practicing statisticians. Some still like to use them in pilot studies, where more conservative methods might eliminate promising groups. However, this author would suggest using the liberal, and simple, Fisher's LSD in such cases.

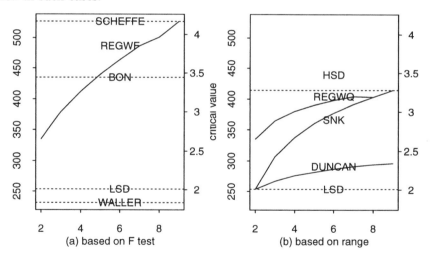

Figure 6.3. *Cloning balanced multiple comparisons: dashed lines indicate 5% critical values for methods independent of the size of the subset of groups being compared. Other methods (solid lines) show increasing values as the number of groups in a subset increases. Values based on range have been multiplied by $\sqrt{2}$ to be comparable. See Figure 6.2.*

Example 6.4 Cloning: The plant pathologist used comparisons based on ranges and found similar results to those shown in Table 6.2. The critical

Clone	Mean	LSD	HSD	SNK	REGWQ	Duncan
a	1358	A	A	A	A	A
l	1068	B	A	B	A	B
f	645	C	B	C	B	C
e	428	CD	B	C	B	CD
k	406	CD	B	C	B	CD
m	329	D	B	C	B	D
i	288	D	B	C	B	D
h	279	D	B	C	B	D
d	265	D	B	C	B	D
Critical Value		254	414	*	*	*

Table 6.3. *Cloning multiple comparisons based on range*

values for range comparisons based on all the data are shown in Figure 6.2(b). However, be somewhat cautious interpreting these since the properties of the statistic are only understood for equal sample sizes. To that end, the plant pathologist compared only the 11 clones that had all five replicates and similar variation. The critical value for range of means comparison is $q(0.05, 8, 36)/\sqrt{2} = 3.30$, or 414 in units of titer, as compared to 254 for the LSD ($t_{0.025;36} = 2.03$) for these nine clones. Again, clones with the same letter are not significantly different at the 5% level using the indicated method. Table 6.3 only shows multiple comparison results based on ranges but Figure 6.3 includes critical values for comparisons based on (a) F tests and (b) ranges. ◊

6.4 Comparison of comparisons

It is difficult to put forth one 'best' recommendation for multiple comparison of means. There are numerous tradeoffs between methods, depending on the type and number of contrasts to be considered. This section indicates some consistent guidelines and remarks on the compromises among certain methods.

The following relationships always hold:

$$\text{LSD} < \text{REGWF} < \text{SCHEFFÉ}$$
$$\text{LSD} < \text{BONFERRONI}$$
$$\text{LSD} < \text{SNK} < \text{REGWQ} < \text{HSD}$$

Example 6.5 Cloning: The BONFERRONI method outperformed the SCHEFFÉ method in this experiment. This is usually the case. Note that one is com-

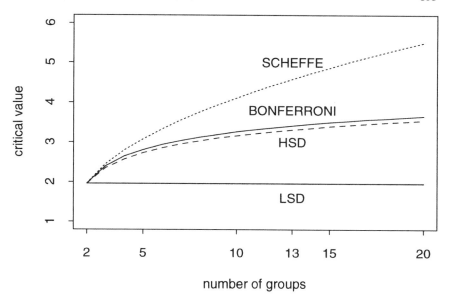

Figure 6.4. *Cloning critical value by number of groups. There is in general no ordering of* SCHEFFÉ, BONFERRONI *and Tukey's* HSD *critical values. However,* SCHEFFÉ *typically is much more conservative, rising substantially with the number of groups.* BONFERRONI *does not appear to be much worse than* HSD.

paring $t_{\alpha/c;n.-a}$ to $\sqrt{(a-1)F_{\alpha;a-1,n.-a}}$, with c the number of comparisons (there are $c = a(a-1)/2 = 78$ comparisons of pairs of means, or $c = 78$ with $a = 13$ means). Figure 6.4 compares the critical values (standardized to $\hat{\sigma} = 1$) for the SCHEFFÉ, BONFERRONI and Tukey's HSD as the number of groups a increases, assuming $n.$ is very large. Notice that BONFERRONI does only slightly worse than HSD! Both of these inflate the critical value over Fisher's LSD value of 1.96 but this value is not doubled until one has at least 25 groups to compare.

The SCHEFFÉ critical value exceeds twice the LSD for comparisons of eight or more groups. Put another way, over 1000 planned comparisons are needed to have the BONFERRONI be worse than SCHEFFÉ for 13 groups; the 78 pairwise comparisons using BONFERRONI correspond to the critical value for SCHEFFÉ for comparing six or seven groups (Figure 6.5).

Tukey's HSD is always better than SCHEFFÉ for comparing pairs of means. However, the SCHEFFÉ critical value can be smaller than the HSD for contrasts involving general contrasts of means. For instance, in the cloning example, if one compares the average of two means to the average of another two means in the reduced set of 11 means, the critical values for SCHEFFÉ is 373, while HSD still uses 395. ◊

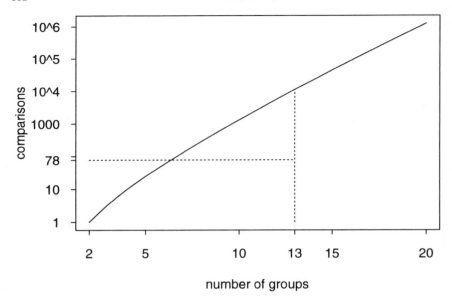

Figure 6.5. *Cloning match of* **BONFERRONI** *to* **SCHEFFÉ** *comparisons: With 13 groups there are 78 paired comparisons. However, the* **SCHEFFÉ** *critical value is roughly twice the* **BONFERRONI** *critical value. The* **BONFERRONI** *critical value for paired comparisons of 13 groups is roughly the same as the* **SCHEFFÉ** *critical value for six or seven groups.*

Similar comparison applies for the REGWQ and REGWF methods as for the HSD and SCHEFFÉ, respectively. Note that REGWQ and REGWF for comparing two means are equal and lie about half-way between the liberal LSD and conservative BONFERRONI methods. These approaches take advantage of the dependencies among comparisons, resulting in more power while controlling the experiment-wise error rate. These two methods also control the experiment-wise error rate under all partial null hypotheses in which only some means are equal.

In general, the Duncan and WALLER methods are related to the LSD as follows:

$$\text{WALLER} < \text{LSD} < \text{Duncan} < \text{SNK}$$

However, this may be altered depending on the prior information used to construct these methods.

The methods based on the multiple-range statistics are not usually consistent with the overall F test. They are appropriate for comparing means one-to-one provided sample sizes are nearly equal. Of these, the SNK is the most powerful method which controls the the experiment-wise error rate.

The SCHEFFÉ and REGWF are the only methods completely compatible

PROBLEMS
103

with the F test. However, the SCHEFFÉ method is much too conservative and the REGWF may be too conservative for many. For a recent discussion of multiple-stage tests, see Wright (1992).

In practice, the multiple-stage tests take some explaining to scientists. Sometimes that effort can interfere with the interpretation of other more pressing issues about data analysis and the technical details cloud the big picture. Usually, the protected LSD – that is, an overall F test followed by multiple comparisons using Fisher's LSD – is perfectly adequate for the needs of the moment.

Keep in mind that any multiple comparison method only provides a method of assigning significance to the ordering of means. Clear-cut differences with one method are usually supported by any approach. The difficulties arise with uncertain separation of means, and the scientist must understand the limitations of taking any particular test too seriously.

6.5 Problems

6.1 Diet: The scientist wanted to conduct several contrasts of means based on the way they were constructed. That is, several ingredients (UIP, SBM, ABP) present in varying amounts merited particular attention.
(a) Conduct two forms of multiple comparisons of the six diets.
(b) Perform the following contrasts selected by the scientist:

contrast/diet	1	2	3	4	5	6
UIP linear	-6	-1	-1	1	1	6
UIP quadratic	2	-1	-1	-1	-1	2
SBM	1	-1	0	0	0	0
ABP linear	0	0	-1	0	0	1
ABP quadratic	0	0	1	0	-2	1

(c) Comment on the comparisons. What cautions would you make about interpretation? What general conclusions, if any, emerge from these data?

6.2 Cloning: You are given information of summary statistics and partial results from an analysis of variance with some LSDs and some contrasts of interest to the scientist. Your job is to write a cogent description of what was found. In your report you may want to comment on the following:
(a) the effect of the missing data value;
(b) any problems with assumptions;
(c) relationship of contrasts to the LSD table.
You may suggest modifications or show how (or whether) a more 'appropriate' analysis might alter results.

6.3 Tree: Reconsider the growth of a soil pathogen in different soils. The scientist really wanted to compare soil from nm-6 with the other three tree soils. This would suggest using Dunnett's many-to-one approach, which is supported by SAS as well as many other packages.

(a) Lay out the overall hypothesis for comparing nm-6 to the tree soils. Be sure to indicate how the non-tree soils are to be handled.

(b) Perform Dunnett's test if it is readily available.

(c) Argue for the use of an overall test of no difference among means as opposed to the null hypothesis in (a). While the argument can be based partly on convenience and easy of interpretation, it should also address the chance of detecting differences (power).

(d) Select two other types of multiple comparisons. You must state why you chose these: what are their strengths and weaknesses (be brief!)? How do they compare in theory and practice? What are the tradeoffs between experiment-wise and comparison-wise error rates?

(e) Follow the overall test by your chosen methods of multiple comparisons and interpret the results with particular attention to the scientist's key question.

(f) Use a multiple comparisons approach to contrast the four controls with the soils of interest. Follow this with a contrast of NM-6 with the other four soils.

PART C

Sorting out Effects with Data

This part covers standard analysis of variance settings in which there are two or more factors of interest in a completely randomized design. The context is set and models are developed, leading to estimates and tests of the effects of factors.

A statistical model based on means is fairly straightforward for a single factor. However, introducing factorial effects into more complicated models allows separation of important concepts of main effects and interaction which later lead to formal questions of inference. Chapter 7 formalizes the means model, motivates the effects models and establishes the general form of estimable functions, with some attention to side conditions.

Factorial arrangements for balanced experiments, developed in Chapter 8, allow comparison of main effects for each factor. However, synergism or antagonism – interaction – among factors leads to interpretation of one factor in reference to the other.

Model selection is addressed in Chapter 9, drawing on ideas from regression. However, model selection in analysis of variance usually follows the hierarchy of factors, including main effects if interactions are considered. Issues of pooling interaction into error and examining evidence for interaction when there is no replication are discussed.

CHAPTER 7

Factorial Designs

The cell means model is an example of a statistical model for a factorial design. As the design becomes more complicated, it is helpful to move from a discussion of means to a focus on effects. This chapter develops factor effects notation for factorial models.

The cell means model in Section 7.1 formalizes the assumption that the response for each combination of factor levels is centered about a mean and has some random variation about that mean. Section 7.2 recasts models in terms of factor effects, showing how this can partition mean response among main effects and interactions for multi-factor experiments. Such models are over-specified, suggesting a discussion of linear constraints and estimable functions (Section 7.3). Finally, since some experiments are incomplete, by design or by chance, Section 7.4 examines connected cells to show how the pattern of empty cells can limit the range of meaningful comparisons.

7.1 Cell means models

Most models in this book can be written as a mean component plus an error component,

$$\text{response} = \text{group mean} + \text{random error} .$$

The **cell means model** can be expressed with notation as

$$y_{ij} = \mu_i + e_{ij} .$$

Independence of the responses y_{ij} necessitates that the random errors e_{ij} are independent. The usual assumptions that the model is correct and variances are equal yield, respectively,

$$E(e_{ij}) = 0 \text{ and } V(e_{ij}) = \sigma^2 .$$

Normality, if assumed, implies that $e_{ij} \sim N(0, \sigma^2)$.

The means μ_i and variance σ^2 are the **parameters**, or unknown constants, of the means model. In the analysis of variance, it is often useful to think about the information in the data, y_{ij}, in terms of **degrees of freedom**. There are $n. = \sum n_i$ degrees of freedom, or $n. - 1$ corrected for the grand mean. The a groups provide a degrees of freedom for the model, or $a-1$ model degrees of freedom when corrected for the grand mean. This leaves $n.-a = \sum(n_i-1)$ degrees of freedom for error. There is a one-to-one correspondence between the a model degrees of freedom and the a group mean parameters μ_i.

The **two-factor means model** can be thought of as a big cell means model. Consider two factors A and B, with levels $A_1, \cdots, A_a$ and $B_1, \cdots, B_b$, respectively. The observed response can be modeled as

$$y_{ijk} = \mu_{ij} + e_{ijk}, \text{ with independent } e_{ijk} \sim N(0, \sigma^2).$$

Each of the ab cells, or factor combination levels (A_i, B_j), $i = 1, \cdots, a$, $j = 1, \cdots, b$, has n_{ij} experimental units. The index for EUs at level (i, j) runs from $k=1$ to $k=n_{ij}$, with the total sample size, or number of EUs, being $n.. = \sum_i \sum_j n_{ij}$. Three-factor and higher-order layouts can be developed in the same manner.

Example 7.1 Design: Suppose six combinations of nutrients and temperature are to be used in an experiment on plant growth in controlled conditions, eventually planned for the space station. There are two nutrient levels, high and low, and three temperatures, low, medium and high. Thus, this experiment involves two factors, A = nutrient with $a = 2$ levels and B = temperature with $b = 3$ levels. The response is y = plant growth. The particular response y_{121} denotes the low nutrient A_1, medium temperature B_2 and 1st replicate. The expected response for this treatment combination is μ_{12}. ◇

Note that if any $n_{ij} = 0$ there are no data in cell (i, j) and no estimator for the corresponding cell mean. This would reduce the number of model parameters, and the number of model degrees of freedom. If there are m cells with no data then only $ab - m$ cell means can be estimated, and there can be at most $ab - m$ model degrees of freedom. For this chapter, assume $n_{ij} > 0$ for all cells. In fact, the next two chapters of this part assume the data are balanced $(n_{ij} = n)$. However, the ideas in the present chapter do not require this balance.

The cell means μ_i, μ_{ij} or μ_{ijk} in the one-, two- or three-factor arrangements are **estimable** provided that there is at least one observation per factor combination. That is, there are unique, unbiased estimates of these quantities which are linear combinations of the response, namely $\bar{y}_{i\cdot}$, $\bar{y}_{ij\cdot}$ and $\bar{y}_{ijk\cdot}$, respectively. Note that any linear combination of model means is also estimable, provided each level has at least one observation.

7.2 Effects models

This section first recasts the cell means model into the equivalent effects model and then considers the two-factor and more complicated factorial ar-

EFFECTS MODELS

rangements. Sometimes a scientist is interested in comparing group means directly in an experiment involving several factors. For instance, if the goal is to find the 'best' combination of two chemicals, then a segment of analysis may focus on a cell means model with multiple comparisons of group means. However, interest usually focuses on separating the effects of factors A and B to determine if either or both have substantial effect on response. Should there be evidence that the factors are not simply additive (a synergistic or antagonistic interaction), then statements about the effect of one factor must be qualified by indicating the level of the other.

The **effects model** has three components, two for model plus one for error:

$$\text{response} = \text{reference} + \text{group effect} + \text{random error} .$$

It is convenient to think of the **reference point** as the population grand mean, but this is not correct. Some packages such as SAS call this the intercept, which is also incorrect. The reference point is arbitrary. The group effect measures how far a group mean is from this reference point,

$$\text{group effect} = \text{group mean} - \text{reference} .$$

With notation this can be written compactly as

$$y_{ij} = \mu + \alpha_i + e_{ij} \text{ with } \alpha_i = \mu_i - \mu .$$

In other words, $\mu_i = \mu + \alpha_i$. The reference or benchmark standard, μ, is arbitrary. It may be useful in some situations to make it the standard (old) treatment. Alternatively, it could be set to the average of all factor levels.

It is difficult to see the value of the effects model over the cell means model until there are two or more factors. The **two-factor effects model** partitions the cell mean into four distinct components, including the effects of both factors and the interaction associated with combinations of factors

$$\mu_{ij} = \mu + \alpha_i + \beta_j + \gamma_{ij} .$$

The three-factor effects model simply has more terms, which are commonly expressed with lots of Greek symbols, such as

$$\mu_{ijk} = \mu + \alpha_i + \beta_j + \gamma_k + \delta_{ij} + \xi_{ij} + \zeta_{jk} + \eta_{ijk}$$

or in a form acknowledging interaction pairs and triplets

$$\mu_{ijk} = \mu + \alpha_i + \beta_j + \gamma_k + (\alpha\beta)_{ij} + (\alpha\gamma)_{ik} + (\beta\gamma)_{jk} + (\alpha\beta\gamma)_{ijk} .$$

Recall that these models can be interpreted for some purposes as cell means models, leading to examination of specific linear contrasts of interest.

The value of partitioning cell means into effects emerges when trying to determine if factor combinations can be explained in terms of the **population marginal means**

$$\bar{\mu}_{i\cdot} = \sum_{j=1}^{b} \mu_{ij}/b , \; \bar{\mu}_{\cdot j} = \sum_{i=1}^{a} \mu_{ij}/a .$$

Consider the following table in which each **cell** represents a distinct combination of factor levels.

Cell and Marginal Means

factor A levels	factor B levels				margin
	1	2	$\cdots$	b	
1	μ_{11}	μ_{12}	$\cdots$	μ_{1b}	$\bar{\mu}_{1\cdot}$
2	μ_{21}	μ_{22}	$\cdots$	μ_{2b}	$\bar{\mu}_{2\cdot}$
$\vdots$	$\vdots$	$\vdots$	$\ddots$	$\vdots$	$\vdots$
a	μ_{a1}	μ_{a2}	$\cdots$	μ_{ab}	$\bar{\mu}_{a\cdot}$
margin	$\bar{\mu}_{\cdot 1}$	$\bar{\mu}_{\cdot 2}$	$\cdots$	$\bar{\mu}_{\cdot b}$	$\bar{\mu}_{\cdot\cdot}$

The population grand mean is the simple average of cell means

$$\bar{\mu}_{\cdot\cdot} = \sum_{i=1}^{a}\sum_{j=1}^{n_i} \mu_{ij}/ab = \sum_{i=1}^{a} \bar{\mu}_{i\cdot}/a = \sum_{j=1}^{n_i} \bar{\mu}_{\cdot j}/b \ .$$

The population marginal means are always related to effects model parameters by the following relations:

$$\bar{\mu}_{i\cdot} = \mu + \alpha_i + \bar{\beta}_{\cdot} + \bar{\gamma}_{i\cdot}$$
$$\bar{\mu}_{\cdot j} = \mu + \bar{\alpha}_{\cdot} + \beta_j + \bar{\gamma}_{\cdot j}$$
$$\bar{\mu}_{\cdot\cdot} = \mu + \bar{\alpha}_{\cdot} + \bar{\beta}_{\cdot} + \bar{\gamma}_{\cdot\cdot} \ .$$

7.3 Estimable functions

The specific estimable functions which are employed in data analysis depend on the questions being addressed. This section specifically identifies the estimable functions for the one-factor and two-factor effects models while introducing the more general concept of estimability in linear models. What linear combinations of parameters in the effects models are estimable? How can this be made explicit?

Questions about cell means can be posed in terms of means or effects with equal ease. However, only the cell means and linear combinations of those cell means are estimable. To see this, consider minimizing the error sum of squares for the one-factor effects model

$$SS = \sum_{i=1}^{a}\sum_{j=1}^{n_i} (y_{ij} - \mu - \alpha_i)^2$$

which is solved by setting partial derivatives of SS with respect to the

ESTIMABLE FUNCTIONS

parameters μ and α_i to 0:

$$0 = \frac{\partial SS}{\partial \mu} = -2\sum_{i=1}^{a}\sum_{j=1}^{n_i}(y_{ij} - \mu - \alpha_i)$$
$$0 = \frac{\partial SS}{\partial \alpha_i} = -2\sum_{j=1}^{n_i}(y_{ij} - \mu - \alpha_i) , \; i = 1, \cdots, a .$$

This leads to the normal equations

$$n.\mu = \sum_{i=1}^{a}\sum_{j=1}^{n_i} y_{ij} - \sum_{i=1}^{a} n_i \alpha_i$$
$$n_i \alpha_i = \sum_{j=1}^{n_i} y_{ij} - n_i \mu , \; i = 1, \cdots, a .$$

These normal equations do not have unique solutions in μ and α_i since the $a+1$ equations are not linearly independent. That is,

$$\mu = \bar{y}.. - \sum n_i \alpha_i / n. \text{ and } \alpha_i = \bar{y}_{i.} - \mu .$$

However, $\mu_i = \mu + \alpha_i$ is uniquely estimated by $\hat{\mu}_i = \bar{y}_{i.}$.

Estimable functions are defined as those functions of parameters which do not depend on the particular solution of the normal equations. An **estimable linear function** is any linear combination of model parameters that can be expressed as a linear combination of expected responses. Linear combinations of estimable functions are estimable.

Example 7.2 Design: Suppose six lakes are to be examined under three different treatment regimes of added nutrients with the goal of comparing water pH. The mean pH for treatment group i can be written as

$$\mu_i = \mu + \alpha_i .$$

It would be possible to define $\mu = 7$ and consider the effects as deviations from a neutral pH, $\alpha_i = \mu_i - 7$. Alternatively, the reference could be defined as the mean pH over the three treatments, $\mu = \bar{\mu}.$. In any event, let $a = 3$ and $n_1 = n_2 = n_3 = 2$ in the effects model. The model can be represented as a product of a design matrix with 1s and 0s and a vector of the parameters. That is, the response vector is the product of the design matrix times the parameter vector plus the error vector:

$$\begin{pmatrix} y_{11} \\ y_{12} \\ y_{21} \\ y_{22} \\ y_{31} \\ y_{32} \end{pmatrix} = \begin{bmatrix} 1 & 1 & 0 & 0 \\ 1 & 1 & 0 & 0 \\ 1 & 0 & 1 & 0 \\ 1 & 0 & 1 & 0 \\ 1 & 0 & 0 & 1 \\ 1 & 0 & 0 & 1 \end{bmatrix} \begin{pmatrix} \mu \\ \alpha_1 \\ \alpha_2 \\ \alpha_3 \end{pmatrix} + \begin{pmatrix} e_{11} \\ e_{12} \\ e_{21} \\ e_{22} \\ e_{31} \\ e_{32} \end{pmatrix} .$$

This is just another way of writing the six model equations

$$y_{11} = \mu + 1 \times \alpha_1 + 0 \times \alpha_2 + 0 \times \alpha_3 + e_{11}$$
$$y_{12} = \mu + 1 \times \alpha_1 + 0 \times \alpha_2 + 0 \times \alpha_3 + e_{12}$$
$$y_{21} = \mu + 0 \times \alpha_1 + 1 \times \alpha_2 + 0 \times \alpha_3 + e_{21}$$
$$y_{22} = \mu + 0 \times \alpha_1 + 1 \times \alpha_2 + 0 \times \alpha_3 + e_{22}$$
$$y_{31} = \mu + 0 \times \alpha_1 + 0 \times \alpha_2 + 1 \times \alpha_3 + e_{31}$$
$$y_{32} = \mu + 0 \times \alpha_1 + 0 \times \alpha_2 + 1 \times \alpha_3 + e_{32} \ .$$

[A matrix approach is briefly developed in Chapter 12, Part D. A concise but accessible review of matrix algebra is available in Healy (1986).] The design matrix has four columns, but is only of rank 3, since the last column can be written as a the first column minus the sum of the middle columns. There is no unique solution for μ and α_i. Thus μ and the α_i are not estimable. ◊

The reference point μ is not estimable although cell means μ_i are. This is important. For instance, a set of data run through different computer packages can yield vastly different solutions for μ and effects parameters due to differing conventions about linear constraints (discussed in the next section). However, they must all agree in estimates of estimable functions such as linear combinations of cell means. Take care to focus attention only on estimable functions since their estimates are the same regardless of the solution.

The cell means μ_i and all linear combinations of cell means $\sum_i c_i \mu_i$ are estimable functions since $\mu_i = \mu + \alpha_i$ are estimated uniquely by $\hat{\mu}_i = \bar{y}_i.$ regardless of the particular solutions for μ and α_i. Therefore, the estimable functions must take the form

$$\sum_i c_i(\mu + \alpha_i) = c.\mu + \sum_i c_i \alpha_i \ .$$

Linear combinations with $c. = \sum_i c_i = 0$ are known as **linear contrasts**. The estimable linear contrasts can be expressed in the same way in terms of cell means or effects as

$$\sum_i c_i \mu_i = \sum_i c_i \alpha_i \text{ with } \sum_i c_i = 0 \ .$$

In the **two-factor model** each cell mean μ_{ij} is naturally estimable. Linear combinations of cell means are estimable since they are linear combinations of estimable functions. Hence all marginal means are estimable, but the parameters for factor effects and interactions are not. This can be seen once again by minimizing the error sum of squares

$$SS = \sum_{i=1}^{a} \sum_{j=1}^{n_i} (y_{ij} - \mu - \alpha_i - \beta_j - \gamma_{ij})^2$$

ESTIMABLE FUNCTIONS

leading to normal equations

$$
\begin{aligned}
n_{..}\mu &= \sum_{ijk} y_{ijk} - \sum_i n_{i.}\alpha_i - \sum_j n_{.j}\beta_j - \sum_{ij} n_{ij}\gamma_{ij} \\
n_{i.}\alpha_i &= \sum_{jk} y_{ijk} - n_{i.}\mu - \sum_j n_{ij}\beta_j - \sum_j n_{ij}\gamma_{ij} \\
n_{.j}\beta_j &= \sum_{ik} y_{ijk} - n_{.j}\mu - \sum_i n_{ij}\alpha_i - \sum_i n_{ij}\gamma_{ij} \\
n_{ij}\gamma_{ij} &= \sum_k y_{ijk} - n_{ij}(\mu + \alpha_i + \beta_j + \gamma_{ij}) \ .
\end{aligned}
$$

There are $ab+a+b+1$ equations but only ab unknowns. Thus these are not linearly independent and all the parameters cannot be estimable. However, $\mu_{ij} = \mu + \alpha_i + \beta_j + \gamma_{ij}$ is uniquely estimated by $\hat{\mu}_{ij} = \bar{y}_{ij.}$.

Questions of estimability suggest what kinds of comparisons make sense for two-factor models. For instance, it may be interesting to contrast population marginal means at two different levels of A or of B

$$
\begin{aligned}
\bar{\mu}_{1.} - \bar{\mu}_{2.} &= (\alpha_1 - \alpha_2) + (\bar{\gamma}_{1.} - \bar{\gamma}_{2.}) \\
\bar{\mu}_{.3} - \bar{\mu}_{.4} &= (\beta_3 - \beta_4) + (\bar{\gamma}_{.3} - \bar{\gamma}_{.4}) \ .
\end{aligned}
$$

These **main effects contrasts** isolate the marginal effects of a factor on response. Note that the effects terms α_i are completely confused with the marginal interactions $\bar{\gamma}_{i.}$. Similar confounding arises for contrasts of main effects for the second factor. This cannot be eliminated, since neither α_i nor $\bar{\gamma}_{i.}$ is estimable.

Evidence of interaction would suggest comparisons of two levels of one factor at particular levels of the other, such as

$$
\begin{aligned}
\mu_{13} - \mu_{23} &= (\alpha_1 - \alpha_2) + (\gamma_{13} - \gamma_{23}) \\
\mu_{13} - \mu_{14} &= (\beta_3 - \beta_4) + (\gamma_{13} - \gamma_{14}) \ .
\end{aligned}
$$

These comparisons confound main effects and interactions. It is possible to isolate the **pure interaction comparisons** by examining the difference in response to A at different levels of B,

$$(\mu_{13} - \mu_{23}) - (\mu_{14} - \mu_{24}) = (\gamma_{13} - \gamma_{23}) - (\gamma_{14} - \gamma_{24}) \ ,$$

which is the same as comparing the response to B at different levels of A,

$$(\mu_{13} - \mu_{14}) - (\mu_{23} - \mu_{24}) = (\gamma_{13} - \gamma_{14}) - (\gamma_{23} - \gamma_{24}) \ .$$

Models without interactions are simpler to interpret. The **two-factor additive effects model** assumes $\gamma_{ij} = 0$:

$$\mu_{ij} = \mu + \alpha_i + \beta_j \ .$$

This model has no pure interaction. The marginal means are

$$\bar{\mu}_{i\cdot} = \mu + \alpha_i + \bar{\beta}.$$
$$\bar{\mu}_{\cdot j} = \mu + \bar{\alpha}. + \beta_j.$$

Effects for each factor can be interpreted independently of the other factor. Further, there are fewer parameters in the model, allowing more power for tests. The drawback is that it is easy to overlook important synergy or antagonism of certain factor combinations by ignoring evidence in the data for interaction.

7.4 Linear constraints

Most models are over-specified in order fully to present the concepts of design structure and treatment structure. Simply put, there are more model parameters than model degrees of freedom. Thus not all model parameters, or linear combinations of model parameters, are estimable. This section examines two commonly used sets of linear constraints. Linear constraints are merely a convenient way to reduce the dimension of the model to solve the normal equations. They do not make model parameters estimable.

The one-factor effects model has a groups and hence a model degrees of freedom, but $a+1$ parameters $(\mu, \alpha_1, \cdots, \alpha_a)$. Placing one linear constraint on α can lead to a unique solution to the normal equations. However, the constraint does not make the parameters estimable! This can be made more explicit by considering particular linear constraints. The two main types of linear constraints are sum-to-zero and set-to-zero. The former are handy for model interpretation while the latter are easier to implement in computer packages. SAS uses a set-to-zero constraint while S-Plus by default uses a different sum-to-zero constraint. Still other conventions can be found in other packages. It can be rather confusing to encounter different solutions from two packages, while the estimated cell means are the same. Understanding linear constraints can usually unravel this mystery.

The **sum-to-zero linear constraints** set the marginal sums of effects parameters to zero. In the one-factor model, $\sum_{i=1}^{a} \alpha_i = 0$ makes the reference equal to the population grand mean ($\mu = \bar{\mu}. = \sum_{i=1}^{a} \mu_i$) and the α_i correspond to the group effects as deviations of group means from the grand mean. This type of constraint is often implicitly assumed by statisticians used to working in a theoretical setting. They can be tricky to build into a statistical package in such a way that they handle all manner of design imbalance. This is the default constraint for S-Plus.

Example 7.3 Design: For the same setup as Example 7.2, consider the sum-to-zero linear constraints,

$$\sum_{i=1}^{3} \alpha_i = 0, \text{ or } \alpha_3 = -(\alpha_1 + \alpha_2) .$$

This can be represented with matrices as

$$\begin{pmatrix} y_{11} \\ y_{12} \\ y_{21} \\ y_{22} \\ y_{31} \\ y_{32} \end{pmatrix} = \begin{bmatrix} 1 & 1 & 0 \\ 1 & 1 & 0 \\ 1 & 0 & 1 \\ 1 & 0 & 1 \\ 1 & -1 & -1 \\ 1 & -1 & -1 \end{bmatrix} \begin{pmatrix} \mu \\ \alpha_1 \\ \alpha_2 \end{pmatrix} + \begin{pmatrix} \epsilon_{11} \\ \epsilon_{12} \\ \epsilon_{21} \\ \epsilon_{22} \\ \epsilon_{31} \\ \epsilon_{32} \end{pmatrix} ,$$

leading to the following relations between mean response and model parameters:

$$\begin{aligned} \mu_1 &= \mu + \alpha_1 \\ \mu_2 &= \mu + \alpha_2 \\ \mu_3 &= \mu - \alpha_1 - \alpha_2 . \end{aligned}$$

Expressing model parameters in terms of group means gives $\mu = \bar{\mu}.$ and $\alpha_i = \mu_i - \bar{\mu}.$ with $\bar{\mu}. = (\mu_1 + \mu_2 + \mu_3)/3$. $\diamond$

The far easier **set-to-zero linear constraints** simply set enough parameters equal to 0 to specify the model completely. This must be done in a careful way, of course. Packages such as SAS (SAS Institute 1992) set the last level of each factor effect to zero (e.g. $\alpha_a = 0$) in the effects model. For more complicated designs it is important to keep track of how many degrees of freedom are available for each model component and set parameters to zero accordingly in a separate fashion for each component. This is readily accomplished in a logical manner but makes interpretation of model parameters difficult.

Example 7.4 Design: For the same setup as in Example 7.2, the set-to-zero linear constraint is $\alpha_a = 0$. The matrix form is

$$\begin{pmatrix} y_{11} \\ y_{12} \\ y_{21} \\ y_{22} \\ y_{31} \\ y_{32} \end{pmatrix} = \begin{bmatrix} 1 & 1 & 0 \\ 1 & 1 & 0 \\ 1 & 0 & 1 \\ 1 & 0 & 1 \\ 1 & 0 & 0 \\ 1 & 0 & 0 \end{bmatrix} \begin{pmatrix} \mu \\ \alpha_1 \\ \alpha_2 \end{pmatrix} + \begin{pmatrix} \epsilon_{11} \\ \epsilon_{12} \\ \epsilon_{21} \\ \epsilon_{22} \\ \epsilon_{31} \\ \epsilon_{32} \end{pmatrix} .$$

The relationship between model parameters and treatment means is a bit different here, with

$$\begin{aligned} \mu_1 &= \mu + \alpha_1 \\ \mu_2 &= \mu + \alpha_2 \\ \mu_3 &= \mu + 0 = \mu . \end{aligned}$$

Model parameters have rather different interpretation with these constraints:

$$\alpha_1 = \mu_1 - \mu_3$$
$$\alpha_2 = \mu_2 - \mu_3$$
$$\mu = \mu_3 \ .$$

Thus the α_i can be interpreted as differences of two factor means. Getting the population grand mean and effects requires some algebra:

$$\bar{\mu}_{.} = \mu + (\alpha_1 + \alpha_2)/3$$
$$\mu_1 - \bar{\mu}_{.} = (2\alpha_1 - \alpha_2)/3$$
$$\mu_2 - \bar{\mu}_{.} = (2\alpha_2 - \alpha_1)/3$$
$$\mu_3 - \bar{\mu}_{.} = -(\alpha_1 + \alpha_2)/3 \ .$$

Be very careful about how model parameters and their estimates are interpreted. Model parameters under set-to-zero linear constraints are rarely of direct interest to the scientist unless one factor level is a natural reference point. ◊

Note that whatever linear constraints are imposed, the population grand mean and group effects have the following relationship to model parameters in the effects model with three groups:

$$\bar{\mu}_{.} = \mu + \bar{\alpha} = \mu + (\alpha_1 + \alpha_2 + \alpha_3)/3$$
$$\mu_1 - \bar{\mu}_{.} = \alpha_1 - \bar{\alpha} = (2\alpha_1 - \alpha_2 - \alpha_3)/3$$
$$\mu_2 - \bar{\mu}_{.} = \alpha_2 - \bar{\alpha} = (2\alpha_2 - \alpha_1 - \alpha_3)/3$$
$$\mu_3 - \bar{\mu}_{.} = \alpha_3 - \bar{\alpha} = (2\alpha_3 - \alpha_1 - \alpha_2)/3 \ .$$

Replacing α_3 by its value under either the sum-to-zero or set-to-zero linear constraints reduces the above representation to that found in Example 7.3 or 7.4, respectively. These ideas can be readily extended to more complicated models by carefully keeping track of terms.

Side constraints for the two-factor model are slightly more complicated. The model cell mean

$$\mu_{ij} = \mu + \alpha_i + \beta_j + \gamma_{ij}$$

has $ab + a + b + 1$ parameters in the effects version, but only ab estimable cell means. There must be at least $a + b + 1$ linear constraints in order for the model to be well determined.

The **sum-to-zero linear constraints**,

$$\alpha_{.} = \beta_{.} = \gamma_{i\cdot} = \gamma_{\cdot j} = \gamma_{..} = 0 \ ,$$

specify $a + b + 3$ linear relations, but only $a + b + 1$ unique constraints on the parameters in the effects model. These provide a nice interpretation of the model parameters, as indicated earlier in this section. The population grand mean is $\mu = \bar{\mu}_{..}$ while the effect of treatment A_i averaged over the levels of factor B is $\alpha_i = \bar{\mu}_{i\cdot} - \bar{\mu}_{..}$, with a similar interpretation for β_j . Any modification of effects (that is, interaction) that might result

GENERAL FORM OF ESTIMABLE FUNCTIONS 117

from the combination of A_i and B_j on the same EU are measured by $\gamma_{ij} = \mu_{ij} - \bar{\mu}_{i\cdot} - \bar{\mu}_{\cdot j} + \bar{\mu}_{\cdot\cdot}$. Put another way, the marginal means for A_i and B_j can be represented in terms of these model parameters as, respectively,

$$\bar{\mu}_{i\cdot} = \mu + \alpha_i + \bar{\beta}_{\cdot} + \bar{\gamma}_{i\cdot} = \mu + \alpha_i$$
$$\bar{\mu}_{\cdot j} = \mu + \bar{\alpha}_{\cdot} + \beta_j + \bar{\gamma}_{\cdot j} = \mu + \beta_j \ .$$

The confounding of α_i and $\gamma_{i\cdot}$ is now hidden, since the latter is zero due to the linear constraints. The sum-to-zero linear constraints would appear to be the natural choice. However, if there are any missing cells ($n_{ij} = 0$), then some of the sum-to-zero constraints are not well defined.

The **set-to-zero linear constraints**,

$$\alpha_a = \beta_b = \gamma_{ib} = \gamma_{aj} = \gamma_{ab} = 0$$

again have are $a+b+3$ constraints specified, but two are redundant. These constraints lead to markedly different interpretations of the model parameters. That is,

$$\mu = \mu_{ab}$$
$$\alpha_i = \mu_{ib} - \mu_{ab}$$
$$\beta_j = \mu_{aj} - \mu_{ab}$$
$$\gamma_{ij} = \mu_{ij} - \mu_{ib} - \mu_{aj} + \mu_{ab} \ .$$

While these may seem odd, they span the space of estimable functions. Thus any linear combinations of these estimable functions can be estimated, and any estimable function can be written as a linear combination of them.

7.5 General form of estimable functions

The general form of estimable functions provides a convenient, general approach to estimable functions and linear constraints. This section recasts the types of models considered in this chapter into the general form of estimable functions to indicate how to interpret these in other situations. The set-to-zero linear constraint case is shown first, since it is easier. The coefficients for the set-to-zero case are denoted by the letter L to conform with the notation used in SAS (see Littell, Freund and Spector 1991, sec 4.3).

Example 7.5 Design: Consider the design matrix for the unconstrained effects model in Example 7.2, but including only one row per group. Each

component	level	effect	coefficient
reference		μ	L_1
factor A	1	α_1	L_2
	2	α_2	L_3
	3	α_3	$L_4 = L_1 - L_2 - L_3$

Table 7.1. *General form of estimable functions for one factor*

column is identified by a coefficient L:

$$\begin{pmatrix} \mu_1 \\ \mu_2 \\ \mu_3 \end{pmatrix} = \begin{matrix} L_1 & L_2 & L_3 & L_4 \\ \begin{bmatrix} 1 & 1 & 0 & 0 \\ 1 & 0 & 1 & 0 \\ 1 & 0 & 0 & 1 \end{bmatrix} & & & \end{matrix} \begin{pmatrix} \mu \\ \alpha_1 \\ \alpha_2 \\ \alpha_3 \end{pmatrix}.$$

Notice that the first column can be written as a linear combination of the other three,

$$\begin{pmatrix} 1 \\ 1 \\ 1 \end{pmatrix} = \begin{pmatrix} 1 \\ 0 \\ 0 \end{pmatrix} + \begin{pmatrix} 0 \\ 1 \\ 0 \end{pmatrix} + \begin{pmatrix} 0 \\ 0 \\ 1 \end{pmatrix}.$$

Thus in order for a linear combination of the model parameters, say $f = L_1\mu + L_2\alpha_1 + L_3\alpha_2 + L_4\alpha_3$, to be estimable the coefficients L_i must have the relationship

$$L_1 = L_2 + L_3 + L_4 \text{ or } L_4 = L_1 - L_2 - L_3.$$

The general form of estimable functions is summarized in Table 7.1. That is, estimable functions f are linear combinations of parameters that have the form

$$f = L_1\mu + L_2\alpha_1 + L_3\alpha_2 + (L_1 - L_2 - L_3)\alpha_3,$$

or equivalently, combining by coefficient L,

$$f = L_1(\mu + \alpha_3) + L_2(\alpha_1 - \alpha_3) + L_3(\alpha_2 - \alpha_3).$$

Thus setting $L_2 = L_3 = 0$ and $L_1 = 1$ yields $\mu_3 = \mu + \alpha_3$ as estimable. Similarly, $L_1 = L_2 = 0$ and $L_3 = 1$ results in the difference of the last two group means $\alpha_2 - \alpha_3 = \mu_2 - \mu_3$. To construct the contrast $\mu_1 - \mu_2 = \alpha_1 - \alpha_2$ set $L_1 = 0$, $L_2 = 1$ and $L_3 = -1$. The population grand mean $\bar{\mu}.$ can be shown to be estimable by setting $L_1 = 1$ and $L_2 = L_3 = 1/3$. However, it is not possible to set the L coefficients to isolate μ or any one of the α_i parameters. ◊

Example 7.6 Design: The set-to-zero linear constraint $\alpha_3 = 0$ simplifies the calculation of estimable functions to

$$f = L_1\mu + L_2\alpha_1 + L_3\alpha_2 \ .$$

Here the reference point is $\mu = \mu_3$ and α_1 and α_2 are differences of the first two group means with the third. The SAS `solution` option to the `model` statement produces estimates of these estimable functions. The general form of estimable functions with the set-to-zero linear constraints can be examined in SAS using the `e` option to the `model` statement in `proc glm`.

```
proc glm;       /* print solution, estimable functions */
   class A;
   model Y = A / solution e;
```

The SAS output is considerably less readable than Table 7.1. ◊

Example 7.7 Design: The sum-to-zero linear constraint $\bar{\alpha}. = 0$ ($\bar{\mu}. = \mu$) could be implemented in a variety of ways. A common choice sets $\alpha_3 = -(\alpha_1 + \alpha_2)$ yielding estimable functions

$$f = L_1\mu + (2L_2 + L_3 - L_1)\alpha_1 + (2L_3 + L_2 - L_1)\alpha_2 \ ,$$

or equivalently, combining by coefficient L,

$$f = L_1(\mu - \alpha_1 - \alpha_2) + L_2(2\alpha_1 + \alpha_2) + L_3(2\alpha_2 + \alpha_1) \ .$$

Note that setting $L_1 = 1$ and $L_2 = L_3 = 0$ yields the third mean $\mu_3 = \mu - \alpha_1 - \alpha_2$ as above, although its relationship to the parameters is different since the reference point is now $\mu = \bar{\mu}.$, the grand mean. Other linear combinations yield exactly the same estimable functions as in the previous example. However, it is not possible to estimate the parameters μ and α_i . In other words, the parameters have different interpretations since the reference is different. ◊

Thus, given a set of constraints, it is possible to establish the general form of estimable functions. These estimable functions are the same regardless of what particular constraints are chosen. While linear constraints ensure unique solutions to the normal equations, they do not make the model parameters estimable. The only **estimable functions** of parameters are those whose estimates are the same regardless of the particular solution of the normal equations.

Example 7.8 Budding: A genetic experiment involving members of the mustard family, varieties of *Brassica rapa*, found strong association between days to budding of flowers and a particular genetic marker (Song, Slocum and Osborn 1995; *cf.* Yandell 1991). Based on 95 lines with no missing data, the means (and standard errors) by genotype were as follows:

genotype		Spring broccoli	Chinese cabbage	hybrid
mean	$\hat{\mu}_i$	85.29	97.75	88.02
standard error	$\hat{\sigma}/\sqrt{n_i}$	2.06	2.32	1.70
sample size	n_i	28	22	41

There were very significant differences among the means ($F = 8.84$, $p = 0.0003$) which could be readily explained (e.g. with Fisher's LSD or other multiple comparisons) as cabbage taking longer to bud than the other two, which were not significantly different.

The SAS package provides a solution for the effects model with set-to-zero linear constraints.

```
proc glm;   /* least squares group means and grand mean */
   class locus;
   model days = locus / solution e;
   lsmeans locus / stderr pdiff;
   estimate 'grand mean' intercept 3 locus 1 1 1
      / divisor=3;
```

The least squares (LS) solutions from SAS are

```
Parameter          Estimate  Test=0  p-value   SE
INTERCEPT          88.02 B   51.82   0.0001   1.70
LOCUS  broccoli    -2.74 B   -1.03   0.3073   2.67
       cabbage      9.75 B    4.03   0.0001   2.87
       hybrid       0    B      .       .       .
```

The coefficient for the INTERCEPT, 88.02, estimates the hybrid LOCUS mean $\mu_3 = \mu + \alpha_3$, corresponding to $L_1 = 1$ and $L_2 = L_3 = 0$. The estimate of the contrast between broccoli and hybrid, $\mu_1 - \mu_3$ is –2.74. The estimate of the population grand mean $\bar{\mu}.$ is

$$88.02 + (-2.74 + 9.75 + 0)/3 = (85.29 + 88.02 + 97.77)/3 = 90.36 .$$

This estimate differs from the sample grand mean (89.54) which weights the group means by sample size. This latter estimate is provided by many packages such as SAS (stated as LOCUS mean on the same line with R-Square and Root MSE), while the appropriate estimator must be computed by hand or with an estimate command!   ◊

Example 7.9 Design: The general form of estimable functions for the **two-factor model** is shown in Table 7.2 when $a = 3$ and $b = 2$. The last coefficient can be found by noticing that $L_{12} = L_6 - L_8 - L_{10}$ and substituting expressions for all three of the terms on the right-hand side.

GENERAL FORM OF ESTIMABLE FUNCTIONS

component	level		effect	coefficient
reference			μ	L_1
factor A	1		α_1	L_2
	2		α_2	L_3
	3		α_3	$L_4 = L_1 - L_2 - L_3$
factor B	1		β_1	L_5
	2		β_2	$L_6 = L_1 - L_5$
combination AB	1	1	γ_{11}	L_7
	1	2	γ_{12}	$L_8 = L_2 - L_7$
	2	1	γ_{21}	L_9
	2	2	γ_{22}	$L_{10} = L_3 - L_9$
	3	1	γ_{31}	$L_{11} = L_5 - L_7 - L_9$
	3	2	γ_{32}	$L_{12} = L_1 - L_2 - L_3 - L_5 + L_7 + L_9$

Table 7.2. *General form of estimable functions for two factors*

The general form of estimable functions f is thus

$$f = L_1\mu + L_2\alpha_1 + L_3\alpha_2 + (L_1 - L_2 - L_3)\alpha_3 + L_5\beta_1 + (L_1 - L_5)\beta_2$$
$$+ L_7\gamma_{11} + (L_2 - L_7)\gamma_{12} + L_9\gamma_{21} + (L_3 - L_9)\gamma_{22}$$
$$+ (L_5 - L_7 - L_9)\gamma_{31} + (L_1 - L_2 - L_3 - L_5 + L_7 + L_9)\gamma_{32} ,$$

which can also be expressed in terms of cell means as

$$f = L_1(\mu_{32}) + L_2(\mu_{12} - \mu_{32}) + L_3(\mu_{22} - \mu_{32}) + L_5(\mu_{31} - \mu_{32})$$
$$+ L_7(\mu_{11} - \mu_{12} + \mu_{32} - \mu_{31}) + L_9(\mu_{21} - \mu_{22} + \mu_{32} - \mu_{31}) .$$

Packages like SAS set this up automatically, but provide no interpretation with the output (but see Littell, Freund and Spector 1991). ◊

Example 7.10 Design: The set-to-zero linear constraints for $a = 3$ and $b = 2$ are

$$\alpha_3 = \beta_2 = \gamma_{3j} = \gamma_{i2} = 0 .$$

Thus in matrix form the relationship between cell means and parameters, after dropping columns corresponding to zero constraints, can be written as

$$\begin{pmatrix} \mu_{11} \\ \mu_{12} \\ \mu_{21} \\ \mu_{22} \\ \mu_{31} \\ \mu_{32} \end{pmatrix} = \begin{bmatrix} \overset{L_1}{1} & \overset{L_2}{1} & \overset{L_3}{0} & \overset{L_5}{1} & \overset{L_7}{1} & \overset{L_9}{0} \\ 1 & 1 & 0 & 0 & 0 & 0 \\ 1 & 0 & 1 & 1 & 0 & 1 \\ 1 & 0 & 1 & 0 & 0 & 0 \\ 1 & 0 & 0 & 1 & 0 & 0 \\ 1 & 0 & 0 & 0 & 0 & 0 \end{bmatrix} \begin{pmatrix} \mu \\ \alpha_1 \\ \alpha_2 \\ \beta_1 \\ \gamma_{11} \\ \gamma_{21} \end{pmatrix} .$$

The corresponding estimable functions are thus

$$f = L_1\mu + L_2\alpha_1 + L_3\alpha_2 + L_5\beta_1 + L_7\gamma_{11} + L_9\gamma_{21} .$$

A package such as SAS obtains the particular solution with the set-to-zero constraints and uses the general form for estimable functions. For instance, the marginal means can be expressed in the general form with a little care. The first marginal for A, $\bar{\mu}_{1.} = (\mu_{11} + \mu_{12})/2$, is derived by setting $L_1 = L_2 = 1$, $L_5 = L_7 = 1/2$ and $L_3 = L_9 = 0$. Fortunately, in SAS it is possible to use the lsmeans or estimate statements rather than calculate directly from the solutions.

```
proc glm;
   class A B;
   model Y = A | B / solution e;
   lsmeans A B / stderr pdiff;      /* marginal means */
   lsmeans A*B / stderr pdiff;      /* cell means */
   estimate 'grand mean' intercept 6 A 2 2 2 B 3 3
       A*B 1 1 1 1 1 1 / divisor = 6;
   estimate 'A1' intercept 1 A 1 0 0 B .5 .5
       A*B .5 .5 0 0 0 0;
   estimate 'A2' intercept 1 A 0 1 0 B .5 .5
       A*B 0 0 .5 .5 0 0;
   estimate 'A3' intercept 1 A 0 0 1 B .5 .5
       A*B 0 0 0 0 .5 .5;
   estimate 'B1' intercept 6 A 2 2 2 B 6 0
       A*B 2 0 2 0 2 0 / divisor = 6;
   estimate 'B2' intercept 6 A 2 2 2 B 0 6
       A*B 0 2 0 2 0 2 / divisor = 6;
```

The entries have been stretched over several lines to make it easier to check that values are properly recorded. It is better to use a divisor to avoid roundoff error (0.3333 is not 1/3), although exact decimal fractions can be used as in the second estimate statement. ◊

Note in the estimate statement that the intercept and all levels of any factor effect or combination of factors (interaction) can be specified. The order of levels is alphanumeric and the order of factors in a combination of

GENERAL FORM OF ESTIMABLE FUNCTIONS   123

factors making up an interaction is determined by the class statement. This can be checked by examining the order in the solution from the model or by specifying the e option to the estimate statement. More details can be found in Littell, Freund and Spector (1991).

Example 7.11 Design: The general form of estimable functions with sum-to-zero linear constraints corresponds to a different relationship between cell means and model parameters. For $a = 3$ and $b = 2$ the constraints

$$\begin{aligned} \alpha_3 &= -\alpha_1 - \alpha_2, & \beta_2 &= -\beta_1, \\ \gamma_{12} &= -\gamma_{11}, & \gamma_{22} &= -\gamma_{21}, \\ \gamma_{31} &= -\gamma_{11} - \gamma_{21}, & \gamma_{32} &= \gamma_{11} + \gamma_{21}, \end{aligned}$$

lead to the matrix form

$$\begin{pmatrix} \mu_{11} \\ \mu_{12} \\ \mu_{21} \\ \mu_{22} \\ \mu_{31} \\ \mu_{32} \end{pmatrix} = \begin{bmatrix} 1 & 1 & 0 & 1 & 1 & 0 \\ 1 & 1 & 0 & -1 & -1 & 0 \\ 1 & 0 & 1 & 1 & 0 & 1 \\ 1 & 0 & 1 & -1 & 0 & -1 \\ 1 & -1 & -1 & 1 & -1 & -1 \\ 1 & -1 & -1 & -1 & 1 & 1 \end{bmatrix} \begin{pmatrix} \mu \\ \alpha_1 \\ \alpha_2 \\ \beta_1 \\ \gamma_{11} \\ \gamma_{21} \end{pmatrix}.$$

Using the notation of the general form of estimable functions this corresponds to

$$f = L_1\mu + (2L_2 + L_3 - L_1)\alpha_1 + (2L_3 + L_2 - L_1)\alpha_2$$
$$+ (2L_5 - L_1)\beta_1$$
$$+ (L_1 - 2L_2 - L_3 - 2L_5 + 4L_7 + 2L_9)\gamma_{11}$$
$$+ (L_1 - L_2 - 2L_3 - 2L_5 + 2L_7 + 4L_9)\gamma_{21}.$$

Some packages such as S-Plus use this form. Other forms are possible, limited only by the imagination. ◇

It can be confusing to find that two packages give very different solutions. In fact, they should if they use different linear constraints on model parameters. What is important to verify is that they give the same estimates of **estimable functions**. Knowing how those are calculated is vital!

In practice scientists tend to fit the over-specified model, letting a statistical package make intelligent choices about what functions are estimable. However, it is useful to understand how these choices can be made and, in some situations, to know how to impose appropriate constraints on a problem. Sometimes this can be done using a statistical package by specifying factors to estimate or comparisons to be tested. At other times, it may be easier to do a few steps by hand with a calculator.

Modern statistical packages can compute the population grand mean and marginal means using package options. However, it can be important when trying to estimate model features of interest to know exactly how to reproduce these relationships. This knowledge can be used to check the accuracy of statistical package, or to check the veracity of different analyses of the same data.

7.6 Problems

7.1 Infer: This problem concerns the normal equations and side constraints. Consider the model

$$y_{ij} = \mu + \alpha_i + \epsilon_{ij}, \ i = 1, 2, \ j = 1, 2, 3.$$

with the usual assumptions on ϵ_{ij}. As it stands, the model is over-specified, with three parameters in the mean piece but only two groups.
(a) Write down the normal equations for the model. Show that $\mu + \alpha_1$ is estimable but α_1 is not. [Hint: You can do this by picking a particular solution.]
(b) Impose a sum-to-zero linear constraint and write out the model for each group ($i = 1, 2$) separately.
(c) Average the model for each group from (b) over the experimental units ($j = 1, 2, 3$). That is, what is the distribution of $\bar{y}_{i\cdot}$ in terms of model parameters under the sum-to-zero constraint?
(d)–(e) Repeat (b)–(c) for set-to-zero constraint.
(f) Write out the general form of estimable functions. Show how it 'simplifies' for both sum-to-zero and set-to-zero constraints.

7.2 Tree: The S-Plus system implicitly establishes contrasts among the levels of each factor in a data frame. These contrasts can be envisioned as the building blocks for columns of the design matrix. Understanding them can help in interpreting estimates of model parameters. Consider the following in terms of the example in Problem 5.3.
(a) Verify that contrasts set by `contr.sum` and `contr.treatment` correspond to the sum-to-zero and set-to-zero linear constraints, respectively.
(b) The default contrasts in S-Plus, called Helmert coding, enforce sum-to-zero constraints in a slightly different way. The first contrast is between the lowest two levels of a factor, the second is between their average and the third, and so on (see Venables and Ripley 1994, sec. 6.2). Verify that these impose the sum-to-zero constraint.
(c) Show how the solutions of the normal equations (the coefficients) are related to the constraints for the three types of contrasts above. [Caution: the Helmert coding leads to a slighly different model formulation.]

CHAPTER 8

Balanced Experiments

This chapter begins the examination of more complicated treatment and design structures, while keeping things balanced for ease of presentation and analysis. These experimental designs are sometimes referred to as **factorial designs** since they are built from several factors, or treatments. Consider an experiment with a number of cells, corresponding to different combinations of factor levels, with experimental units assigned at random to cells in a completely randomized design. The two-factor case allows segregation of the effects of factor combination into a component for each factor (main effects) and a component measuring the synergy or antagonism among factors (interactions).

There are several ways to interpret experiments involving two or more factors. Recall that a complicated treatment structure can be viewed as a single factor by examining cell means, which opens up multiple comparisons as in the previous chapter. However, it is more efficient to consider multiple comparisons in the context of main effects and interactions when examining two or more factors. That is, overall tests for model components that yield significant results can be followed by contrasts and multiple comparisons for those components.

Section 8.1 presents some of the new issues that arise when considering more than one factor. The balanced two-factor effects model with an equal number of observations per factor combination is discussed in detail in Section 8.2. Additive models, containing main effects but no interactions, are developed in Section 8.3. Section 8.4 briefly considers higher-order models. Finally, interaction plots are encouraged in Section 8.5.

8.1 Additive models

The additive model assumes that each factor influences the mean response separately. For two factors, the **additive model with no replication** of factor combinations is simply

$$\text{response} = \text{mean} + \text{factor } A \text{ effect} + \text{factor } B \text{ effect} + \text{error} ,$$

Source	df	SS
A	$a-1$	$\sum_i b(\bar{y}_{i\cdot} - \bar{y}_{\cdot\cdot})^2$
B	$b-1$	$\sum_j a(\bar{y}_{\cdot j} - \bar{y}_{\cdot\cdot})^2$
error	$(a-1)(b-1)$	$\sum_i \sum_j (y_{ij} - \bar{y}_{\cdot j} - \bar{y}_{i\cdot} + \bar{y}_{\cdot\cdot})^2$
total	$ab-1$	$\sum_i \sum_j (y_{ij} - \bar{y}_{\cdot\cdot})^2$

Table 8.1. *Two-factor additive model*

or as a mathematical model,

$$y_{ij} = \mu_{ij} + e_{ij} = \mu + \alpha_i + \beta_j + e_{ij}$$

with $i = 1, \cdots, a$, $j = 1, \cdots, b$, and independent errors $e_{ij} \sim N(0, \sigma^2)$. Interest focuses upon questions about these **main effects**. In fact, they can be discussed separately. That is, the effect of factor A is the same regardless of the level of factor B. This is precisely the meaning of **no or negligible interaction** among factors. The overall hypothesis for factor A main effect is

$$H_0 : \alpha_i = 0 \text{ or } H_0 : \mu_{ij} = \mu_j = \mu + \beta_j \text{ for } i = 1, \cdots, a .$$

Similarly, the overall hypothesis for B is $H_0 : \beta_j = 0$ or $H_0 : \mu_{ij} = \mu_i = \mu + \alpha_i$ for $j = 1, \cdots, b$.

The total variation is partitioned into sums of squares for factors A and B plus the unexplained variation:

$$SS_{\text{TOTAL}} = SS_A + SS_B + SS_{\text{ERROR}} .$$

The analysis of variance table summarizes this relation (Table 8.1). Expected mean squares are similar to the one-factor model:

$$\begin{aligned} E(MS_A) &= \sigma^2 + b \sum_i (\alpha_i - \bar{\alpha}_\cdot)^2 / (a-1) \\ E(MS_B) &= \sigma^2 + a \sum_j (\beta_j - \bar{\beta}_\cdot)^2 / (b-1) \\ E(MS_E) &= \sigma^2 \end{aligned}$$

Formal tests follow naturally as well. For factor A, the F test for the hypothesis $H_0 : \alpha_i = 0$ is

$$F = MS_A / MS_E \sim F_{(a-1),(a-1)(b-1);\delta_A^2}$$

with non-centrality parameter $\delta_A^2 = b \sum_i (\alpha_i - \bar{\alpha}_\cdot)^2 / \sigma^2$.

ADDITIVE MODELS

Source	df	SS
A	$a-1$	$bn\sum_i(\bar{y}_{i..}-\bar{y}_{...})^2$
B	$b-1$	$an\sum_j(\bar{y}_{.j.}-\bar{y}_{...})^2$
error	$abn-a-b+1$	$\sum_{ijk}(y_{ijk}-\bar{y}_{.j.}-\bar{y}_{i..}+\bar{y}_{...})^2$
total	$abn-1$	$\sum_{ijk}(y_{ijk}-\bar{y}_{...})^2$

Table 8.2. *Two-factor additive model with replication*

Least squares estimates are developed by minimizing the residual sum of squares $SS = \sum_{ij}(y_{ij} - \mu - \alpha_i - \beta_j)^2$ leading to normal equations

$$ab\mu = \sum_{ij} y_{ij} - b\sum_i \alpha_i - a\sum_j \beta_j$$
$$b\alpha_i = \sum_j y_{ij} - b\mu - \sum_j \beta_j$$
$$a\beta_j = \sum_i y_{ij} - a\mu - \sum_i \alpha_i$$

which have no unique solutions. However, the following parameters, among others, are estimable by appealing to the previous chapter:

model parameter		estimator
$\bar{\mu}_{..}$	$= \mu + \bar{\alpha}_{.} + \bar{\beta}_{.}$	$\bar{y}_{..}$
$\bar{\mu}_{i.}$	$= \mu + \alpha_i + \bar{\beta}_{.}$	$\bar{y}_{i.}$
$\bar{\mu}_{.j}$	$= \mu + \bar{\alpha}_{.} + \beta_j$	$\bar{y}_{.j}$
μ_{ij}	$= \mu + \alpha_i + \beta_j$	$\bar{y}_{i.} + \bar{y}_{.j} - \bar{y}_{..}$
$\bar{\mu}_{i.} - \bar{\mu}_{..}$	$= \alpha_i - \bar{\alpha}_{.}$	$\bar{y}_{i.} - \bar{y}_{..}$
$\bar{\mu}_{.j} - \bar{\mu}_{..}$	$= \beta_j - \bar{\beta}_{.}$	$\bar{y}_{.j} - \bar{y}_{..}$

Additive models with replication can be written as

$$y_{ijk} = \mu + \alpha_i + \beta_j + e_{ijk}$$

with $k = 1, \cdots, n$, leading to an anova table of similar form (Table 8.2). Expected mean squares are also similar, with for instance

$$E(MS_A) = \sigma^2 + bn \sum_i \alpha_i^2/(a-1) \ .$$

trt	bact	temp	run 1	run 2
1	ctrl	cold	39.77	40.23
2	myco	cold	39.19	38.95
3	ctrl	warm	40.37	41.71
4	myco	warm	40.21	40.78

Table 8.3. *Bacteria mean bill length*

source	df	SS	MS	F	p-value
bact	1	1.08	1.08	4.31	0.093
temp	1	3.02	3.02	12.0	0.018
error	5	1.26	0.25		
total	7	5.36			

Table 8.4. *Bacteria additive anova*

Formal tests follow naturally as well. Normal equations simply involve averaging over replications to get estimates as follows:

model parameter		estimator
$\bar{\mu}_{..}$	$= \mu + \bar{\alpha}_{.} + \bar{\beta}_{.}$	$\bar{y}_{...}$
$\bar{\mu}_{i.}$	$= \mu + \alpha_i + \bar{\beta}_{.}$	$\bar{y}_{i..}$
$\bar{\mu}_{.j}$	$= \mu + \bar{\alpha}_{.} + \beta_j$	$\bar{y}_{.j.}$
μ_{ij}	$= \mu + \alpha_i + \beta_j$	$\bar{y}_{i..} + \bar{y}_{.j.} - \bar{y}_{...}$

Example 8.1 Bacteria: An experiment performed at the National Wildlife Health Research Center under the supervision of Michael Samuel (Samuel et al. 1995) examined the effect of a certain bacteria strain (mycoplasma) on the development of birds. Unfortunately, due to the danger of aerial infection, they had to isolate treatment groups in different rooms. A very conservative approach would take each chamber as an experimental unit, regardless of the number of birds per room. There were four treatments: cold, cold+myco, warm and warm+myco. In addition, the experiment was run in two runs since there were only four rooms. The data for bill length are summarized in Table 8.3. For the present, these are all the data available.

Suppose the scientist believed ahead of time that interaction of bacteria and temperature was negligible. In this case the simpler additive model would be appropriate. Notice how the tests for main effects are more significant in Table 8.4. Since the F test for interactions in Example 8.3, was less than 1, the estimate of error variance in the additive model is smaller. This, combined with the extra degree of freedom for error, leads to smaller p-values. The marginal means for the two factors are the same as before although the standard error has decreased from 0.39 to 0.25. The results are qualitatively similar to those for the full two-factor model, but a strict reading (e.g. rejecting the null hypothesis if $p > 0.1$) would give contradictory interpretation for bacteria between the two models. A more pragmatic reading is that the results for bacteria are borderline in either case.   ◊

8.2 Full models with two factors

The additive model imposes strong restrictions on the cell means μ_{ij} which may be unwarranted. It is often worthwhile considering the **two-factor cell means model**

$$y_{ijk} = \mu_{ij} + e_{ijk}$$

with $i = 1, \cdots, a$, $j = 1, \cdots, b$, $k = 1, \cdots, n$, and independent errors $e_{ij} \sim N(0, \sigma^2)$. Solutions of the normal equations lead directly to cell mean and variance estimates

$$\hat{\mu}_{ij} = \bar{y}_{ij\cdot} = \sum_{k=1}^{n} y_{ijk}/n$$
$$\hat{\sigma}^2 = \sum_{i=1}^{a} \sum_{j=1}^{b} \sum_{k=1}^{n} \frac{(y_{ijk} - \bar{y}_{ij\cdot})^2}{ab(n-1)},$$

with $\hat{\mu}_{ij} \sim N(\mu_{ij}, \sigma^2/n)$ and $ab(n-1)\hat{\sigma}^2/\sigma^2 \sim \chi^2_{ab(n-1)}$. That is, there are ab degrees of freedom in the model means, leaving $ab(n-1)$ degrees of freedom to estimate the error variance. Tests of hypotheses or confidence intervals for means can be developed in the same manner as for the one-factor cell means model. Alternatively, it may be useful to partition the cell means into main effects and interactions with a **two-factor effects model**

$$y_{ijk} = \mu + \alpha_i + \beta_j + \gamma_{ij} + e_{ijk}$$

introduced in the previous chapter. The balanced two-factor experiment has least squares estimates of cell means and marginal population means as follows:

model parameter		estimator	standard error
μ_{ij}	$= \mu + \alpha_i + \beta_j + \gamma_{ij}$	$\bar{y}_{ij\cdot}$	$\sigma/\sqrt{n}$
$\bar{\mu}_{i\cdot}$	$= \mu + \alpha_i + \bar{\beta}_{\cdot} + \bar{\gamma}_{i\cdot}$	$\bar{y}_{i\cdot\cdot}$	$\sigma/\sqrt{bn}$
$\bar{\mu}_{\cdot j}$	$= \mu + \bar{\alpha}_{\cdot} + \beta_j + \bar{\gamma}_{\cdot j}$	$\bar{y}_{\cdot j\cdot}$	$\sigma/\sqrt{an}$
$\bar{\mu}_{\cdot\cdot}$	$= \mu + \bar{\alpha}_{\cdot} + \bar{\beta}_{\cdot} + \bar{\gamma}_{\cdot\cdot}$	$\bar{y}_{\cdot\cdot\cdot}$	$\sigma/\sqrt{abn}$

Notice that when there is no replication ($n = 1$), it is not possible to make inference about all cell means or about full interaction. Alternatives to the

130   BALANCED EXPERIMENTS

additive model considered in the previous section are developed toward the end of the next chapter.

What are the key questions one wishes to ask in such an experiment? And what questions can be addressed with data analysis? Sometimes a scientist is interested in comparing these means directly. For instance, if the goal is to find the 'best' combination of two chemicals, then part of the analysis may focus on the means model and a one-factor interpretation with means comparisons. However, usually it is helpful to separate the effects of factors A and B to determine if either or both have substantial effect on response. Evidence of interaction would suggest qualifying statements about the effect of one factor by indicating the level of the other.

Example 8.2 Infer: Consider an experiment with three varieties (B_1, B_2, B_3) and two fertilizers (A_1, A_2) laid out in 12 plots with a completely randomized design. In other words, the six treatment combinations were assigned at random to two plots each. Here are some questions and appropriate hypotheses. If the key question is 'Which variety is best?', attention should focus on

$$H_0 : \bar{\mu}_{.1} = \bar{\mu}_{.2} = \bar{\mu}_{.3} \text{ or } H_0 : \bar{\mu}_{.j} = \bar{\mu}_{..}$$

or the hypothesis that the $\beta_j + \bar{\gamma}_{.j}, j = 1, 2, 3$, are constant. For the question 'Are fertilizers the same?' the hypothesis of interest is

$$H_0 : \bar{\mu}_{1.} = \bar{\mu}_{2.} \text{ or } H_0 : \bar{\mu}_{i.} = \bar{\mu}_{..}$$

or that the $\alpha_i + \bar{\gamma}_{i.}, i = 1, 2$, are constant. The question 'Does fertilizer matter when comparing varieties?' concerns interactions

$$H_0 : \mu_{11} - \mu_{21} = \mu_{12} - \mu_{22} = \mu_{13} - \mu_{23} \text{ or } H_0 : \mu_{ij} = \bar{\mu}_{i.} + \bar{\mu}_{.j} - \bar{\mu}_{..}$$

or the hypothesis that the γ_{ij}, $i = 1, 2$, $j = 1, 2, 3$, are constant. ◊

The centered response can be partitioned into components for main effects, pure interaction and residual:

centered mean = A effect   + B effect   + interaction + residual

$$y_{ijk} - \bar{\mu}_{..} = (\bar{\mu}_{i.} - \bar{\mu}_{..}) + (\bar{\mu}_{.j} - \bar{\mu}_{..}) + (\mu_{ij} - \bar{\mu}_{i.} - \bar{\mu}_{.j} + \bar{\mu}_{..}) + e_{ijk} \; .$$

This suggests a corresponding **partition of the total sum of squares** as shown in Table 8.5. This partition can be justified by arguments similar to those for the partition for the one-factor model in Chapter 5.

Expected mean squares can be computed directly. For instance, expand the sum of squares for factor A as

$$\begin{aligned} SS_A &= nb \sum_i (\bar{\mu}_{i.} - \bar{\mu}_{..} + \bar{e}_{i..} - \bar{e}_{...})^2 \\ &= nb \sum_i [(\bar{\mu}_{i.} - \bar{\mu}_{..})^2 + 2(\bar{\mu}_{i.} - \bar{\mu}_{..})(\bar{e}_{i..} - \bar{e}_{...}) + (\bar{e}_{i..} - \bar{e}_{...})^2] \; . \end{aligned}$$

FULL MODELS WITH TWO FACTORS

Source	df	Sum of Squares	
A	$a-1$	SS_A	$= nb\sum_i(\bar{y}_{i\cdot\cdot} - \bar{y}_{\cdot\cdot\cdot})^2$
B	$b-1$	SS_B	$= na\sum_j(\bar{y}_{\cdot j\cdot} - \bar{y}_{\cdot\cdot\cdot})^2$
$A*B$	$(a-1)(b-1)$	SS_{A*B}	$= n\sum_{ij}(\hat{\mu}_{ij} - \bar{y}_{i\cdot\cdot} - \bar{y}_{\cdot j\cdot} + \bar{y}_{\cdot\cdot\cdot})^2$
error	$ab(n-1)$	SS_E	$= \sum_{ijk}(y_{ijk} - \hat{\mu}_{ij})^2$
total	$nab-1$	SS_T	$= \sum_{ijk}(y_{ijk} - \bar{y}_{\cdot\cdot\cdot})^2$

Table 8.5. *Two-factor anova table*

Source	$E(MS)$
A	$\sigma^2 + nb\sum_i(\bar{\mu}_{i\cdot} - \bar{\mu}_{\cdot\cdot})^2/(a-1)$
B	$\sigma^2 + nt\sum_j(\bar{\mu}_{\cdot j} - \bar{\mu}_{\cdot\cdot})^2/(b-1)$
$A*B$	$\sigma^2 + n\sum_{ij}(\mu_{ij} - \bar{\mu}_{i\cdot} - \bar{\mu}_{\cdot j} + \bar{\mu}_{\cdot\cdot})^2/(a-1)(b-1)$
error	σ^2

Table 8.6. *Two-factor expected mean squares*

The first term is constant. The third term is proportional to a χ^2 variate, as seen earlier in the one-factor model,

$$\sum_i(\bar{e}_{i\cdot\cdot} - \bar{e}_{\cdot\cdot\cdot})^2 \sim \sigma^2\chi^2_{a-1}/nb ,$$

since it is the centered sum of squares of normal random variates with variance σ^2/nb. The cross-products in the second term of SS_a have zero mean. Combining this information yields

$$E(SS_A) = nb\sum_i(\bar{\mu}_{i\cdot} - \bar{\mu}_{\cdot\cdot})^2 + (a-1)\sigma^2 .$$

The expected means squares are presented in Table 8.6. For balanced designs, the SS and other parts of the anova table do not change value regardless of order. However, order is extremely important for unbalanced designs, as seen in Part D.

It is certainly possible to conduct an overall F test for the combination of factors, which is essentially the one-factor model test of equal means

$$F = \frac{(SS_A + SS_B + SS_{A*B})/(ab-1)}{SS_E/ab(n-1)} \sim F_{ab-1, ab(n-1); \delta}$$

with non-centrality parameter

$$\delta^2 = n \sum_{ij} (\mu_{ij} - \bar{\mu}_{..})^2/\sigma^2 \ .$$

In practice this is seldom employed, particularly since the tests for main effects and for pure interaction shown below are more powerful than the overall test. It is advisable first to test the interaction with the statistic

$$F = MS_{A*B}/MS_E \sim F_{(a-1)(b-1), ab(n-1); \delta^2_{AB}}$$

with $MS = SS/df$ and non-centrality parameter

$$\delta^2_{AB} = n \sum_{ij} (\gamma_{ij} - \bar{\gamma}_{..})^2/\sigma^2 \ ,$$

which is zero if there are no interactions. The main effects test statistics follow in a similar fashion,

$$\begin{aligned} F &= MS_A/MS_E \sim F_{a-1, ab(n-1); \delta^2_A} \\ F &= MS_B/MS_E \sim F_{b-1, ab(n-1); \delta^2_B} \end{aligned}$$

with non-centrality parameters similar to the one-factor models,

$$\begin{aligned} \delta^2_A &= nb \sum_i (\bar{\mu}_{i.} - \bar{\mu}_{..})^2/\sigma^2 \\ \delta^2_B &= na \sum_j (\bar{\mu}_{.j} - \bar{\mu}_{..})^2/\sigma^2 \ . \end{aligned}$$

These comprise three orthogonal tests for the two main effects and the interaction. As in the one-factor model, it is important first to conduct overall tests before investigating multiple comparisons. A significant result for interactions suggests multiple comparisons among levels of one factor at each level of the other factor. Lack of significance of interactions followed by rejection of the null hypothesis for a main effect would lead to multiple comparisons among marginal means for that factor.

Example 8.3 Bacteria: The anova table for the evaluation of the effects of bacteria and temperature is presented in Table 8.7. Since the interaction is not significant ($p = 0.65$), multiple comparisons across all four means are not recommended. Instead, it makes sense to compare the marginal means. Here, control chicks had longer bills (40.52) than those exposed to bacteria (39.78). Further, chicks in warm rooms had longer bills (40.77) than those in cold rooms (39.54). The standard error for these four marginal means is 0.39 since they all involve four values in simple averages. This analysis confirms the pattern shown in the interaction plot of Figure 8.2. ◊

source	df	SS	MS	F	p-value
bact	1	1.08	1.08	3.65	0.13
temp	1	3.02	3.02	10.19	0.033
bact*temp	1	0.0704	0.0704	0.24	0.65
error	4	1.19	0.297		
total	7	5.36			

Table 8.7. *Bacteria two-factor anova*

8.3 Interaction plots

The presence of interactions compromises interpretation of main effects, depending on the magnitude of interaction. This can be depicted by **interaction plots** of response means by factor combinations against one factor. That is, assign the levels of one factor (say A) across the horizontal axis, with the vertical axis being in the units of the response. For each level A_i, plot the b means $\hat{\mu}_{ij}$, $j = 1, \cdots, b$ above that assigned position. Now draw a line to connect the a means $\hat{\mu}_{ij}$, $i = 1, \cdots, a$ which are at level B_j.

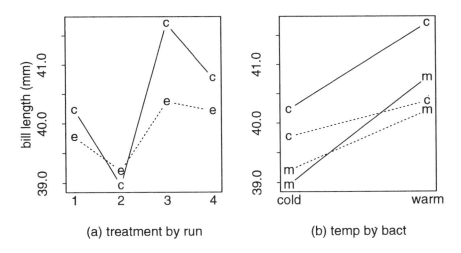

(a) treatment by run   (b) temp by bact

Figure 8.1. *Bacteria interaction plots. (a) Horizontal axis is arbitrary treatment number, which is not helpful. (b) Interaction plot shows interaction between* temperature *and* bacteria *(c = ctrl, m = myco) in their effect on* bill length. *Solid lines for* chick *(c) run and dashed lines for* egg *(e) run.*

Example 8.4 Bacteria: Figure 8.1 shows (a) a scatter plot of these data by treatment number and (b) an interaction plot by the two factors. In both cases, values from the same run are connected by lines (c = chick solid, e = egg dashed). However, Figure 8.1(a) requires a lookup of what the treatment numbers signify. Figure 8.1(b) shows a strong temperature effect, and control (c = ctrl) birds with longer bills than mycoplasma (m) birds. There appears to be some evidence in both figures for differences between runs, but this is of secondary interest. It is the primary feature of Figure 8.1(a), while Figure 8.1(b) emphasizes more appropriately the effects of temperature and bacteria. There is a suggestion in Figure 8.1(b) that the effect of bacteria is less pronounced at higher temperature. ◇

If there is no interaction, then the lines on an interaction plot which connect points at the same level of the second factor B are parallel, that is separated by a constant amount. Such a graphical pattern indicates that the effect of changing the level of B is to raise or lower the mean response by a certain amount, regardless of the level of A. Sometimes patterns are more apparent with one or the other factor along the horizontal axis. It pays to experiment by switching the roles of the two factors.

Any deviation from parallel is evidence of interaction, although this should be verified with a formal test against the estimated unexplained variation $\hat{\sigma}^2$. Placing a small vertical bar the width of $\hat{\sigma}$ (or of a standard error $\hat{\sigma}/\sqrt{n}$) can help others use interaction plots for graphical tests, provided the bar and all lines are properly labelled.

There are many other ways to lay out factor levels along the horizontal axis. With only two levels, the choice is elementary. However, some care is needed with three or more levels. While alphabetical or numerical order may seem reasonable, it does not necessarily reflect the evidence in the data. More useful plots often arise when ordering factor levels by increasing marginal mean response (e.g. $\hat{\mu}_i$ for A_i). This can be taken another step by using those marginal means for exact positioning of factor levels along the horizontal axis as a **margin plot**. (Milliken and Johnson (1989) introduced these as Type II interaction plots.) Ordering factor levels by marginal means can be helpful in tables as well.

Example 8.5 Bacteria: The interaction plots in Figure 8.2 contain a vertical bar to show the size of the LSD = 1.51 ($= t_{0.05;4} \times 0.545\sqrt{2/2}$). Figure 8.2(a) is similar to Figure 8.1(b), but averages over the runs. Figure 8.2(b) shows a margin plot with the horizontal spacing based on the estimated main effects of temperature, adding a dashed identity line. Both figures show only mild evidence for interaction, but strong evidence for both main effects of temperature and bacteria. ◇

Interaction plots can get too busy. Usually one should limit any plot to 3–5 lines (e.g. levels of B), separating levels into two or more plots as needed. Sometimes there are natural breakdowns suggested by the design

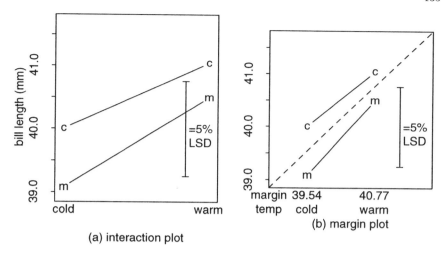

Figure 8.2. *Bacteria two-factor interaction plots. (a) Standard interaction plot with* temperature *levels along the horizontal axis and* bacteria *levels for each line (*c = ctrl, m = myco*). (b) Margin plot is shown square to emphasize the marginal means of* bill *length by level of* temperature *which establish horizontal axis. The* LSD *bar is based on full interaction model.*

of the experiment. In other situations, switching the role of A and B, as mentioned above, can do the trick. Still other situations might suggest averaging over several levels of B for presentation, particularly if this is supported by results of contrasts or by the design of the scientist's key questions.

Other questions still persist for interactions. What if interactions are present but there is no evidence of main effects? Sometimes a log or other **transform** of the data will lead to an additive model in this case. Residual plots may provide valuable insights into such issues.

One new feature introduced in higher-order arrangements is the **three-factor interaction**. If present, the interaction of $A * B$ depends on the level of factor C. For instance, nitrogen (B) affects yield (y) of variety (A) differently under very wet conditions than under dry conditions (C). Depending on recent rainfall, added nitrogen (B_1 versus B_2) could improve yield or have the reverse effect. Thus one must discuss the three factors in the experiment together when interpreting results.

Example 8.6 Bacteria: The runs differed in an important way. The first run involved inoculation of bird eggs while the second run, done several weeks later, inoculated young chicks. Figure 8.3 shows separate interaction plots for the two runs; this is essentially Figure 8.1(b) split into two figures. Notice that bill lengths were shorter for run 1. This could reflect

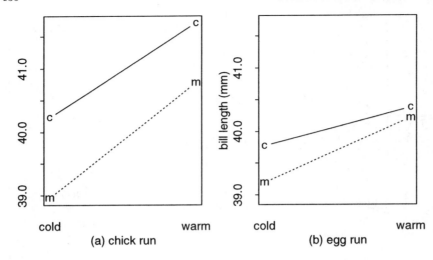

Figure 8.3. *Bacteria three-factor interaction plots: interaction plots of* temperature *by* bacteria *for each* run. *Runs differed by whether (a)* chicks *or (b)* eggs *were inoculated.*

source	df	SS
temp	1	3.023
bact	1	1.083
inoc	1	0.564
temp:inoc	1	0.351
bact:inoc	1	0.270
temp:bact	1	0.070
temp:bact:inoc	1	0.001
total	7	5.362

Table 8.8. *Bacteria three-factor sums of squares*

seasonal differences or be related to the method of inoculation; these are completely confounded by design.

Table 8.8 shows the sums of squares for the three main effects and four interactions. It is not possible to perform tests since there are no degrees of freedom left for error. However, one or more interactions could be assumed to be significant. Notice that if only the three-factor interaction were deemed neglible, most other interactions and main effects would be significant. This would actually reflect the small value of the three-factor interaction. ◊

Sometimes it is valuable to use plots to examine all effects even if the model is saturated. (Here for convenience consider the sum-to-zero linear constraints to define effects as deviations from the grand mean, etc.) A **half-normal plot** is very handy for factors with only two levels. That is, if there are n residuals, the ith largest absolute effect is plotted against the $q_i = (i-s)/(n+1-2s)$ quantile of the square root of the χ_1^2 distribution, or equivalently the $(1+q_i)/2$ quantile of the standard normal, with offset $s = 3/8$ if $n < 10$ and $s = 1/2$ otherwise.

Effect plots are an alternative which work well with more than two levels of factors. Basically, model effects are rescaled by noticing a clever decomposition of mean squares (see Hoaglin, Mosteller and Tukey 1991). For instance, in the two-factor experiment, the mean square for main effect A can be written as

$$MS_A = \sum_i n_{i\cdot}(\bar{y}_{i\cdot\cdot} - \bar{y}_{\cdots})^2/(a-1) = \sum_i r_i(A)^2/a \;,$$

in which $r_i(A) = (\bar{y}_{i\cdot\cdot} - \bar{y}_{\cdots})\sqrt{an_{i\cdot}/(a-1)}$ is the effect adjusted for mean square. Similarly, the residuals could be rescaled as

$$MS_E = \sum_{ijk}(y_{ijk} - \hat{\mu}_{ij})^2/(n_{\cdot\cdot} - ab) = \sum_i r_{ijk}^2/n_{\cdot\cdot} \;,$$

with $r_{ijk} = (y_{ijk} - \hat{\mu}_{ij})\sqrt{n_{\cdot\cdot}/(n_{\cdot\cdot} - ab)}$. These adjusted residuals $r_i(A)$, $r_j(B)$, $r_{ij}(AB)$ and r_{ijk} have mean 0 and variance $\hat{\sigma}^2$ if there are no factor effects, although they are not independent. Plots with these residuals, or the residuals standardized by $\hat{\sigma}$, can reveal unusual patterns in model fit. For instance, side-by-side dot-plots of main effects, interactions and a box-plot of the residuals show graphically the amount of variation explained by each model term.

Example 8.7 Bacteria: Consider two ways to examine all seven effects shown in Table 8.8. Figure 8.4 shows (a) a half-normal plot and (b) an effect plot for comparison. Note that temperature is higher and the three-factor interaction is smaller than expected in Figure 8.4(a). These are also at the extremes in Figure 8.4(b). The half-normal plot suggests that the two-factor interactions may be negligible, but this is less apparent in Figure 8.4(b). In any event, with only eight observations, it seems prudent to consider an additive model, only main effects, as in Table 8.9. Here temperature is significant, bacteria less so and inoculation method probably not important. ◇

Another aspect is the potential of **three sets of two-factor interactions**. Some apparent three-factor interactions can be decomposed into more tractable combinations of two-factor interactions. Suppose that ni-

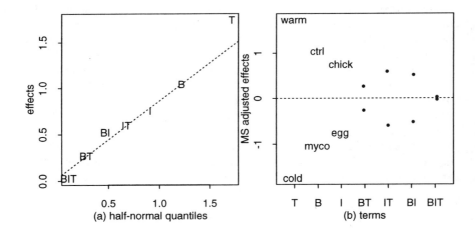

Figure 8.4. *Bacteria half-normal and effect plots. (a) Effects are square root of sum of squares. Expected values are square roots of quantiles of the χ_1^2 distribution. (b) Mean square adjusted effects rescaled so that the SDs are the square root of MS for that effect. Symbols are combined to indicate main effects or interactions with* B = bacteria, I = inoculation *and* T = temperature.

source	df	SS	MS	F	p-value
temp	1	3.023	3.023	14.57	0.014
bact	1	1.083	1.083	6.25	0.067
inoc	1	0.564	0.564	3.26	0.15
error	4	0.693	0.173		
total	7	5.362			

Table 8.9. *Bacteria additive anova*

trogen (B) level affects ordering of varieties (A), rainfall (C) affects order of varieties (A) and rainfall (C) affects availability of nitrogen (B) in soil. Suppose there is no three-factor interaction. Then one can consider factor pairs when interpreting effects. It is difficult to sort this out in the field by just looking at factor means. Even with statistical analysis, it can be tricky. If the design is unbalanced, then interpretation can be very messy and possibly inconclusive.

Three-factor interactions may be viewed with sets of interaction plots. For each level of one factor, draw a separate interaction plot for the other two (with a common SD or SE bar). If there are not too many factor levels, it is possible to present all three factors on one plot, viewing a two-factor combination as one factor. However, this can lead to plots which are too busy and confusing. If these interaction plots are basically the same, only shifted up or down from each other, there is no visual evidence of interaction. If the plots differ it is still not clear whether one has three-factor interaction or some two-factor interactions. Sometimes switching the positions of the three factors in the plots can sort this out. .

To examine the two-factor interactions directly, draw interaction plots using estimated means for factor pairs, averaged over the other factor. If patterns of interaction are still unclear, one may want to fit models and examine interaction plots based on residuals. For instance, one may fit a model with two two-factor interactions and then examine interaction plots for further evidence of three-factor interactions.

8.4 Higher-order models

Experiments involving three or more factors can become quite large. There are potentially many distinct components for main effects and interactions to be considered. These can consume a large number of degrees of freedom. Prudent assumptions about negligible interactions can increase the power to detect real differences. However, excessive assumptions can hide important interactions.

Higher-order layouts are used extensively in **quality improvement** where it may be reasonable to suppose that there are many main effects and most interactions are negligible. Usually only a fraction of the factor combinations are employed in a balanced fractional factorial design. The aim may be to find the few factors of interest and investigate interactions in subsequent experiments. Such an approach typically involves many small experiments which evolve based on results of the previous ones. Fractional factorial designs are considered briefly in Chapter 11, Part D, and in more detail in Box, Hunter and Hunter (1978).

Ideas from balanced two-factor models extend directly to three factors and models of higher order. The **three-factor model** is

$$y_{ijkl} = \mu_{ijk} + e_{ijkl}$$

with $i = 1, \cdots, a$, $j = 1, \cdots, b$, $k = 1, \cdots, c$, and $l = 1, \cdots, n$. The mean value can be partitioned as

response = mean + main effects + two-factor interactions
+ three-factor interaction + error .

Recall that this model can be interpreted for some purposes as a one-factor means model, leading to examination of specific linear contrasts of interest.

Mathematical notation for higher-order **effects models** can get rather involved. For convenience, consider using a single Greek letter for main effects and pair up those letters for interactions:

$$\mu_{ijk} = \mu + \alpha_i + \beta_j + \gamma_k + (\alpha\beta)_{ij} + (\alpha\gamma)_{ik} + (\beta\gamma)_{jk} + (\alpha\beta\gamma)_{ijk} \ .$$

This avoids the need to introduce more Greek letters and makes it easier to keep track. Main effect and interaction comparisons may be addressed in a similar manner to the two-factor model discussed earlier in this chapter. However, there are more **estimates** to consider, such as the marginal means

$$\begin{aligned}
\text{sample grand mean} &= \bar{y}.... \\
A \text{ marginal mean} &= \bar{y}_{i}... \\
B \text{ marginal mean} &= \bar{y}._{j}.. \\
AB \text{ marginal mean} &= \bar{y}_{ij}.. \\
&\cdots \\
ABC \text{ cell mean} &= \bar{y}_{ijk}.
\end{aligned}$$

For simplicity, consider the sum-to-zero constraints

$$\bar{\alpha}. = \bar{\beta}. = \bar{\gamma}. = \cdots = \overline{(\alpha\beta\gamma)}_{ij.} = \overline{(\alpha\beta\gamma)}_{i\cdot k} = \overline{(\alpha\beta\gamma)}_{.jk} = 0$$

and formally define the estimates of main effects and interactions

$$\begin{aligned}
\hat{\mu} &= \bar{y}.... \\
\hat{\alpha}_i &= \bar{y}_{i}... - \bar{y}.... \\
\hat{\beta}_j &= \bar{y}._{j}.. - \bar{y}.... \\
\hat{\gamma}_{ij} &= \bar{y}_{ij}.. - \bar{y}_{i}... - \bar{y}._{j}.. + \bar{y}.... \\
&\cdots \\
(\widehat{\alpha\beta\gamma})_{ijk} &= \bar{y}_{ijk}. - \bar{y}_{ij}.. - \bar{y}_{i\cdot k}. - \bar{y}._{jk}. + \bar{y}_{i}... + \bar{y}._{j}.. + \bar{y}._{\cdot k}. - \bar{y}.... \ .
\end{aligned}$$

In fact, these are non-unique solutions of the normal equations, depending heavily on the linear constraints on model parameters.

The **three-factor anova table** is much more involved but follows the same general approach (Table 8.10). Expected mean squares are similarly developed. For instance,

$$E(MS_{B*C}) = \sigma^2[1 + \delta^2_{B*C}/(b-1)(c-1)] \ ,$$

with $\delta^2_{B*C} = nt\sum_{jk}(\beta\gamma)^2_{jk}/\sigma^2$ and so on. Formal F tests follow the usual form for balanced designs, with degrees of freedom taken from the anova in Table 8.10.

The SS correspond to squared distances in sample n-space for a series of nested models. There are nine orthogonal spaces spanned by $\hat{\mu}$, $\{\hat{\alpha}_i\}$, $\{\hat{\beta}_j\}$, $\cdots$ and $\{y_{ijkl} - \hat{\mu}_{ijk}\}$. These can be combined in various ways. For instance, consider the factor combinations involving A and B. Collapsing the corresponding three spaces $\{\hat{\alpha}_i\}$, $\{\hat{\beta}_j\}$ and $\{(\widehat{\alpha\beta})_{ij}\}$ into one results in a two-factor model for factors C and $A*B$, partitioning n-space into four orthogonal spaces instead of nine. In other words, the three orthogonal spaces for B, A and $A*B$ form a subspace which is orthogonal to the rest. Similar statements can be made about interactions of these three terms

Source	Degrees of freedom	Sums of squares	
A	$a-1$	SS_A	$= nbc \sum_i \hat{\alpha}_i^2$
B	$b-1$	SS_B	$= nct \sum_j \hat{\beta}_j^2$
C	$c-1$	SS_C	$= nbt \sum_k \hat{\gamma}_k^2$
$A*B$	$(a-1)(b-1)$	SS_{A*B}	$= nc \sum_{ij} (\hat{\alpha\beta})_{ij}^2$
$A*C$	$(a-1)(c-1)$	SS_{A*C}	$= nb \sum_{ik} (\hat{\alpha\gamma})_{ik}^2$
$B*C$	$(b-1)(c-1)$	SS_{B*C}	$= nt \sum_{jk} (\hat{\beta\gamma})_{jk}^2$
$A*B*C$	$(a-1)(b-1)(c-1)$	SS_{A*B*C}	$= n \sum_{ijk} (\hat{\alpha\beta\gamma})_{ij}^2$
error	$abc(n-1)$	SS_{ERROR}	$= \sum_{ijkl}(y_{ijkl} - \hat{\mu}_{ijk})^2$
total	$abcn-1$	SS_{TOTAL}	$= \sum_{ijkl}(y_{ijkl} - \bar{y}....)^2$

Table 8.10. *Three-factor anova*

Source	df	SS
$A*B$	$ab-1$	$nc \sum_{ij} \left[\hat{\alpha}_i^2 + \hat{\beta}_j^2 + (\hat{\alpha\beta})_{ij}^2 \right]$
C	$c-1$	$nbt \sum_k \hat{\gamma}_k^2$
$C*(A*B)$	$(ab-1)(c-1)$	$n \sum_{ijk} \left[(\hat{\alpha\gamma})_{ik}^2 + (\hat{\beta\gamma})_{jk}^2 + (\hat{\alpha\beta\gamma})_{ijk}^2 \right]$
error	$abc(n-1)$	$\sum_{ijkl}(y_{ijkl} - \hat{\mu}_{ijk})^2$
total	$abcn-1$	$\sum_{ijkl}(y_{ijkl} - \bar{y}....)^2$

Table 8.11. *Three-factor anova collapsed into two*

with C. In other words, it is possible to test for the AB factor combination as a single factor, as shown in Table 8.11.

The **three-factor additive model** can be developed by a similar argument. The model

$$y_{ijkl} = \mu + \alpha_i + \beta_j + \gamma_k + e_{ijkl}$$

leads to the anova in Table 8.12. Tests for main effects follow naturally.

Source	Degrees of freedom	Sums of Squares
A	$a-1$	$nbc\sum_i(\bar{y}_{i...} - \bar{y}_{....})^2$
B	$b-1$	$nct\sum_j(\bar{y}_{.j..} - \bar{y}_{....})^2$
C	$c-1$	$nbt\sum_k(\bar{y}_{..k.} - \bar{y}_{....})^2$
error	$abcn - a - b - c + 2$	$\sum_{ijkl}(y_{ijkl} - \bar{y}_{i...} - \bar{y}_{.j..} - \bar{y}_{..k.} + 2\bar{y}_{....})^2$
total	$abcn-1$	$\sum_{ijkl}(y_{ijkl} - \bar{y}_{....})^2$

Table 8.12. *Three-factor additive anova*

Notice the marked increase in error degrees of freedom. However, inference is very sensitive to the assumption that interactions are negligible.

If $n=1$ (no replication), the last space (error space) is empty and one must alias some interactions with error. That is, one must assume some interactions are negligible in order to estimate error variance for inference purposes. One could employ various approaches to investigate interaction in this case, such as interaction plots or Tukey's one-degree-of-freedom test presented in Section 9.4.

8.5 Problems

8.1 Bacteria: Reproduce the analysis of this experiment. That is, review the various examples, lay out a model, conduct a formal analysis and present results. Be sure to document your steps.

8.2 Running: This study (Myers *et al.* 1993) examined how the metabolic cost of locomotion varied with speed, stride frequency and body mass. Cost was determined by measuring oxygen consumption (vo2), analyzing the oxygen content in air inhaled and exhaled by the subjects through a mask. The rate of oxgen consumption was measured for three **subjects** locomoting at all combinations of low and high levels of the three factors — running **speed**, **stride** frequency, and **mass** distribution in the leg. The first factor was set using a treadmill, the second by synchronizing the subjects' pace with a metronome, and the third by varying the positions of weights strapped onto the legs or waist. In addition to these three design factors, other variables were measured by filming each trial with a high-speed motion camera. Ignore these other measurements for this problem.

The **order** of the eight combinations was randomly assigned for each test session. For each test combination, **subjects** continued running until their rate of oxygen consumption levelled off, between eight and 15 minutes into the run, signalling that steady state had been reached. Subjects were given daily sessions for an initial training period, during which the rate of oxygen consumption at each test combination decreased to a relatively constant level. Subjects were considered trained when their oxygen consumption for all test combinations were consistent and repeatable between test sessions. All subjects achieved the trained state within two weeks. The data used in this study were collected in single sessions from trained subjects. However, there still might be some concern about incomplete training causing systematic differences in consumption during the session. How does the metabolic cost of locomotion vary as a function of speed, stride frequency and mass distribution? Analyze this as a completely randomized design with three replicates for each of the eight treatment combinations. Your analysis should include discussion of appropriate overall tests, steps leading to a 'parsimonious' model (one that is simple enough to explain the data, but still maintains the key design features), and any contrasts or other comparisons that you feel are important. Include at least one interaction plot with appropriate bar(s) to indicate precision.

CHAPTER 9

Model Selection

Model selection is the process of finding the 'best' model supported by data. In practice, this involves compromise. On the one hand, a model should fit the data well, avoiding bias from being too simple. On the other hand, an over-specified model may be too complicated, including factor combinations that are negligible. The aim is a **parsimonious model** that balances bias and over-fit.

Model selection is traditionally associated with regression. Therefore, this chapter draws heavily upon ideas from that realm (see, for instance, Draper and Smith 1981). However, some aspects are unique to analysis of variance. Most scientists prefer to maintain the **hierarchy** of factorial models. That is, if an interaction is kept in a model, then the corresponding main effects are also retained. In addition, main effects and interactions typically have several degrees of freedom while regressors have only one.

Formal F tests focus attention primarily on comparing nested models among a collection of partially ordered models. That is, two models are formally comparable if one is a reduced form of the other, allowing for instance the use of the general F test of the full against the reduced model. Model selection, on the other hand, involves comparing models that may not be nested. The general approach is to evaluate one or more statistics for each model and pick the 'best' model based on those statistics.

Section 9.1 examines the question of when to pool interactions. However, these ideas can be applied to experiments with replication as well. Several approaches to selecting the 'best' model appear in Section 9.2. While these ideas are presented for balanced experiments, they are applicable in general to unbalanced experiments and other linear models. Finally, some simpler forms of interactions are considered in Section 9.3 in the context of experiments with only one observation per cell.

9.1 Pooling interactions

Formal statistical theory is largely developed to handle one question in isolation. The formal concern in a broad sense is that the results of one test (for interaction) may influence the results of other tests (for main effects). A similar dilemma arose in Part B concerning multiple comparisons. Pooling interactions is another realm where formal, deductive inference must be balanced against informed, inductive judgement.

Pooling interactions into error may inflate MS_{ERROR} if the interaction is real. On the other hand, it deflates MS_{ERROR} if the F statistic is less than one. In general, the practice of testing then pooling affects the results of all subsequent tests (e.g. the p-values for tests of main effects) in non-obvious ways. Whenever possible, it is important to use results of previous experiments to guide in model selection *before* conducting a new experiment. Information about negligible interactions can help determine on which side of the philosophical dilemma to begin. For the future, tests of interaction can suggest if they need be part of models in subsequent experiments. Often these concerns are most important for experiments of modest sample size. This is all the more reason to discuss the importance of interaction prior to designing a new experiment.

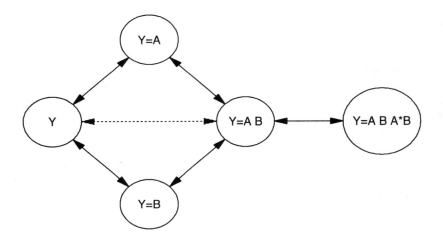

Figure 9.1. *Hierarchical models for two factors. Usually compare along solid lines, but consider null to full additive model (dotted) even if neither factor A nor B is significant on its own. Forward selection is left to right; backward elimination is right to left.*

Pragmatically, in the present situation it might be best to consider both approaches to main effects, under models with and without interactions. If they give essentially the same results, then the question of which to use is moot. However, if they do not, then it is important to investigate further. Suppose factor differences which were anticipated to be significant fall slightly short, while expected interactions did not emerge. Results after pooling could change dramatically ($p = 0.12$ to $p = 0.001$) or very slightly

($p = 0.056$ to $p = 0.048$). The former is likely only when there are many degrees of freedom in the questionable interaction, while the latter situation indicates a borderline result either way. It may be best to report both analyses, with a view toward future experiments to clarify issues raised in the present study.

Example 9.1 Infer: With two factors, there are five hierarchical models to consider. Figure 9.1 presents these in a path diagram in terms of generic factors A, B and response Y. The question often arises whether to pool interaction into error if the test for interaction is not significant when testing main effects. Suppose the scientist believed there was no interaction but decided to test it to make sure. If interaction proved not to be significant, should inference proceed directly to main effects in the additive model? Suppose instead that there is a suspicion of interaction but a formal test finds none. Should inference remain within the full interactive model, or is it appropriate to use the new evidence to test main effects in the additive model? ◊

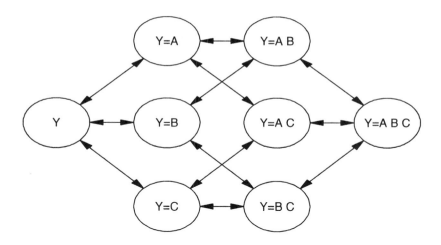

Figure 9.2. *Hierarchical additive models for three factors. The single step comparisons of nested models are indicated by lines. Now imagine adding interactions!*

9.2 Selecting the 'best' model

This section examines some issues in selecting among an array of anova models. The collection of possible models to be examined might include all possible subsets. However, it is usually most appropriate to consider only hierarchical models. Since there may be a large number of models, most selection procedures involve comparing two models that differ only by one term, either an interaction or a main effect. However, it is sometimes important to look ahead a few steps to double-check the process. Criteria for model selection are presented in the next section.

Example 9.2 Infer: Selection for a three-factor additive model would involve only the eight model choices indicated in Figure 9.2. The resulting model would allow separate interpretation of the main effects since there are no interactions to begin with. **Three-factor models** that allow interactions are more numerous. There are 18 hierarchical models listed below:

```
Y = A              Y = A B C
Y = B              Y = A B C A*B
Y = C              Y = A B C A*C
Y = A B            Y = A B C B*C
Y = A C            Y = A B C A*B A*C
Y = B C            Y = A B C A*B B*C
Y = A B A*B        Y = A B C A*C B*C
Y = A C A*C        Y = A B C A*B A*C B*C
Y = B C B*C        Y = A B C A*B A*C B*C A*B*C
```

In other words, non-hierarchical models such as Y = A*B and Y = A B A*C would not be considered. Recall that in addition there is the empty model in which no factor considered adequately explains any portion of the variation.

Here is a suggested method of analysis. If three-factor interactions are present and significant, pick one factor (of lesser interest) and analyze the two-factor model for each level of this factor and compare these analyses. If there is no three-factor interaction, look for 2-factor interactions. If only one two-factor interaction is found, analysis is pretty straight-forward. For instance, with A*B consider this factor combination as a new factor D = A+B+A*B and examine the two-factor model with C and D. If two or more two-factor interactions are present, analyses from several perspectives may be helpful. Consider separate analysis at levels of one factor as suggested for three-factor interactions. Remember that if any interactions are present interpretation of main effects must be modified appropriately. ◇

These comments so far do not deal directly with how to move among the possible models. While all models for three factors can be considered, it represents a great deal of work. Consideration of m factors leads to 2^m additive models plus many additional models with interactions. Most of these model comparisons are likely to be uninteresting. Therefore, in general investigators do not undertake selection among all models, even among all hierarchical models.

Stepwise procedures involve comparing models that differ by exactly one term. **Forward selection** starts by including the most important factor, ascertained by an F test or one of the criteria in the next section. It then proceeds for additive models in a similar fashion to regression but with attention to factor degrees of freedom. Forward selection with interaction terms proceeds in a natural way, being mindful of hierarchy. Terms are added until criteria are no longer satisfied or until the full model is reached.

Backward elimination begins with the full model and proceeds to drop terms (absorbing them into the error) one at a time. It is not uncommon for complicated situations to arrive at different models with backward elimination and forward selection. Sometimes this can be resolved with a stepwise compromise, checking at each step whether terms can be selected or eliminated. Often it is beneficial to perform a general F test to compare two models that differ by two or more terms, as indicated by the dotted line in Figure 9.1.

Hoaglin *et al.* (1991) suggest following the **rule of 2** in which interactions are pooled into error if the F test is less than 2. Their approach involves 'sweeping down' from the main effects onward through two-factor and higher-order interactions. They suggest combining simpler terms with higher-order terms (e.g. main effect A with interaction $A * B$) if the ratio of mean squares (MS_A/MS_{A*B}) is less than 2. Highest-order interactions would be similarly pooled with error if their F test were less than 2. This approach emphasizes the interpretation of combinations of factors where appropriate. For instance, it avoids the misleading reporting of non-significant main effects when there are significant interactions. However, the formal properties of this approach, such as its effects on p-values, is not well understood.

Occasionally, different model selection procedures return two disparate models which support the data equally well but are mutually incompatible. There may be no way to resolve this paradox with the current experiment. This can happen when an experimental design is overly ambitious, including too many factors for the sample size. It can also arise when anticipated differences are more subtle than expected. In any event, it can highlight important questions that can help design future experiments.

There are few automated tools for model selection in statistical analysis of variance packages. Factors with only two levels can be recast as regressors and used in a stepwise regression procedure. Factors with more levels could be split into orthogonal or non-orthogonal contrasts and handled similarly. However, it is important to consider whether all contrasts for a factor should be included if any one contrast is selected. It is certainly possible to take the results from a stepwise regression procedure as a starting point for subsequent analysis. Beyond this, judicious choices of model subsets can provide valuable information and save considerable time.

9.3 Model selection criteria

For the purposes of this section, suppose there are n observations and a particular model under evaluation has p parameters. This section primarily considers the explained variation $R^2(p)$, mean square error $MS_{\text{ERROR}}(p)$ and Mallow's $C(p)$ statistic. However, there is no 'best' procedure. In practice, there is not only a compromise between bias and over-fit, but also a compromise in the choice of procedure!

Plots can provide insight into the model selection process. While it is not feasible to examine residual plots with symbols for all models, they can help unravel paradoxes such as those discussed above. Half-normal plots of orthogonal contrasts for full model effects and residuals can be used to suggest quickly what main effects and interactions might be important. Path diagrams such as Figures 9.1 and 9.2 can be augmented with the values of selection criteria to provide a graphical summary along the way. They may not be publishable, but can shed important light on what is important. In addition, selection criteria discussed below can be plotted against the number of model parameters p as an aid. This is particularly useful for Mallow's $C(p)$ as discussed below. Once the 'final' model has been selected, interaction plots and tables of means provide important summaries of the key results supported by the data. If there are many factors, it may be useful to consider stem-and-leaf diagrams of cell means as an alternative to large tables (Hoaglin et al. 1991, ch. 11).

The explained variation for a model is defined as the ratio of the model sum of squares to the total sum of squares $R^2 = SS_{\text{MODEL}}/SS_{\text{TOTAL}}$. As the model becomes more complicated, the explained variation continues to rise, at least for nested models. Therefore, $R^2(p)$ can provide a heuristic guide to model selection, but would always suggest the most complicated model on its own. However, it can be instructive to consider the range of $R^2(p)$ among the models being considered.

Unfortunately, the distribution of the $R^2(p)$ depends on the error degrees of freedom $(n - p)$. The explained variation has been adjusted in common practice to give

$$R_a^2 = 1 - (1 - R^2)(n - 1)/(n - p) = 1 - \frac{MS_{\text{ERROR}}(p)}{SS_{\text{TOTAL}}/(n - 1)}.$$

That is, $1 - R_a^2$ is the ratio of the the model mean square error to the total mean square error of the responses. In other words, examining the adjusted R^2 is equivalent to considering the model $MS_{\text{ERROR}}(p)$. Again, this is a heuristic guide for informal model selection. It may be worth noting if and when model $MS_{\text{ERROR}}(p)$ changes drastically, either up or down, between two models. However, it should not be used alone.

Mallow (1973) suggested a statistic $C(p)$ which roughly balances between bias from an incompletely specified model and increased variance from an over-specifed model. The criterion is

$$C(p) = \frac{SS_{\text{ERROR}}(p)}{\hat{\sigma}^2} - (n - 2p)$$

with $\hat{\sigma}^2$ the estimated variance from the fullest model. If p parameters are adequate, then $E(SS_{\text{ERROR}}(p)) = (n-p)\sigma^2$. In this case, at least approximately, $C(p) = p$. Models with $C(p) > p$ tend to have large bias due to an under-specified model. Models with $C(p) < p$ have no evidence of bias. Thus, it would be preferred to select the 'smallest' model with $C(p) \leq p$.

Notice that Mallow's $C(p)$ is sensitive to the estimate of σ^2. In particular, it cannot be used on a saturated full model in which there are no degrees of freedom for error. In this case, some choice of reduced model must be made 'artfully', perhaps based on $MS_{\text{ERROR}}(p)$. Further reduction can then proceed using this criterion. However, if the estimate of σ^2 is not very accurate, based on only a few degrees of freedom, this criterion should be interpreted with caution.

9.4 One observation per cell

Balanced designs with one observation per cell raise interesting dilemmas about model selection. Consider the two-way model

$$y_{ij} = \mu + \alpha_i + \beta_j + \gamma_{ij} + e_{ij} ,$$

with independent normal errors, $e_{ij} \sim N(0, \sigma^2)$, for $i = 1, \cdots, a$, and $j = 1, \cdots, b$. As it stands, interaction is confounded with unexplained variation. If all the usual degrees of freedom are devoted to interaction, there is no way to estimate error variance. There must be some compromise.

Example 9.3 Tukey: A plant geneticist wanted to compare the growth characteristics of two inbred lines A and B with a hybrid combination H. The plants were grown at field stations at three latitudes, 25, 35, 45, and a number of measurements were taken. Unfortunately, the records were burned and the scientist only has the mean values from each station. Figure 9.3 shows the plant height plotted by genotype and latitude. This is in effect an **interaction plot**, but there is no replication. The data suggest that genotype and latitude are not additive. All three lines appear to perform equally well at latitude 25. Line A grows shorter while line B and the hybrid H grow taller as latitude increases. The hybrid shows evidence of 'hybrid vigor', doing much better than either parent line at higher latitudes. It might make sense in this situation to regress plant height on latitude, but the relationship with genotype would be nonlinear even if it were recoded, say, as the number of A alleles for height ($A = 2$, $H = 1$, $B = 0$). However, what could be done if there were three locations rather than three latitudes? Three unrelated lines rather than three genetically related lines?   ◊

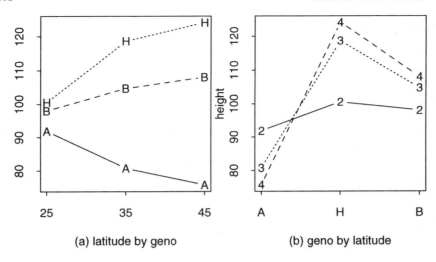

Figure 9.3. *Tukey interaction plots. These suggest an interaction, particularly in (b). However, with only one observation per factor combination, further assumptions are needed.*

One approach is to assume no interaction, which reduces consideration to the additive model as presented in the previous chapter. Alternatively, it is possible use one or a few degrees of freedom to investigate special forms of interaction. John Tukey (1949; see Scheffé 1959, sec. 4.8) suggested considering interaction of the form

$$\gamma_{ij} = G\alpha_i\beta_j ,$$

with G fixed but unknown, leading to a one-degree-of-freedom test of the interaction hypothesis $H_0 : G = 0$.

The motivation behind this involves interactions which enhance (or inhibit for $G < 0$) the effect of A by the effect of B in more than an additive fashion. That is, this particular form of synergy (or inhibition) supposes that the difference in mean response among two levels (say 1 and 2) of factor A for B at level j is $(\alpha_1 - \alpha_2)(1 + G\beta_j)$. This dependence on B_j may be natural in many practical settings. The approach is symmetric: the difference in effects of two levels of B (say 3 and 4) when A is at level i is $(\beta_3 - \beta_4)(1 + G\alpha_i)$. The model can be written as

$$y_{ij} = \mu + \alpha_i + \beta_j + G\alpha_i\beta_j + e_{ij}$$

with the usual assumptions. This so-called **Tukey interaction model** is not linear in the model parameters. However, this pattern of interaction can easily be detected with a margin plot. Further, Tukey's one-degree-of-freedom test can be performed in a few steps with most statistical packages, as shown in the next section.

A **margin plot** is an interaction plot of cell mean $\hat{\mu}_{ij}$ on a marginal mean, say $\bar{y}_{i\cdot}$ for factor A. One purpose of such a plot is to provide a natural order for factor levels. Those levels are further positioned by their marginal means. Levels with similar marginal effect are placed beside each other. In addition, margin plots provide a way to interpret factors as regressors. That is, imagine regressing the response on the marginal means for A

$$y_{ij} = \beta_0 + \beta_1 x_{ij} + e_{ij}$$

with $x_{ij} = \bar{y}_{i\cdot}$. The margin plot is a scatter plot of y_{ij} against x_{ij} using plot symbols or connected lines for levels of factor B. This is conceptually reasonable, but has certain theoretical problems, since the response is on both sides of the equation.

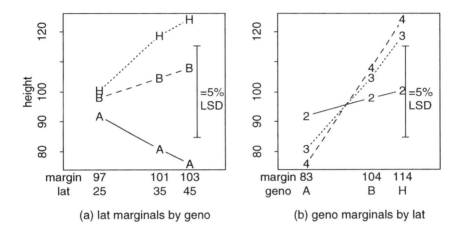

Figure 9.4. *Tukey margin plots of raw data. Tukey's one-degree-of-freedom test corresponds to straight lines on margin plots. Notice the* **Hybrid** *in (b) has higher response, but the points are now on a line. The evidence suggests that* **latitude** *25 is associated with profoundly different responses across* **genotypes**. *LSD bar based on additive model.*

Example 9.4 Tukey: Figure 9.4 displays the plant **height** data against the marginal means in a set of margin plots. The unequal spacing of **latitude** levels corresponds to deviations from a strict linear ordering. The **genotype** levels have been reordered to reflect the marginal mean **heights**. Now a linear relationship for the interactions is fairly evident. However, it is difficult to say much formally at this point. The LSD bar corresponds to the mean square error estimate from the additive model of Table 9.1, which is clearly inadequate. ◇

source	df	SS	MS	F	p-value
geno	2	1494.1	747.1	5.32	0.075
lat	2	76.1	38.0	0.27	0.78
error	4	562.0	140.5		
total	8	2132.2			

Table 9.1. *Tukey additive anova*

For convenience, constrain the effects parameters to sum to zero and formally consider estimates $\hat{\alpha}_i$ and $\hat{\beta}_j$ respectively of the main effects $\alpha_i = \bar{\mu}_{i\cdot} - \bar{\mu}_{\cdot\cdot}$ and $\beta_j = \bar{\mu}_{\cdot j} - \bar{\mu}_{\cdot\cdot}$ for the additive model

$$y_{ij} = \mu_{ij} + e_{ij} = \mu + \alpha_i + \beta_j + e_{ij} .$$

Now solve for G which minimizes the sum of squared residuals

$$SS = \sum_{ij} (y_{ij} - \mu - \alpha_i - \beta_j - G\hat{\alpha}_i\hat{\beta}_j)^2 ,$$

conditioned on $\hat{\alpha}_i$ and $\hat{\beta}_j$. This is very close to the 'proper' residual sum of squares for the Tukey interaction model when G is small. When $G = 0$, it reduces to the additive model. Thus an F test for the hypothesis of zero slope ($H_0 : G = 0$) tests the null hypothesis that the model is additive against an alternative that is very close to the Tukey interaction model.

This author learned a nice model-fitting trick from Peter M Crump (College of Agriculture and Life Sciences, UW–Madison) which simplifies the development of margin plots and inferential tools for the Tukey and related interaction models. Rather than solve for the effects as indicated above, use the square of the estimated cell means from the additive model as a **covariate** x_{ij} for a second fit of the additive model,

$$x_{ij} = \hat{\mu}_{ij}^2/2 = (\bar{y}_{\cdot\cdot}^2 + \bar{y}_{i\cdot}^2 + \bar{y}_{\cdot j}^2)/2 + \hat{\alpha}_i\hat{\beta}_j .$$

The first three terms contribute to the constant, A factor effect and B factor effect, respectively, while the last term captures the Tukey interaction. The Tukey interaction model can be written as

$$\begin{aligned} y_{ij} &= \mu^* + \alpha_i^* + \beta_j^* + Gx_{ij} + e_{ij} \\ &= [\mu^* + G\bar{y}_{\cdot\cdot}^2/2] + [\alpha_i^* + G\bar{y}_{i\cdot}^2/2] + [\beta_j^* + G\bar{y}_{\cdot j}^2/2] + G\hat{\alpha}_i\hat{\beta}_j + e_{ij} \\ &= \mu + \alpha_i + \beta_j + G\hat{\alpha}_i\hat{\beta}_j + e_{ij} . \end{aligned}$$

Example 9.5 Tukey: The margin plot in Figure 9.5 highlights the linear relationship fundamental to Tukey's interaction. The LSD bar is based on the mean square error estimate developed in the next section. It is much

TUKEY'S TEST FOR INTERACTION

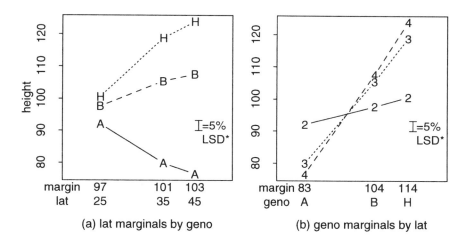

Figure 9.5. *Tukey margin plots of fitted model. Fit of Tukey's one-degree-of-freedom interaction yields straight lines on margin plots. Smaller LSD bar than Figure 9.4 provides strong evidence of interaction.*

shorter than the LSD bar in Figure 9.4 because it takes advantage of the one-degree-of-freedom interaction to reduce the error. ◊

9.5 Tukey's test for interaction

The following argument formally justifies such an F test for the Tukey one-degree-of-freedom test for interaction. It is of necessity a local approximation and should be interpreted in that light. However, its main use is to explore the possibility of interaction when there is only one observation per cell. It makes it possible in some cases to guard against blithely assuming additivity when the data suggest otherwise. This approach, introduced by John Tukey, can be used for more complicated models and with replication (see Milliken and Johnson 1989). The basic idea is to find the extra variation explained by the interaction given that the true model is additive. Essentially this is a score test for the parameter G (Rao 1965; see Scheffé 1959, sec. 4.2). It is a standard idea in nonlinear methods which works nicely in the present setting.

The least squares estimate of G could be found by differentiating the residual sum of squares,

$$SS = \sum_{ij}(y_{ij} - \mu - \alpha_i - \beta_j - G\alpha_i\beta_j)^2,$$

with respect to G to obtain the (nonlinear) estimator of G, conditional on estimators of α_i and β_j,

$$\hat{G} = \frac{\sum_{ij} \alpha_i \beta_j y_{ij}}{\sum_i \alpha_i^2 \sum_j \beta_j^2}.$$

It is nonlinear, since the denominator includes unknown parameters. There is no simple form for a particular solution μ, α_i, β_j and G which minimizes SS. The 'best' solution can be found by iteration. That is, (1) start with initial solutions $\alpha_i = \hat{\alpha}_i$ and $\beta_i = \hat{\beta}_j$ from the additive model with $G = 0$; (2) solve for G given $\hat{\alpha}_i$ and $\hat{\beta}_j$; (3) update the solution of the normal equations for μ, α_i and β_j given G; (4) repeat (2) and (3) until the solution converges in some sense.

However, there is no need to iterate in order to test $G = 0$. The conditional form of the estimator of G above suggests how to construct the sum of squares for interaction. Consider the $(a-1)(b-1)$-dimensional residual space spanned by $(y_{ij} - \bar{y}_{i\cdot} - \bar{y}_{\cdot j} + \bar{y}_{\cdot\cdot})$ in the additive model. Given any α_i and β_j, the estimator $\hat{G}$ is a linear combination of y_{ij} which lies in this residual space and forms a one-dimensional subspace. Thus for any α_i and β_j, the residual space can be partitioned into two orthogonal spaces with interaction sum of squares

$$SS_{G|A,B} = \sum_{ij} \hat{G}^2 \alpha_i^2 \beta_j^2 = \frac{(\sum_{ij} \alpha_i \beta_j y_{ij})^2}{\sum \alpha_i^2 \sum \beta_j^2}$$

and residual sum of squares

$$SS_{E|A,B} = \sum_{ij} (y_{ij} - \bar{y}_{i\cdot} - \bar{y}_{\cdot j} + \bar{y}_{\cdot\cdot})^2 - SS(G|A,B).$$

These are both quadratic forms in y_{ij} for any given α_i and β_j. Thus

$$SS_{G|A,B} \sim \sigma^2 \chi_1^2$$
$$SS_{E|A,B} \sim \sigma^2 \chi_{ab-a-b}^2.$$

Now substitute $\hat{\alpha}_i = \bar{y}_{i\cdot} - \bar{y}_{\cdot\cdot}$ for α_i and $\hat{\beta}_j = \bar{y}_{\cdot j} - \bar{y}_{\cdot\cdot}$ for β_j in these sums of squares, leading to the Tukey's one-degree-of-freedom F test for interaction

$$F = \frac{SS_{G|A,B}}{SS_{E|A,B}/(ab-a-b)}$$

which is distributed as $F_{1,ab-a-b}$ under $H_0: G = 0$. Notice that this approach partitions the total sum of squares as

$$SS_{\text{TOTAL}} = SS_A + SS_B + SS_{G|A,B} + SS_{E|A,B}.$$

Example 9.6 Tukey: The anova table for plant height with the one-degree-of-freedom interaction test is presented in Table 9.2. The test for

source	df	SS	MS	F	p-value
geno	2	1494.1	747.1	143.4	0.0011
lat	2	76.1	38.0	7.3	0.070
inter	1	546.4	546.4	104.9	0.0029
error	3	15.6	5.21		
total	8	2132.2			

Table 9.2. *Tukey interaction anova*

interaction is highly significant ($p = 0.003$). However, the tests for main effects are not really appropriate and should be ignored. In any event, significant interactions argue for interpretation of genotypes in the context of latitude. The mean square dropped dramatically from 140.5 for the additive model to 5.21 for the Tukey interaction model. This is reflected in the LSD bars shown on Figures 9.4 and 9.5. ◊

More complicated interaction models have been proposed along similar lines (see Milliken and Johnson 1989). The Tukey model forces all the lines in a margin plot such as Figure 9.5 to intersect at one point. The Mandel model allows for separate lines for one factor by introducing an asymmetry as follows:

$$\mu_{ij} = \mu + \alpha_i + \beta_j + \alpha_i \gamma_j \ .$$

Details of this and other approaches can be found in Milliken and Johnson (1989). It is possible to test $H_0 : \gamma_j = 0$ and to develop margin plots for the Mandel interaction model in an analogous way to the Tukey model. The trick here is to fit the additive model $\mu_{ij} = \mu + \alpha_i + \beta_j$, let $x_{ij} = \bar{y}_{i\cdot}$ and fit the model

$$y_{ij} = \mu + \alpha_i + \beta_j + \gamma_j x_{ij} \ ,$$

much as was done above. Tests can be developed in a similar fashion as well.

9.6 Problems

9.1 Infer: Verify that Mallow's $C(p)$ is p for correctly specified or overspecified models when the true variance σ^2 is used instead of $\hat{\sigma}^2$.

9.2 Infer: Show that Tukey's one-degree-of-freedom test for interaction can be used when there are replicates. That is, how does the argument for the test statistic need to be modified.

9.3 Company: A manufacturer wants to measure the 'quality' of a product by using the ratio of the maximum to minimum diameter of a particular hole punched in plastic (the ratio should ideally be 1.000 ± 0.001). Rather than record this value, which is costly to measure, each product line is given a visual score between 0 and 100 as to its quality.

The parts in question are of several types (A,B,C) and are manufactured using steel from several companies (1–5). In addition, the hole punch may occur at one of two sites in the manufacturing plant. Four replicates were collected for each combination of type, company and site in a completely randomized design for a total of 120 measurements.

(a) Write down a model for the quality score. Be sure to define everything carefully.

(b) Fit the full model with all two-factor and three-factor interactions. Summarize the findings for formal F tests.

(c) Based on the results of (b), construct interaction plots and/or tables of means to show the key features. Include measures of precision such as standard errors, LSD bars and the like. Plots and tables should be clearly labeled!

(d) Use ideas from model selection to reduce the full model to a 'parsimonious' model supported by the data. If you use a stepwise approach, clearly state how you proceed. It is not enough to state 'the answer'.

(e) Conduct a general F test between the full model and your parsimonious model. Interpret results.

(f) Plot predicted values against residuals to check for lack of fit or problems with assumptions. Do not attempt any formal analysis, but merely show the data. Always use plot symbols to identify factors or factor combinations.

PART D

Dealing with Imbalance

This part covers data analysis when factorial arrangements are not nicely balanced. Some care is needed in estimating effects and in determining their significance. In some ways, it is easier to consider unbalanced designs as a special case of a general linear model.

Any imbalance in the design structure calls for caution interpreting main effects and interactions. The type of tests for main effects must be chosen properly, as developed in Chapter 10. Otherwise comparisons may depend on sample sizes as well as on characteristics of the population under study.

Some comparisons of effects are meaningless or impossible if there are missing factor combinations, or cells. Chapter 11 examines approaches to balanced comparisons of cell means which address questions about main effects and interactions in this case. Dropping higher-order interactions, if appropriate, can lead to easier analysis with missing cells. Certain designs such as Latin squares and fractional factorials specifically miss factor combinations in order to reduce the size of experiments.

The general linear model provides a useful framework to discuss designs considered so far, as well as setting the stage for more complicated designs. The factorial arrangement is placed in matrix form in Chapter 12 as a more advanced topic for the adventurous. Some basic results are stated, to set the stage for results for more complicated designs used in later parts of this book. Factorial arrangements are shown by example to be special cases of a larger class of linear models.

CHAPTER 10

Unbalanced Experiments

Many experiments begin with a well-thought-out plan but run into problems during execution. Plants or animals may die or be taken off study, perhaps for reasons related to a combinations of factors. Materials may be accidentally destroyed, or data misrecorded in ways that cannot be later deciphered. These and other problems, including limited planning or loss of resources, can upset the perfect balance of an experimental design. In many other experiments the experimenter cannot control the balance. Sometimes experiments are intentionally unbalanced. Whatever the reason, unbalanced experiments are more difficult to analyze than balanced ones, requiring special care for estimation and testing.

There are two main types of imbalance, one more serious than the other. Both concern the number of samples per factor combination, or cell. Unbalanced design structures have unequal sample sizes across factor combinations (Section 10.1). Imbalance requires careful consideration of how to compare means, which becomes clearer after examining the additive model in Section 10.2. The four different types of hypotheses and their corresponding partitioning of sums of squares are examined in Section 10.3. Experiments with unbalanced treatment structures, or missing cells, are deferred to the next chapter.

10.1 Unequal samples

Consider an experiment which yields observations on all factor combinations but may have differing numbers of measurements for each factor combination. Thus the **treatment structure** is balanced but the **design structure** is not. For the two-factor model

$$y_{ijk} = \mu_{ij} + e_{ijk}$$

with $i = 1, \cdots, a$, $j = 1, \cdots, b$, but possibly unequal sample sizes $k = 1, \cdots, n_{ij}$. For this chapter, $n_{ij} > 0$. If the imbalance is slight, say due to the loss of one or two observations, it might seem appropriate to substitute an estimate of the missing observation ($\hat{\mu}_{ij}$ for y_{ijk}) and proceed with analysis as in a balanced experiment. This is supported by theory, provided that the degrees of freedom for error are reset to reflect the actual number of observations. However, when the imbalance is more pronounced, it is better to address its effect on inference directly and carefully.

The estimators for μ_{ij} and σ^2 are the same as for the one-factor model with unequal sample sizes:

$$\hat{\mu}_{ij} = \bar{y}_{ij.} = \sum_{k=1}^{n_{ij}} y_{ij}/n_{ij} \quad \sim N(\mu_{ij}, \sigma^2/n_{ij})$$
$$\hat{\sigma}^2 = \sum_{ijk}(y_{ijk} - \hat{\mu}_{ij})^2/(n_{..} - ab) \quad \sim \sigma^2 \chi^2_{n_{..}-ab}/(n_{..} - ab)$$

Testing for all means equal, $H_0 : \mu_{ij} = \bar{\mu}_{..}$, could proceed as in the one-factor model, ignoring details of treatment structure. This can be followed by inference for planned and unplanned multiple comparisons as discussed in Chapter 6, Part B. Note that comparisons based on F tests (*e.g.* Fisher's LSD, BONFERRONI, SCHEFFÉ) are appropriate for unequal sample sizes but those based on the range of means (*e.g.* Tukey's HSD and SNK) are generally not since they need equal sample sizes. Again, the one-factor approach may not be very powerful if the main questions concern model effects.

The two-factor treatment structure can be employed to examine main effects and interactions. However, care is needed. The natural hypotheses for main effects,

$$H_{0A}: \quad \bar{\mu}_{i.} = \bar{\mu}_{..} ,$$
$$H_{0B}: \quad \bar{\mu}_{.j} = \bar{\mu}_{..} ,$$

are not addressed by the naïve SS and F tests from balanced experiments. However, interaction hypotheses

$$H_{0B*A} : \mu_{ij} = \bar{\mu}_{i.} + \bar{\mu}_{.j} - \bar{\mu}_{..}$$

are tested in the same way as before, comparing the fit for the full model with interactions to the reduced, additive model.

The implications of unequal sample size on questions of inference can be understood by looking at estimation of **population marginal means**. The marginal population means are the averages of the population cell means across all other factors. That is, for the two-factor model

$$\bar{\mu}_{i.} = \sum_j \mu_{ij}/b$$
$$\bar{\mu}_{.j} = \sum_i \mu_{ij}/a$$
$$\bar{\mu}_{..} = \sum_i \sum_j \mu_{ij}/ab .$$

The unbiased estimators of these are, respectively,

$$\hat{\bar{\mu}}_{i.} = \sum_j \hat{\mu}_{ij}/b = \sum_j \sum_k y_{ijk}/bn_{ij}$$
$$\hat{\bar{\mu}}_{.j} = \sum_i \hat{\mu}_{ij}/a = \sum_i \sum_k y_{ijk}/an_{ij}$$
$$\hat{\bar{\mu}}_{..} = \sum_i \sum_j \hat{\mu}_{ij}/ab = \sum_i \sum_j \sum_k y_{ijk}/abn_{ij}$$

UNEQUAL SAMPLES 163

with corresponding variances

$$
\begin{aligned}
V(\hat{\bar{\mu}}_{i\cdot}) &= (\sigma/b)^2 \sum_j 1/n_{ij} \\
V(\hat{\bar{\mu}}_{\cdot j}) &= (\sigma/a)^2 \sum_i 1/n_{ij} \\
V(\hat{\bar{\mu}}_{\cdot\cdot}) &= (\sigma/ab)^2 \sum_i \sum_{j=1}^{n_i} 1/n_{ij} \ .
\end{aligned}
$$

Hypothesis tests and confidence intervals for questions involving marginal population means can be constructed with these estimates. Our aim is to develop appropriate overall F tests based on sums of squares, showing along the way that some approaches which seem natural may in fact be rather far from what was desired.

Note that in unbalanced experiments, the above estimates of marginal population means do not correspond to the **marginal sample means**

$$
\begin{aligned}
\bar{y}_{i\cdot\cdot} &= \sum_j \sum_k y_{ijk}/n_{i\cdot} \\
\bar{y}_{\cdot j\cdot} &= \sum_i \sum_k y_{ijk}/n_{\cdot j} \\
\bar{y}_{\cdot\cdot\cdot} &= \sum_i \sum_j \sum_k y_{ijk}/n_{\cdot\cdot} \ .
\end{aligned}
$$

whose expectations are weighted averages of population cell means,

$$
\begin{aligned}
E(\bar{y}_{i\cdot\cdot}) &= \sum_j n_{ij}\mu_{ij}/n_{i\cdot} &\neq \bar{\mu}_{i\cdot} \\
E(\bar{y}_{\cdot j\cdot}) &= \sum_i n_{ij}\mu_{ij}/n_{\cdot j} &\neq \bar{\mu}_{\cdot j} \\
E(\bar{y}_{\cdot\cdot\cdot}) &= \sum_i \sum_j n_{ij}\mu_{ij}/n_{\cdot\cdot} &\neq \bar{\mu}_{\cdot\cdot} \ .
\end{aligned}
$$

The marginal sample means would only be appropriate for estimating weighted marginal means with weights $w_{ij} = n_{ij}/n_{\cdot\cdot}$ or for sampling proportionally from sub-populations.

Example 10.1 Oocyte: Rose-Hellekant and Bavister (1996) has been studying the chemistry of oocyte (egg) development. In one experiment, she examined the effect of two agents, forskolin and h89. Initial sample sizes were six for each combination of three levels of these agents, but one sample was lost (Table 10.1). While the sample means $\hat{\mu}_{ij} = \bar{y}_{ij\cdot}$ are unbiased estimates of μ_{ij}, the marginal sample means depend on sample size. For instance, the marginal sample mean for 100 units of h89 is

$$E(\bar{y}_{3\cdot\cdot}) = (5\mu_{31} + 6\mu_{32} + 6\mu_{33})/17 = \bar{\mu}_{3\cdot} + (\bar{\mu}_{3\cdot} - \mu_{31})/17 \ .$$

Questions concerning the population marginal means,

$$
\begin{aligned}
H_{0A} &: \quad \bar{\mu}_{1\cdot} = \bar{\mu}_{2\cdot} = \bar{\mu}_{3\cdot} \ , \\
H_{0B} &: \quad \bar{\mu}_{\cdot 1} = \bar{\mu}_{\cdot 2} = \bar{\mu}_{\cdot 3} \ ,
\end{aligned}
$$

h89/fors	0	0.5	1.0	total
0	6	6	6	18
50	6	6	6	18
100	5	6	6	17
total	17	18	18	53

Table 10.1. *Oocyte sample sizes by factor combination*

do not correspond to questions about the expected sample marginal means

$$H_{0A}: \quad E(\bar{y}_{1..}) = E(\bar{y}_{2..}) = E(\bar{y}_{3..}),$$
$$H_{0B}: \quad E(\bar{y}_{.1.}) = E(\bar{y}_{.2.}) = E(\bar{y}_{.3.}).$$

These latter hypotheses can be written as

$$H_{0A}: \quad \bar{\mu}_{1.} = \bar{\mu}_{2.} = \bar{\mu}_{3.} + (\bar{\mu}_{3.} - \mu_{31})/17,$$
$$H_{0B}: \quad \bar{\mu}_{.1} + (\bar{\mu}_{.1} - \mu_{31})/17 = \bar{\mu}_{.2} = \bar{\mu}_{.3}.$$

In this case, the problem may not be that bad, as the hypotheses are only off by $(\bar{\mu}_{3.} - \mu_{31})/17$ because the imbalance is very slight. The worse the imbalance, the farther the expectations of the sample marginal means from the population marginal means.

The data can be balanced artificially by filling in the missing value as the cell mean ($y_{316} = \bar{y}_{31.}$) and reducing the residual degrees of freedom by 1 at the end of the analysis ($df_{\text{ERROR}} = 53 - 9 = 44$). This was a standard approach for hand calculation (see Searle 1987) but is unnecessary with proper use of most statistical packages. ◊

The hypotheses for expected marginal sample means depend on the sample sizes. This may be desired in some experiments where by design the cell sample sizes are proportional to population sizes by factor combination. In general, key questions should be independent of sample sizes such as those concerning the marginal population means.

10.2 Additive model

Estimation for the **additive model**,

$$y_{ijk} = \mu_{ij} + e_{ijk} = \mu + \alpha_i + \beta_j + e_{ijk},$$

requires some attention for unbalanced experiments. In the balanced experiment, everything works out nicely. That is, the estimates for main effects are the differences between sample marginal means and the sample grand

ADDITIVE MODEL

mean ($\bar{y}_{i..} - \bar{y}_{...}$ and $\bar{y}_{.j.} - \bar{y}_{...}$) for both the full and additive models because the variation explained by main effects always has the same form. These deviations are still the estimates of main effects in the full model with interaction for unbalanced experiments. However, for the additive model the objective involves minimizing the error sum of squares

$$SS_{\text{ERROR}} = \sum_{ijk}(y_{ijk} - \mu - \alpha_i - \beta_j)^2$$

or equivalently maximizing the model sum of squares

$$\begin{aligned}SS_{\text{MODEL}} &= SS_{\text{TOTAL}} - SS_{\text{ERROR}} \\ &= \sum_{ijk}(y_{ijk} - \mu)^2 - \sum_{ijk}(y_{ijk} - \mu - \alpha_i - \beta_j)^2 \\ &= R(\mu, \alpha, \beta) - R(\mu) = R(\alpha, \beta|\mu)\end{aligned}$$

with $R(\mu) = n_{..}\bar{y}_{...}^2$ the grand mean variation. The notation $R()$ follows Searle (1987) and stands for the 'reduction' in sum of squares. The objective leads to the following normal equations for model parameters:

$$\begin{aligned}\mu : \quad \sum_{ijk} y_{ijk} &= n_{..}\mu + \sum_i n_{i.}\alpha_i + \sum_j n_{.j}\beta_j \\ \alpha_i : \quad \sum_{jk} y_{ijk} &= n_{i.}\mu + n_{i.}\alpha_i + \sum_j n_{ij}\beta_j \\ \beta_j : \quad \sum_{ik} y_{ijk} &= n_{.j}\mu + \sum_i n_{ij}\alpha_i + n_{.j}\beta_j \;.\end{aligned}$$

Solving for β_j in terms of μ and α_i yields

$$\beta_j = \sum_i (\bar{y}_{ij.} - \alpha_i)n_{ij}/n_{.j} - \mu$$

which can be used to solve for α_i resulting in a system of a equations in a unknowns:

$$\sum_j n_{ij}\left[\alpha_i - \sum_\ell n_{\ell j}\alpha_\ell/n_{.j}\right] = \sum_j n_{ij}\left[\bar{y}_{ij.} - \sum_\ell n_{\ell j}\bar{y}_{\ell j.}/n_{.j}\right].$$

However, there are only $a - 1$ degrees of freedom. While it is possible to use the sum-to-zero side constraint $\bar{\alpha}_{.} = 0$, the set-to-zero side constraint $\alpha_a = 0$ is much easier. Solving for the other $a-1$ levels of α is then straightforward but somewhat tedious. Similar arguments obtain for the estimation of β_j with eventual estimation of μ.

Example 10.2 Oocyte: The effects of the two chemicals on oocyte development could be independent. That is, the cell mean lines by h89 level would be parallel, as in Figure 10.1(a, b). To a certain degree this is true but Figure 10.1(c, d) shows that some sort of interaction is evident. In particular, the response with neither drug is unusually low. The figures suggest possible additivity ignoring the control. However, there would be a missing cell!

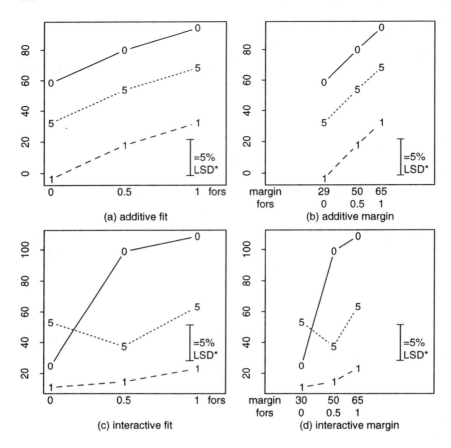

Figure 10.1. *Oocyte additive and interactive models. Top figures show additive model fit as (a) interaction plot and (b) margin plot. Bottom figures do likewise with full interaction model. LSD bars based on (a,b) additive or (c,d) full interaction model. Plot symbols represent levels of* h89 *(0 = none, 5 = 50, 1 = 100). Note that margin plots (b,d) are square.*

Note that the marginal means in Figure 10.1(b, d) were computed under the additive model, using the methods indicated above. Their precision is indicated by the number of digits relative to the standard errors of the marginal means for the additive model. The LSD bar uses the harmonic mean, which is flagged by the asterisk. Table 10.2 shows the marginal means for forskolin and h89 levels under successively fuller models. Notice the dramatic change in standard errors for the full interaction model. The marginal means remain stable because the imbalance in the design is very slight. This raises suspicions that interaction is important. ◊

forskolin model		$\hat{\hat{\mu}}_{i.}$(SD)		
	0.0	0.5	1.0	$\hat{\sigma}$
μ	49.0(5.3)	49.0(5.3)	49.0(5.3)	38.5
$\mu + \alpha_i$	30.8(8.9)	50.3(8.6)	65.0(8.6)	36.5
$\mu + \alpha_i + \beta_j$	28.9(6.4)	50.3(6.2)	65.0(6.2)	26.2
$\mu + \alpha_i + \beta_j + \gamma_{ij}$	29.8(4.8)	50.3(4.6)	65.0(4.6)	19.7
h89 model		$\hat{\hat{\mu}}_{.j}$(SD)		
	0	50	100	$\hat{\sigma}$
μ	49.0(5.3)	49.0(5.3)	49.0(5.3)	38.5
$\mu + \beta_j$	77.5(7.0)	51.2(7.0)	16.6(7.2)	29.8
$\mu + \beta_j + \alpha_i$	77.5(6.2)	51.2(6.2)	15.5(6.4)	26.2
$\mu + \beta_j + \alpha_i + \gamma_{ij}$	77.5(4.8)	51.2(4.6)	16.3(4.8)	19.7

Table 10.2. *Oocyte marginal means by model*

Main effects of A can be tested by comparing the fit between the one-factor model with B present and the two-factor additive model. That is, compare

$$SS_{A|B} = R(\mu, \beta, \alpha) - R(\mu, \beta) = R(\alpha|\mu, \beta)$$

using an F test with $(a-1)$ and $(n_{..} - b - a + 1)$ degrees of freedom

$$F = \frac{R(\alpha|\mu, \beta)/(a-1)}{SS_{\text{ERROR}}/(n_{..} - b - a + 1)} .$$

An analogous test can be developed for testing B in the presence of A.

Thus estimation and testing of main effects in the additive model with unequal sample sizes are rather complicated, while in the full model with interaction, estimation of main effects is fairly direct. This hints at some of the problems of testing main effects in the two-way model. Is it better to take a sequential approach, building from one-way to additive to fully interactive models, or to consider the importance of each part conditional on the rest of the full model? To resolve these issues, the next section considers the strengths and limitations of several alternative approaches.

10.3 Types I, II, III and IV

The focus on questions involving population cell means and marginals leads naturally to consideration of partitioning the two-factor model as

$$y_{ijk} = \hat{\hat{\mu}}_{..} + (\hat{\hat{\mu}}_{i.} - \hat{\hat{\mu}}_{..}) + (\hat{\hat{\mu}}_{.j} - \hat{\hat{\mu}}_{..}) + (\hat{\mu}_{ij} - \hat{\hat{\mu}}_{i.} - \hat{\hat{\mu}}_{.j} + \hat{\hat{\mu}}_{..}) + (y_{ijk} - \hat{\mu}_{ij})$$

which suggests a partition of the total sum of squares as in Table 10.3. The individual sums of squares (SS) terms in this table are the appropriate

source	df	Type III SS
A	$a-1$	$\sum_i n_{i\cdot}(\hat{\bar{\mu}}_{i\cdot} - \hat{\bar{\mu}}_{\cdot\cdot})^2$
B	$b-1$	$\sum_j n_{\cdot j}(\hat{\bar{\mu}}_{\cdot j} - \hat{\bar{\mu}}_{\cdot\cdot})^2$
$B*A$	$(a-1)(b-1)$	$\sum_{ij} n_{ij}(\hat{\mu}_{ij} - \hat{\bar{\mu}}_{i\cdot} - \hat{\bar{\mu}}_{\cdot j} + \hat{\bar{\mu}}_{\cdot\cdot})^2$
error	$n_{\cdot\cdot}-ab$	$\sum_{ijk}(y_{ijk} - \hat{\mu}_{ij})^2$
total	$n_{\cdot\cdot}-1$	(does not add up in general)

Table 10.3. *Type III adjusted sum of squares*

ones for testing hypotheses concerning main effects and interactions. Unfortunately, the SS_{TOTAL} is *not* the sum of the parts unless the experiment is balanced (or analysis involves weighted sums with cell weights proportional to sample sizes). In fact, there are several ways to extract sums of squares from the total to address hypotheses. Four types have been developed which have distinct uses in testing, although they are all equivalent for the balanced experiment.

The **Type I** approach fits a sequence of models. Following the same development as in Chapter 8 for the balanced case, it is possible to consider progressively more complicated models as shown in Table 10.4. This examines models **sequentially**, reducing the residual sum of squares between the model mean estimate and the response by adding more terms into the model to explain more of the variation. In other words, begin with the total sum of squares

$$SS_{\text{TOTAL}} = \sum_{ijk}(y_{ijk} - \bar{y}_{\cdots})^2 = \sum_{ijk} y_{ijk}^2 - R(\mu) \ .$$

The second model explains the effect of factor A

$$SS_A = R(\alpha|\mu) = \sum_i n_{i\cdot}(\bar{y}_{i\cdots} - \bar{y}_{\cdots})^2$$

with the marginal means $\bar{y}_{i\cdots}$ being the least squares (LS) estimates for the estimable parameters $\mu + \alpha_i$. The third model in Table 10.4 is the additive model, with LS estimates derived as discussed in the previous section. The extra variation explained by B after A was derived for the additive model as

$$SS_{B|A} = R(\beta|\mu,\alpha)$$

which cannot be easily written down in closed form for unbalanced experi-

model equation	Type I SS
$y_{ijk} = \mu + e_{ijk}$	—
$y_{ijk} = \mu + \alpha_i + e_{ijk}$	SS_A
$y_{ijk} = \mu + \alpha_i + \beta_j + e_{ijk}$	$SS_{B|A}$
$y_{ijk} = \mu + \alpha_i + \beta_j + \gamma_{ij} + e_{ijk}$	$SS_{B*A|A,B}$

Table 10.4. *Type I sequential two-factor models*

ments. The full model with interactions further explains a part which is not additive

$$SS_{B*A|A,B} = R(\gamma|\mu,\alpha,\beta) = \sum_{ij}(\bar{y}_{ij\cdot} - \bar{y}_{i\cdot\cdot} - \bar{y}_{\cdot j\cdot} + \bar{y}_{\cdot\cdot\cdot})^2 \ .$$

Thus Type I sums of squares partition the full model sum of squares,

$$SS_{\text{MODEL}} = R(\alpha,\beta,\gamma|\mu) = SS_A + SS_{B|A} + SS_{B*A|A,B} \ .$$

Geometrically, the total sum of squares SS_{TOTAL} corresponds to the length of the vector of the centered data $\{y_{ijk} - \bar{y}_{\cdot\cdot\cdot}\}$ in a $(n_{\cdot\cdot} - 1)$-dimensional sample space. The model and error spaces are orthogonal, and hence their sums of squares add up to the total. In addition, the Type I sums of squares add up precisely because they correspond to orthogonal spaces. While this is a nice property, it does not always lead to the appropriate hypotheses in the unbalanced data case, as is shown below.

A new twist arises in the unbalanced case in which $SS_{B|A}$, the extra variation explained by B adjusted for A, does not necessarily coincide with SS_B, the unconditional variation explained by B. That is, switching the order of B and A in the model fit changes the partition of orthogonal spaces and of the model sum of squares.

Example 10.3 Oocyte: Table 10.5 shows Type I sums of squares for the fit beginning with F = forskolin or beginning with H = h89. The full model is the same in either case, but the sums of squares (and their associated tests) of main effects differ depending on the order. Notice that the interaction and error sums of squares are the same for both sequences. After fitting the two main effects the two models are identical. ◊

The Type I hypotheses arising from these sequential sums of squares concern weighted means. Let tildes ($\tilde{\mu}$) signify weighted marginal means as shown below:

$$\begin{aligned}
\tilde{\mu}_{i\cdot} &= \sum_j n_{ij}\mu_{ij}/n_{i\cdot} \\
\tilde{\mu}_{\cdot j} &= \sum_i n_{ij}\mu_{ij}/n_{\cdot j} \\
\tilde{\mu}_{\cdot\cdot} &= \sum_i \sum_j n_{ij}\mu_{ij}/n_{\cdot\cdot} \\
\tilde{\mu}_{\cdot\cdot} &= \sum_i n_{i\cdot}\tilde{\mu}_{i\cdot}/n_{\cdot\cdot} = \sum_j n_{\cdot j}\tilde{\mu}_{\cdot j}/n_{\cdot\cdot} \ .
\end{aligned}$$

Model	SS	Model	SS
fors	10218	h89	32527
h89|fors	33741	fors|h89	11432
h89*fors|fors,h89	15982	fors*h89|h89,fors	15982
error	17054	error	17054
total	76994	total	76994

Table 10.5. *Oocyte Type I sums of squares*

The Type I hypotheses with A fitted first are

$$H_{0A}: \quad \tilde{\mu}_{i\cdot} = \tilde{\mu}_{\cdot\cdot}$$

$$H_{0B|A}: \quad \tilde{\mu}_{\cdot j} = \sum_i n_{ij}\tilde{\mu}_{i\cdot}/n_{\cdot j}$$

$$H_{0B*A|A,B}: \quad \mu_{ij} = \bar{\mu}_{i\cdot} + \bar{\mu}_{\cdot j} - \bar{\mu}_{\cdot\cdot}.$$

The hypothesis H_{0A} and sum of squares SS_A correspond to the null hypothesis and SS for the one-factor model. However, they are not equivalent to the desired ones for marginal population means being equal, since they concern weighted averages across the levels of factor B.

The conditional hypothesis $H_{0B|A}$ and sum of squares $SS_{B|A}$ come from the additive model, adding B last. This can be seen by examining the solution of the normal equations for the additive model discussed in the previous section. The conditional hypothesis for B in the presence of A states that the weighted marginal means must correspond to doubly weighted marginal means,

$$\sum_i \frac{n_{ij}\mu_{ij}}{n_{\cdot j}} = \sum_{ik} \frac{n_{ij}n_{ik}\mu_{ik}}{n_{i\cdot}n_{\cdot j}}.$$

This is rather bizarre. The right-hand side does not in general coincide with either the marginal population grand mean ($\bar{\mu}_{\cdot\cdot}$) or the expectation of the marginal sample grand mean ($\tilde{\mu}_{\cdot\cdot}$).

Switching the order of A and B results in another set of Type I hypotheses

$$H_{0B}: \quad \tilde{\mu}_{\cdot j} = \tilde{\mu}_{\cdot\cdot}$$

$$H_{0A|B}: \quad \tilde{\mu}_{i\cdot} = \sum_j n_{ij}\tilde{\mu}_{\cdot j}/n_{i\cdot}$$

$$H_{0B*A|B,A}: \quad \mu_{ij} = \bar{\mu}_{i\cdot} + \bar{\mu}_{\cdot j} - \bar{\mu}_{\cdot\cdot}.$$

Note the asymmetry and the importance of the order of fit of factors. Both hypotheses for main effects, with either order, depend heavily on sample sizes. However, the interaction hypothesis is as desired, the same as that considered for the balanced experiment.

The **Type II** hypotheses and associated sums of squares are similar to Type I except that they no longer depend on the order of entry of terms into the model. The Type II approach is hierarchical, considering a factor conditional on prior entry of all other model terms which do not include that factor. For the two-factor unbalanced model the hypotheses are

$$H_{0A|B} : \quad \tilde{\mu}_{i\cdot} = \sum_j n_{ij}\tilde{\mu}_{\cdot j}/n_{i\cdot}$$

$$H_{0B|A} : \quad \tilde{\mu}_{\cdot j} = \sum_i n_{ij}\tilde{\mu}_{i\cdot}/n_{\cdot j}$$

$$H_{0B*A|B,A} : \quad \mu_{ij} = \bar{\mu}_{i\cdot} + \bar{\mu}_{\cdot j} - \bar{\mu}_{\cdot\cdot} \ .$$

The hypotheses for main effects still depend on cell sample sizes. The corresponding SS are those extracted from Type I SS by switching positions of A and B. Note that Type II SS do not in general add up for unbalanced experiments

$$SS_{\text{TOTAL}} \neq SS_{A|B} + SS_{B|A} + SS_{B*A|A,B} \ .$$

The Type II approach can be useful if the intent is model building, deciding which terms should be in the model. Type II hypotheses and SS are used regularly in stepwise regression, when considering the question of adding one of several regressors last. However, they are potentially very misleading for questions concerning main effects and interactions in factorial arrangements unless an additive model is assumed at the outset.

Example 10.4 Oocyte: Suppose one wanted to build a model as in stepwise regression, in a hierarchical way such that main effects are included if any interactions appear in the model. The forward selection approach would use Type I sums of squares and hypotheses to find the best fit of one factor, then the second and so on. The backward elimination would use Type II to decide whether a term could be dropped. For the oocyte example, the Type I tests for first effect are

$$F_{\texttt{fors}} = (10218/2)/(66776/50) = 3.83 \ (p = 0.028) \ ,$$

$$F_{\texttt{h89}} = (32527/2)/(44467/50) = 18.3 \ (p < 0.0001) \ .$$

which implies including h89 first. The sequential test for a second main effect is actually the Type II test,

$$F_{\texttt{fors}|\texttt{h89}} = (11432/2)/(33036/48) = 8.31 \ (p = 0.0008) \ ,$$

$$F_{\texttt{h89}|\texttt{fors}} = (33741/2)/(33036/48) = 24.5 \ (p < 0.0001) \ .$$

Thus forskolin has a more significant contribution to explaining variation after accounting for h89 than before. Finally, the test of interaction,

$$F_{\texttt{h89}*\texttt{fors}|\texttt{h89},\texttt{fors}} = \frac{15982/4}{17054/44} = 10.3 \ (p < 0.0001) \ ,$$

implies that one cannot simply consider main effects. One concludes that there is significant interplay between the two factors, which is evident in Figure 10.1(c, d). The interaction makes interpretation of the tests for main effects suspect. However, the figure indicates that h89 effects are likely to be much stronger than forskolin effects. The Type III approach below shows how to interpret these main effects in the presence of real interactions. ◊

The **Type III** hypotheses are the natural ones considered at the beginning of this chapter:

$$H_{0A|B,B*A}: \quad \bar{\mu}_{i\cdot} = \bar{\mu}_{\cdot\cdot}$$

$$H_{0B|A,B*A}: \quad \bar{\mu}_{\cdot j} = \bar{\mu}_{\cdot\cdot}$$

$$H_{0B*A|B,A}: \quad \mu_{ij} = \bar{\mu}_{i\cdot} + \bar{\mu}_{\cdot j} - \bar{\mu}_{\cdot\cdot}$$

The Type III SS are those suggested naturally by comparing the estimates of marginal population means (see Table 10.3). These SS are computed by fitting all other terms in the model except the ones relevant to the hypothesis of interest. For example, the main effects hypothesis for A, $H_{0A|B,B*A}$, is examined by adjusting response by the common value, main effect for B and pure interaction $B*A$, before determining the adjusted SS explained by A. Operationally, this involves comparing the fit of the 'full' model to that of the 'reduced' model with all terms except those of interest and assessing the difference in explained variation between these two models with a general F test.

In essence, the conditional SS for the factor of interest, say $SS_{A|B,B*A}$, is the variation explained by the fit of α_i in the direction orthogonal to the fit of the reduced model piece $\mu + \beta_j + \gamma_{ij}$ (with appropriate side conditions to ensure the γs measure only pure interaction). The latter explains the effect of B averaged across levels of A plus the pure interaction of levels of B and A but contains no effect of A averaged across levels of B. This is placed in the broader context of **linear models** in Chapter 12.

Type III hypotheses and SS represent the correct approach for assessing importance of model effects (main effects and interactions) when there are unequal numbers of observations per factor combination. This measures the Euclidean distance between the full model mean estimate and its projection onto a reduced model space which does not have the terms in question.

Example 10.5 Oocyte: Now it is possible formally to test the main effects of forskolin and h89. (The test for interactions is the same as before.)

$$F_{\text{fors|h89,h89*fors}} = (10862/2)/(17054/44) = 14.0 \ (p < 0.0001),$$

$$F_{\text{h89|fors,h89*fors}} = (32726/2)/(17054/44) = 42.2 \ (p < 0.0001).$$

Both main effects are strongly significant, with somewhat more evidence for

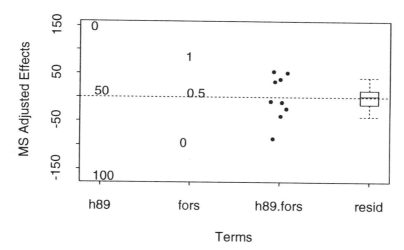

Figure 10.2. *Oocyte mean square adjusted effects: mean square adjusted effects and residuals rescaled so that the SDs are the square root of MS for that effect. Plot symbols are levels of a factor. Box-plot used for residuals.*

Source	df	Type III SS	MS	F	p-value
fors	2	10862	5431	14.0	0.0001
h89	2	32726	1636	42.2	0.0001
fors*h89	4	15982	3996	10.3	0.0001
error	44	17054	388	--	--
total	52	76994	--	--	--

Table 10.6. *Oocyte Type III anova*

h89. However, Figure 10.1(c) suggests a consistent pattern of main effects only after at least one of the chemicals is applied. In summary, Table 10.6 shows the Type III anova. Again, note that the sums of squares do not add up since the experiment is unbalanced. Figure 10.2 shows the mean square adjusted effects, giving a pictorial representation. ◊

Example 10.6 Bacteria: This experiment on the effect of `mycoplasma` and `temperature` on chick development, in particular on `bill` length, has been described up until now as if there were only eight measurements, one per room. However, measurements were taken on almost 200 chicks in these rooms. It is possible to think of the experiment as having chicks (or actually eggs) randomly assigned to the rooms, and to consider the sample

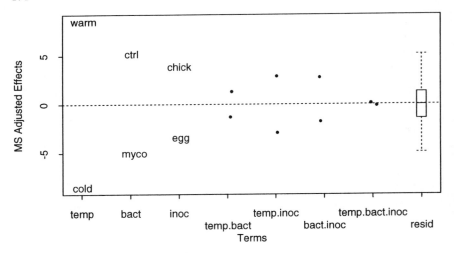

Figure 10.3. *Bacteria effects plot. Mean square adjusted effects and residuals rescaled so that the SDs are the square root of MS for that effect. Plot symbols are levels of a factor. Box-plot used for residuals.*

source	df	MS	F	p-value
inoc	1	10.5	2.66	0.10
bact	1	20.2	5.10	0.025
temp	1	56.3	14.24	0.00022
inoc*bact	1	5.04	1.27	0.26
inoc*temp	1	6.54	1.65	0.20
bact*temp	1	1.31	0.33	0.57
inoc*bact*temp	1	0.01	0.003	0.95
error	184	3.95		

Table 10.7. *Bacteria unbalanced anova*

in each room as a random sample from a population of chicks exposed to that environment (combination of temperature and presence or absence of mycoplasma bacteria). For the present, suppose there were eight such rooms, ignoring the fact that the experiment was conducted in two runs separated by some time. Further, suppose that the eggs and later chicks act independently in these rooms.

Thus there are three factors, temperature (cold, hot), bacteria (ctrl, myco) and inoculaton (egg, chick), arranged in all possible combinations. The number of chicks measured is not the same across rooms for a variety of minor reasons that the scientist does not believe are related to the factors.

source	df	MS	F	p-value
inoc	1	8.33	2.10	0.15
bact	1	16.35	4.13	0.044
temp	1	65.06	16.41	0.000075
error	188	3.96		

Table 10.8. *Bacteria unbalanced additive anova*

The full anova summary for this unbalanced experiment presented in Table 10.7, using adjusted (Type III) sums of squares, shows little evidence for any interactions. Figure 10.3, however, suggests that some interactions may be worth some contemplation. (Compare this figure to Figure 8.4, page 138.)

The additive model fit in Table 10.8, again with Type III sums of squares, can only explain 12.4% of the variation in bill length. However, there is very strong evidence for temperature effects, which were expected. The modest evidence for bacteria effects ($p = 0.025$ or 0.044) gets even slimmer (0.12) if inoculation is dropped. However, the interaction plots examined earlier (Figure 8.1, page 133) suggest there may be a 0.5–1 mm, or 1–2%, shortening of bill length in the presence of mycoplasma, which may be an important indicator of other developmental problems. It is now up to the scientist to weigh this evidence in the light of other evidence about bird development. ◊

The last, or **Type IV**, approach develops hypotheses about the effect of interest such that cell mean coefficients are balanced. When there are no missing factor combinations ($n_{ij} > 0$ for all cells), this approach coincides exactly with Type III. However, certain marginal means are not estimable when some cells are empty. In this situation, questions must revolve around comparing certain levels of factor A averaged across only those levels of factor B for which there are responses. The problem of inference when some cells are missing is examined in further detail in the next chapter.

10.4 Problems

10.1 Oocyte: Reproduce the analysis presented in this chapter as a report to the scientist. That is, carefully state the key questions, lay out a well-defined mathematical and/or word model, conduct analysis and report results in a graphical manner. Be sure to check assumptions.

10.2 Hardy: Scientists in horticulture have been interested in how plants adapt to cold climates such as we find in Wisconsin. They have noticed that plants which have been previously 'acclimatized' suffer less damage. Palta and Weiss (1993) conducted a controlled laboratory experiment to examine two species of **pot**atoes (1, 2) subjecting plants to one of two acclimatization **regimes** (1 kept in cold room; 0 kept at room temperature) for several days. Plants were later subjected to one of two cold **temperature** treatments (1 just below freezing, $-4°C$; 2 way below freezing, $-8°C$). Two responses were measured, both being damage scores for **phot**osynthesis and ion **leak**age.

Initially there were 80 plants, but some were lost during the experiment. Your analysis should take care to address the imbalance due to unequal sample sizes for the different factor combinations.

Your job is to sort out the effects of species, acclimatization and cold treatment on **phot**osynthesis and ion **leak**age. Be sure to include the following:
(a) Lay out models and assumptions.
(b) Justify your process of model selection.
(c) State and perform appropriate formal tests, indicating briefly their value and limitations.
(d) Include both interaction and diagnostic plots, as needed.
(e) Interpret your results, with reference to relevant plots, tables and other summaries of analysis.

10.3 Infer: Show why the regression trick for polynomial contrasts in Example 5.4, page 75 does not work exactly for unbalanced designs, even with only one factor. You may show this mathematically, or demonstrate it with an example (say, using the **Diet** data).

CHAPTER 11

Missing Cells

Suppose some factor combinations are not observed. If all levels of factor B are missing for a particular level of factor A, then simply reduce the number of levels of A. However, what if only a few levels of B are missing? Such an experiment has unbalanced treatment structure and usually an unbalanced design structure.

Section 11.1 considers the nature of missing cells. Connected cells are developed in Section 11.2. Section 11.3 examines the different approaches to analyzing main effects and interactions when there are missing cells, concluding that great caution is in order. Latin square designs and fractional factorial designs, discussed respectively in Sections 11.4 and 11.5, have missing cells by design.

11.1 What are missing cells?

A designed experiment with missing cells, or missing factor combinations, has $n_{ij} = 0$ for at least one cell (A_i, B_j). If $n_{ij} = 0$ for some j and all levels A_i, $i = 1, \cdots, a$, it makes sense to reduce the levels of B (and similarly, interchanging the roles of A and B). This section only considers experiments with $n_{ij} = 0$ for some cells but $n_{ij} > 0$ for at least one cell for every level of each factor.

Example 11.1 Growth: This research concerned growth medium for cloning plants (Stieve, Stimart and Yandell 1992). Plant tissue (a small part of a leaf) was placed in a Petri dish which had a 'food preparation' on it. The dish was sealed and placed in a controlled chamber with adequate light and proper temperature. The 'food preparation', or medium, was designed to encourage growth of new plant shoots from the tissue. The response of interest here was the number of 'adventitious shoots' (`advplt`), or new plants growing out of the tissue. Basically, the more adventitious shoots produced, the better the growing media. The adventitious plants can be separated and grown into numerous individuals for greenhouse cultivation, saving tremendous space and cost for nurseries and florists. The key question is: what is the best growing condition for the shoots?

BA/TDZ	0	0.2	2.0	20	total
0	16	15	16	16	63
0.44	15	16	16	16	63
4.4	13	16	16	0	45
total	44	47	48	32	171

Table 11.1. *Growth sample sizes by factor combination*

There were 12 factor combinations (trt) of interest, coming from a two-factor layout with added control (BHTA, trt = 20). One factor combination was missing; the plant material seemed to be burned by high levels of both chemicals together. Ignoring for the present the added control, the sample sizes for the whole data set are shown in Table 11.1. Most factor combinations have roughly 15 samples but the missing cell is problematic. The cell mean μ_{34} is undefined, as are marginal means $\bar{\mu}_3.$ and $\bar{\mu}_{.4}$. Thus it is not exactly clear yet how to formulate hypotheses. ◊

All cell means where data are present can be estimated in the usual way,

$$\hat{\mu}_{ij} = \bar{y}_{ij.} \sim N(\mu_{ij}, \sigma^2/n_{ij}) \text{ if } n_{ij} > 0 \ .$$

The variance can be estimated using contributions from cells with at least two observations much as before,

$$\hat{\sigma}^2 = \sum_{ijk}(y_{ijk} - \hat{\mu}_{ij})^2/(n.. - c) \sim \sigma^2 \chi^2_{n..-c}/(n.. - c) \ ,$$

with $c < ab$ the number of cells with $n_{ij} > 0$. However, the interpretation of **population marginal means** depends on averaging over the levels of other factors in the model. Some population marginal means are not estimable. That is $\bar{\mu}_{i.}$ and $\bar{\mu}_{.j}$ are not estimable if $n_{ij} = 0$ for a particular i and j. While one could choose to define population marginal means as averages over the available levels of all other factors, comparisons involving such marginal means for A averaged over different levels of B may be slightly misleading or totally ludicrous.

Example 11.2 Growth: Table 11.2 shows the mean of the log of the number of adventitious plants, with or without zeros, by combination of the two chemicals BA and TDZ. Overall tests were moderately significant (that is, $p = 0.052$ for all the data, $p = 0.063$ for the non-zero data), indicating that multiple comparisons should proceed with some caution. Letters in the tables indicate the grouping by Fisher's LSD method. Notice that the best combination for average yield of adventitious plants is TDZ = 0.2 with no added BA, while the combination giving the highest mean

WHAT ARE MISSING CELLS?

		with zeros		
BA/TDZ	0	0.2	2.0	20
0	0.2c	0.51a	0.18bc	0.20bc
0.44	0.23bc	0.23bc	0.29ab	0.11bc
4.4	0.34ab	0.12bc	0.25bc	--
control	0.14bc			LSD=0.26

		without zeros		
BA/TDZ	0	0.2	2.0	20
0	0.30c	0.96ab	0.48c	0.65abc
0.44	0.57bc	0.75abc	0.67abc	0.42c
4.4	1.09a	0.66abc	0.75abc	--
control	0.64abc			LSD=0.48

Table 11.2. *Growth multiple comparisons*

yield when plants survived was BA = 4.4 with no added TDZ. The added control fell in the middle of the range. Since one factor combination was missing, and there was an added control, it is not immediately obvious how to characterize main effects and interactions. Further, the interrelationship of these chemicals seems rather complicated. Thus a one-factor comparison of means seems to be a reasonable first approach. ◇

The two-factor effects model,

$$y_{ijk} = \mu + \alpha_i + \beta_j + \gamma_{ij} + e_{ijk},$$

requires $ab + a + b + 1 - c$ **linear constraints** to achieve a unique solution of the normal equations. In other words, the sum-to-zero or set-to-zero constraints developed earlier are not enough. Further, some of the sum-to-zero constraints involving i or j for which $n_{ij} = 0$ are not well defined. Set-to-zero constraints can be augmented usually by setting $\gamma_{ij} = 0$ if $n_{ij} = 0$. This is reasonable in practice, since one has no information about those interactions. As a consequence, some degrees of freedom for pure interaction are lost when there are missing cells.

Statistical package results for unbalanced experiments with missing treatment combinations can be misleading. Be very sure about what is being done before reporting such results. Some packages may try to estimate missing data, say by assuming an additive model, effectively setting interaction effects for empty cells to 0 (does the package properly adjust degrees of freedom?). Some packages may warn the user of missing data, while others plow ahead without any indication of choices.

Example 11.3 Design: Consider an experiment with two years (first factor) of field trials, using varieties (second factor) D, E, F in year 1 and varieties D, F, G in year 2. The year population means are

$$\bar{\mu}_{1\cdot} = (\mu_{1D} + \mu_{1E} + \mu_{1F})/3$$
$$\bar{\mu}_{2\cdot} = (\mu_{2D} + \mu_{2F} + \mu_{2G})/3 \, ,$$

but they are not strictly comparable. Instead consider only varieties D and F and the balanced hypothesis for years

$$H_0 : (\mu_{1D} + \mu_{1F})/2 = (\mu_{2D} + \mu_{2F})/2 \, .$$

That is, differences between variety E and variety G are confounded with year to year differences. ◇

Thus in an experiment with some missing cells, the marginal means averaged over all levels of the other factor are not all estimable. What are the reasonable approaches to such a problem?

Sometimes the best course of action is to compare all means as in the one-way model. One could perform an overall test of difference and then, if significant, examine linear combinations or contrasts of interest. This approach may have less resolving power than testing main effects hypotheses. However, it can always be done.

11.2 Connected cells and incomplete designs

Sometimes it is useful to consider connected subsets of cells in a factorial arrangement when asking key questions about an experiment with missing cells. At times, this may be the only way to define the question properly. Empty cells may arise by accident. However, they may be part of the design in an effort to economize on scarce resources. Designs with empty cells in a balanced pattern can be quite appropriate for certain questions if the scientist is willing to assume that certain interactions are negligible. Examples of such designs are briefly discussed in the last section of this chapter.

A subset of cells consists of all cells associated with subsets of levels of all factors. A **connected subset** is a subset for which all contrasts of main effects levels are estimable. For two factors, this simply means that it is possible to move from any filled cell to any other filled cell by only horizontal and vertical movements; that is, by changing the level of only one margin each step. Definitions for more than two factors are more complicated, and not really that useful in practice. Instead, consider some examples to fix the idea.

Example 11.4 Design: Consider a factorial arrangement with $a = 3$ and $b = 2$ but with the first cell empty. That is, $n_{11} = 0$ and hence μ_{11} is not

component	level		effect	coefficient
reference			μ	L_1
factor A	1		α_1	L_2
	2		α_2	L_3
	3		α_3	$L_4 = L_1 - L_2 - L_3$
factor B	1		β_1	L_5
	2		β_2	$L_6 = L_1 - L_5$
combination AB	1	2	γ_{12}	$L_7 = L_2$
	2	1	γ_{21}	L_8
	2	2	γ_{22}	$L_9 = L_3 - L_8$
	3	1	γ_{31}	$L_{10} = L_5 - L_8$
	3	2	γ_{32}	$L_{11} = L_1 - L_2 - L_3 - L_5 + L_8$

Table 11.3. *General form of estimable functions with a missing cell*

estimable. This implies that the interaction parameter γ_{11} is undefined. It is not obvious how to define the marginal means $\bar{\mu}_{1\cdot}$ and $\bar{\mu}_{\cdot 1}$ since it is not possible to average across all levels of the other factor. Thus these marginal means are not estimable as well. The general form of estimable functions can be derived by dropping the levels for factor combinations that have missing cells, as shown in Table 11.3. Notice that 11 parameters remain but α_1 and γ_{12} are confounded, which makes sense because there are data for only one level of B ($=2$) when $A = 1$. The estimable functions are thus

$$f = L_1\mu + L_2\alpha_1 + L_3\alpha_2 + (L_1 - L_2 - L_3)\alpha_3 + L_5\beta_1 + (L_1 - L_5)\beta_2$$
$$+ L_2\gamma_{12} + L_8\gamma_{21} + (L_3 - L_8)\gamma_{22}$$
$$+ (L_5 - L_8)\gamma_{31} + (L_1 - L_2 - L_3 - L_5 + L_8)\gamma_{32} ,$$

or, expressed in terms of cell means and combining by Ls,

$$f = L_1(\mu_{32}) + L_2(\mu_{12} - \mu_{32}) + L_3(\mu_{22} - \mu_{32}) + L_5(\mu_{31} - \mu_{32})$$
$$+ L_8(\mu_{21} - \mu_{22} + \mu_{32} - \mu_{31}) .$$

Notice that the missing cell implies that the pure interaction contrast $\mu_{11} - \mu_{12} + \mu_{32} - \mu_{31}$ is not estimable.

The set-to-zero linear constraints

$$\alpha_3 = \beta_2 = \gamma_{3j} = \gamma_{i2} = 0$$

lead to expression of the estimable functions as

$$f = L_1\mu + L_2\alpha_1 + L_3\alpha_2 + L_5\beta_1 + L_8\gamma_{21} .$$

What is a reasonable way to compare different levels of factor A when some marginals are not well defined? It is possible to compare levels of A across selected levels of B in such a way as to balance the comparison. For instance $A = 1$ and $A = 2$ can be compared at level 2 of B. ◊

Example 11.5 Design: Consider the following factorial arrangement of two factors, in which the ×s represent cells with data and the empty cells have no data ($n_{ij} = 0$).

$A \backslash B$	1	2	3	4
1			×	×
2		×	×	×
3	×	×	×	
4	×	×	×	×

Comparing $A = 1$ with $A = 4$ using all levels of B would confound the measured difference in response to A by the different levels of B:

$$(\mu_{13} + \mu_{14})/2 \text{ vs. } (\mu_{41} + \mu_{42} + \mu_{43} + \mu_{44})/4 \;.$$

Instead, $A = 1$ and $A = 4$ could be compared for levels 3 and 4 of B. How could one compare other levels of A in a balanced way? Certain interactions can be investigated in such a design. For instance, pure interaction involving levels 3 and 4 of A with levels 1, 2 and 3 of B can be examined.

Another approach is to consider an additive model,

$$y_{ijk} = \mu + \alpha_i + \beta_j + e_{ijk} \;.$$

It is straightforward to show that all cell means are now estimable, since each can be written as a linear combination of means for nonempty cells. For instance,

$$\mu_{11} = \mu_{13} + \mu_{31} - \mu_{33} \;.$$

Further, all marginal means are estimable. Thus questions can be asked about main effects, provided interactions are assumed to be negligible. ◊

It is good practice to determine connected subsets of the treatment structure when there are missing cells. Balanced subsets of treatment combinations could be viewed as several interrelated experiments. It can be instructive to analyze every balanced subset, although that may be rather time consuming. There is usually a clear choice of subsets and contrasts which span most non-empty cells.

Sometimes cells are not connected to other cells. These cells cannot be compared in a way which can help sort out effects of factors. Therefore, it may be appropriate to rely on comparing cell means for differences, or to make further assumptions about the interrelationship of factors. For

CONNECTED CELLS AND INCOMPLETE DESIGNS 183

instance, the additive model introduced earlier in this chapter can lead to a well-determined problem.

Example 11.6 Design: Consider the following two-factor factorial arrangement with one cell which is not connected to the other five non-empty cells.

$A\backslash B$	1	2	3
1			×
2		×	×
3		×	×
4	×		

Level 4 of factor A and level 1 of factor B stand alone. Any comparison of level $A = 4$ with other levels of A is confounded with factor B, and vice versa. One could analyze the experiment without $(A = 4, B = 1)$, and then consider the experiment as a one-factor layout with five groups.

The additive model

$$y_{ijk} = \mu + \alpha_i + \beta_j + e_{ijk}$$

for this design has $3 \times 4 - 6 = 6$ non-empty cells, but there are only five degrees of freedom. For instance, one can express μ_{22} in terms of its surrounding cells,

$$\mu_{22} = \mu_{23} + \mu_{32} - \mu_{33} .$$

Further, the cell mean μ_{12} is estimable even though its cell is empty, since it can be written as a linear combination of estimable cell means,

$$\mu_{12} = \mu_{13} + \mu_{31} - \mu_{33} .$$

None of the marginal means $\bar{\mu}_{i.}$ or $\bar{\mu}_{.j}$ are estimable, since α_4 and β_1 are completely confounded. However, consider the following subset:

$A\backslash B$	2	3
1		×
2	×	×
3	×	×

The marginal means are now estimable, although their meaning has changed. Note that one could relax the additive model assumption and look at one degree of freedom for interaction.

This experimental design, with one cell unconnected, may seem rather bizarre. However, it is possible to imagine a scenario where it might arise. Factors A and B could be food additives that need to have a fairly constant combined contribution. Thus A could be at a high level (4) only with a low level (1) of B, and B at a medium level (3) must be with A at the intermediate levels (2, 3), and so on. Is an additive model appropriate in such a situation? What if the pattern of empty cells arose as a result of the experiment rather than by design? ◇

The important feature to stress about connectedness is the need for balance in comparisons. The marginal effect of one factor is defined in terms of averages across levels of all other factors. If some factor level combinations are missing, exercise extreme caution. Subset analysis or assumptions of additivity may be necessary in order to draw sensible conclusions to questions of inference.

11.3 Type IV comparisons

Testing main effects must be done carefully. As shown in the previous section, Type I and II hypotheses are not appropriate with unequal sample sizes in the full model with interaction. However, Type II hypotheses should be used if one assumes the additive model. The results in Section 10.2 on the additive model carry over directly if some $n_{ij} = 0$, provided each marginal count ($n_{i\cdot}$ and $n_{\cdot j}$) is positive.

More surprisingly, Type III hypotheses are flawed, since their enforced orthogonality depends on the pattern of missing cells. These hypotheses do not usually correspond to comparisons of marginal population means but to weighted averages of cell means. As a result, Type III hypotheses for main effects can be confounded with pure interaction and rarely agree with one's perception of reasonable comparisons. As before Type I SS depend on order of fit, while Type II and Type III SS do not. None of these would be recommended in practice.

Type IV hypotheses for factor A are composed of balanced contrasts which are averages of cell means across non-empty levels of factor B. The SAS package automatically selects $a-1$ such contrasts but they are usually not orthogonal. SAS compares the last level of A with each other level to develop these contrasts,

$$H_0 : \sum_{j \in S_{ia}} \mu_{aj} = \sum_{j \in S_{ia}} \mu_{ij} , \ k = 1, \cdots, a-1 ,$$

with S_{ia} the set of levels B_j for which both $n_{ij} > 0$ and $n_{aj} > 0$. While the resulting Type IV SS for these balanced contrasts do not depend on the order of fit, they do depend on the order of the treatment labels for A and B. There is no unique 'best' choice of contrasts, leading to an appropriate, unique Type IV SS. Thus Type IV hypotheses and SS for main effects are flawed. However, one may find it useful to enlarge on the Type IV approach by explicitly considering some or all balanced contrasts for main effects.

Example 11.7 Growth: Consider the propagation experiment in which levels of two chemicals were applied and all experimental units at the high-

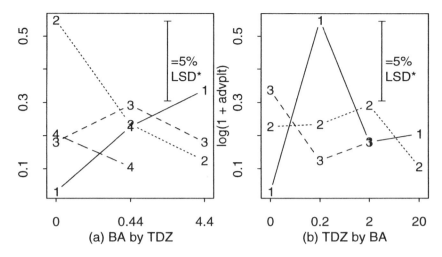

Figure 11.1. *Growth interaction plots with a missing cell: effect of chemicals* BA *and* TDZ *on the number of adventitious plants* (advplt). *Plot symbols represent levels of (a)* TDZ *(1 = 0, 2 = 0.2, 3 = 2, 4 = 20) or (b)* BA *(1 = 0, 2 = 0.44, 3 = 4.4). Not that the combination of the highest levels of both chemicals is missing.*

est level combination were destroyed. Certain marginal means are not well defined and certain contrasts are impossible without further assumptions. Note first that assuming an additive model would suggest examining main effects with a Type II approach:

$$F_{\text{BA}|\text{TDZ}} = 0.01\ (p = 0.99)$$
$$F_{\text{TDZ}|\text{BA}} = 1.12\ (p = 0.34)$$

This is not very satisfying, since a glance at the interaction plot (Figure 11.1) shows evidence of significant interactions. Further, the test of interactions is significant:

$$F_{\text{BA}*\text{TDZ}|\text{BA},\text{TDZ}} = 3.15\ (p = 0.009)\ .$$

It may still be worthwhile to test main effects in some cases, allowing qualitative statements about main effects even if interactions are significant. Contrasts which seem appropriate for main effects of BA, since μ_{34} is not well defined, are

```
 * 0    vs. 0.44:    μ̄₁. = μ̄₂.
   0    vs. 4.4 :    μ₁₁ + μ₁₂ + μ₁₃ = μ₃₁ + μ₃₂ + μ₃₃
 * 0.44 vs. 4.4 :    μ₂₁ + μ₂₂ + μ₂₃ = μ₃₁ + μ₃₂ + μ₃₃
```

The two asterisked contrasts are selected for Type IV hypotheses in SAS when using the actual concentrations as the levels of BA (0, 0.44, 4.4).

However, if number codes were used insted (1, 2, 3), the latter two contrasts would be chosen. Appropriate contrasts for TDZ could be constructed in a similar fashion. Notice again that packages such as SAS which automatically construct Type IV hypotheses tend to select contrasts by a rule which depends on the labelling of the factor levels! With the actual levels, the Type IV contrasts for TDZ chosen are each of 0, 2, 20 against 0.2. Using number codes, the contrasts chosen compare 0, 0.2, 2 with 20. Note that TDZ level 0.2 (0.44 for BA) appears last in the alphanumeric ordering of actual levels because it views the period (.) as the first character, placing it after numerals 0-9. The Type IV tests for these main effects are as follows:

$$F_{BA} = 0.19 \ (p = 0.83)$$
$$F_{TDZ} = 1.93 \ (p = 0.13) \ ,$$

or, using number codes,

$$F_{BA} = 0.02 \ (p = 0.98)$$
$$F_{TDZ} = 2.90 \ (p = 0.036) \ .$$

Which are correct? Wrong question. It is better knowingly to consider particular contrasts, either by explicitly constructing one-degree-of-freedom contrasts or by selecting balanced subsets of the design with no missing cells. The latter approach suggests two balanced subsets, dropping either the highest level of BA or of TDZ. The anova tables with Type III sums of squares after dropping BA = 4.4 or TDZ = 20 are shown in Table 11.4. The results of these subsets are confusing and contradictory. That is in large part because this problem cannot be easily decomposed into main effects and interactions. In fact, the most appropriate analysis here seems to be a one-way analysis to determine the best treatment combination. ◊

It can be beneficial to determine **balanced subsets** of the treatment structure when there are missing cells. Balanced subsets of factor combinations could be viewed as several interrelated experiments. Consider analyzing one or more balanced subsets. There is usually a clear choice of subsets and contrasts which span all non-empty cells.

The Test for **interaction** in a two-factor model is the same regardless of the approach (Type I, II, III or IV). This general F test measures the deviation between the full model with interaction and the additive model. The Type IV hypotheses make explicit the limitation of testing for interaction in the case of missing cells. The only pure interaction contrasts that can be examined are balanced contrasts involving cells with observed responses. Those pure interactions involving empty cells cannot be examined, since there is no information about such possible interaction. The balanced pure interaction contrasts are mutually orthogonal and span the pure interaction space, hence yielding the same SS and hypotheses as for Type III (and I and II), as demonstrated in Example 11.7.

LATIN SQUARE DESIGNS

source	df	dropping BA = 4.4 SS	MS	F	p-value
BA	1	0.027	0.027	0.23	.63
TDZ	3	1.059	0.353	2.93	.036
BA*TDZ	3	1.001	0.334	2.77	.045
error	132	15.931	0.121		
total	139	18.160			

source	df	dropping TDZ = 20 SS	MS	F	p-value
BA	2	0.009	0.004	0.03	.97
TDZ	2	1.148	0.074	0.53	.59
BA*TDZ	4	1.996	0.499	3.54	.009
error	143	20.146	0.141		
total	151	22.501			

Table 11.4. *Growth anova for balanced subsets*

No overall automatic approach is entirely satisfactory when some cells are empty. The basic problem is that one does not have any information about some interaction effects. This suggests one **logical approach**. First test for interaction. If there is no evidence of interaction, examine an additive model using Type II hypotheses and SS. If interaction is in evidence, consider a one-way model with all observed treatment combinations. In either case, it is wise to examine some or balanced contrasts involving main effects.

Given the problems of using a test of the data to determine subsequent action and the tricky nature of interaction, another approach suggests itself. Consider connected subsets of the experiment which have balanced treatment structure (no missing treatment combinations) and analyze these using the Type III approach. Usually at least two balanced subsets present themselves. Compare analysis using all the data with these balanced analyses. If all analyses are consistent, then interpretation of results is easy. If not, then results may be inconclusive, suggesting subsequent experiments to resolve unanswered questions.

11.4 Latin square designs

There are many experiments which have missing cells by design. This arises in situations where examining all factors at all combinations of levels would be too costly or time consuming. Compromises to save time and money involve assuming that some or all of the possible interactions are negligible.

The most condensed design of this nature is the **Latin square**. A scientist may wish to examine three factors with the same number n of levels but may only be ably to use n^2 experimental units, instead of the full design with n^3. Thus there is no replication and insufficient degrees of freedom for any full interactions. However, under an additive model

$$y_{ijk} = \mu + \alpha_i + \beta_j + \xi_k + e_{ijk}$$

there are $3n - 2$ model degrees of freedom, with the remainder for error. Note that this design is balanced even though cells are missing. Typically Latin square design are analyzed using an additive model.

The idea of connected subsets which are balanced (all levels of one factor having data across all levels of another) has been used to construct efficient experimental designs when resources of time and/or materials are scarce. The Latin square design allows examination of (or control of) three factors with the resources normally alloted to two factors.

Example 11.8 Design: Suppose a scientist examines the effect of two factors, temperature and pressure, on the manufacture of a part using three different batches of plastic. However, there can only be nine runs. A possible design is to use each batch of plastic at each of three temperature and pressure levels, but to not repeat any combination of temperature and pressure across the batches. In compact form, with batch identified by letter (a, b, c),

batch by temp and pres

temp\pres	1	2	3
1	a	b	c
2	c	a	b
3	b	c	a

It is important to randomize the labelling of batches, leading to six possible assignments for this Latin square. There is another, complementary, Latin square, with assignments to the pressure levels for the middle and high temperature reversed. Thus there are a total of 12 possible $3 \times 3 \times 3$ Latin squares. The connectedness by batch could be portrayed as:

	batch a			batch b			batch c		
temp\pres	1	2	3	1	2	3	1	2	3
1	×				×				×
2		×				×	×		
3			×	×				×	

Viewed as a three-factor layout, there are no connected cells. Assuming that one of the factors could be ignored (such as batch), the cells for the other two factors are connected, and opening the possibility of a full two-factor

FRACTIONAL FACTORIAL DESIGNS   189

effects model,
$$y_{ijk} = \mu + \alpha_i + \beta_j + (\alpha\beta)_{ij} + e_{ijk} \ .$$
Alternatively, assuming that all interactions are negligible, the additive model,
$$y_{ijk} = \mu + \alpha_i + \beta_j + \gamma_k + e_{ijk} \ ,$$
would have all cell means and all marginal means estimable.   ◊

Despite popular opinion, it is possible to investigate interactions in Latin square designs. One approach is to use Tukey's or Mandel's interaction model, as discussed in the Chapter 9. See Milliken and Johnson (1989) for some interesting developments in this area. In addition, Chapter 27, Part I, examines replicated Latin squares in which selected interactions can sometimes be examined.

11.5 Fractional factorial designs

The Latin square design is an example of a fractional factorial design in which there is a balanced subset of all possible factor level combinations. The design issues and many of the unique features of such designs are beyond the scope of this text. However, the methods developed here can be used on such designs.

Fractional factorial designs typically consider several factors at once in a small experiment by sacrificing high-order interactions. These are most commonly used with two levels per factor, although three-level designs are employed as well. An excellent reference for this area is Box, Hunter and Hunter (1978). Many fractional factorial designs can be analyzed with methods developed earlier in this book.

Example 11.9 Design: Consider an experiment with three factors at two levels. The table below has the eight possible factor combinations, coded generically as low (-) and high (+). Interactions are determined by multiplying the signs as if they were attached to one (-1 or +1).

unit	A	B	C	AB	AC	BC	ABC
1	-	-	-	+	+	+	-
2	-	+	+	-	-	+	-
3	+	-	+	-	+	-	-
4	+	+	-	+	-	-	-
5	+	+	+	+	+	+	+
6	+	-	-	-	-	+	+
7	-	+	-	-	+	-	+
8	-	-	+	+	-	-	+

Suppose that only four of the possible eight combinations can be run. The first four in the table all have two high and two low levels for each factor.

That is, the contrast for factor A compares 3 and 4 with 1 and 2. However, some things are lost. The three-factor interaction at the low level. Further, each main effect is confounded with a two-factor interaction (A with BC, etc.). Nevertheless, with only three degrees of freedom after correcting for the mean, this is the best that can be hoped.

Now suppose that all eight combinations can be run, but only four at a time. For instance, these may be plants to be placed in a growth chamber that can only hold four pots. The experiment is then blocked into two sets of four, as indicated by the separation in the table above. Notice that the three-factor interaction is totally confounded with block, but the added runs relieve the confounding of main effects and two-factor interactions. Of course, the assignment of experimental units should be randomized within each set of four. ◊

It is possible in fact to examine seven factors in eight runs, provided that interactions can be considered negligible. This can be done quite easily using the previous example by matching each of the interactions with another factor (AB with D, AC with E, etc.). Note, however, that this is a **saturated design**, with every degree of freedom associated with a main effect.

For saturated designs, there are no degrees of freedom for error. A common approach to assessing the importance of factors involves graphical analysis: construct all contrasts and construct a normal plot. Actually, it is probably better to remove the sign and construct a half-normal plot instead, which removes some visual bias. Large contrasts are considered significant if they do not follow a straight line with the rest. On the other hand, small contrasts lying along a straight line probably represent negligible effects.

A formal post-hoc analysis could be done after the graphical investigation. However, fractional factorial experiments are often done in situations where quick results are desired to focus attention for future experiments on factors that are more important. That is, each experiment is part of a larger series of experiments. The process of planning a series of experiments to find the best combination of factors is sometimes called evolutionary operations, or 'evop'. This is an aspect of a larger area known as response surface methodology (Box and Draper 1987).

Some researchers have become quite enamored with central replicate designs in which most experimental units are assigned a combination of high and low levels of factors, but a small number have intermediate levels of all factors. These 'central replicate' designs can be quite useful for assessing nonlinear response. However, it is important to consider the whole objective of the experiment carefully to get the most of a study, as illustrated in the following example.

Example 11.10 Design: Consider an experiment with two factors which can be set at low (-), medium (0) or high (+) levels. With only six experi-

FRACTIONAL FACTORIAL DESIGNS   191

mental units, it is not possible to examine nonlinear relationships among
the factors in detail. However, assigning two central replicates allows a
measure of error and of nonlinearity.

factor	1	2	3	4	5	6
A	−	−	+	+	0	0
B	−	+	−	+	0	0

Unfortunately, it is not possible to tell from such an experiment whether
any evidence of nonlinearity points to factor A or factor B.

Now suppose that 12 experimental units were available, say to be run
in two blocks of six each. Why not simply repeat the above? Actually, the
scientist would get more information by including a completely new set:

factor	7	8	9	10	11	12
A	−	+	0	0	0	0
B	0	0	−	+	0	0

This fills out the set of 3 × 3 possible combinations of A and B at low,
medium and high, while adding another central replicate.

		B		
		−	0	+
	+	3	8	4
A	0	9	*	10
	−	1	7	2

Here there are four center replicates (* = 5, 6, 11, 12) which allow measures of block effect, error and some contribution to nonlinear information.
It is now possible to separate quadratic effects of the two factors.   ◇

A number of industries have adopted the Taguchi method for designing experiments, perhaps out of ignorance of its serious flaws (Box, Bisgaard and Fung 1988). Much simpler, more conventional designs can usually achieve research goals and lead to more straightforward analysis. The example below illustrates some of the ambiguities which can arise if one makes unwarranted assumptions.

Example 11.11 Product: A company wanted to examine the response of a product to five design factors. For various reasons, the company wanted to examine three levels of some factors and only two of others. Further, it could only afford 18 experimental units. Thus numerous interactions must be assumed negligible with any design. The design chosen was Taguchi's L^{18}, which confounds some main effects with interactions. Main effects for factors A, B and E were found significant with a variety of approaches, while factor C had a negligible effect on its own or in combination with other factors. If one assumes an additive model, then factor D is highly significant. However, if one allows interactions, then factor D is replaced by

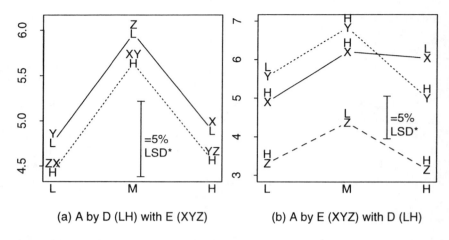

Figure 11.2. *Product interaction plots for model selection. (a) shows the significant ($p = 0.011$) but modest effect of factor* D. *as compared to factor* A *($p = 0.0027$). Notice the confounding of factor* E *from levels printed above those of* D. *(b) shows the* A∗E *interaction ($p = 0.044$), but notice that the lower response at level* H *of factor* D *(printed above those of* E*).*

the interaction A∗E. The models in contention are

$$y = A + B + D + E + \text{error} \quad (R^2 = 94\%)$$
$$y = A + B + E + A*E + \text{error} \quad (R^2 = 98\%) \ .$$

Parsimony would suggest the additive model, which is assumed in Taguchi's method. However, the interaction plots in Figure 11.2 suggest a curious relationship. Added symbols indicate levels of the third factor (E or D); the effect of factor B has been removed for convenience. The left figure suggests the high level of D leads to lower response. However, the added symbols on the right indicate that this relationship holds only at high and low levels of A but is reversed in the middle. Further, if the company is interested in high response, then it may wish to set aside the low level of E entirely. With this done, factor D is totally confounded with the interaction of A and E. Thus it is unclear at the end of the experiment how to attribute effects. The experiment probably should have been done with a few more experimental units and/or only two levels of most factors. As it is, the company may have to redo the experiment to ascertain the importance of D. ◊

11.6 Problems

11.1 Growth: Consider the experiment introduced in Example 11.1. The design structure was a bit complicated. However, for the purposes of this problem, consider the design to be completely randomized. An added problem is that `advplt` is a count, and many values are zero (0). The suggested transformation to stabilize variance is `log10(1+advplt)`.
(a) Present one or more key graphs to summarize the data. Comment on any problems with assumptions.
(b) Find means and standard deviations for each treatment group.
(c) Construct an overall test for equal means. If differences are found, order the means and then perform pairwise t tests to determine significant differences.
(d) Repeat (a)–(c) without the zeros.
(e) Comment on multiple comparison issues. That is, select two methods and briefly discuss their merits and short-comings for this problem.
(f) Comment on the comparison of group means, with and without zeros, noting that the `trt` levels represent three levels of treatment (TDZ = 0.2, 2, 20) plus two controls (TDZ = 0 saline water; TDZ = −1 BHTA).

11.2 Growth: (a) Test whether the data show evidence of differences among the treatment groups (that is, the three different levels of TDZ and the two controls).
(b) Construct rough-and-ready confidence intervals for the five treatment means of the form

$$\text{mean} \pm 2 \times \text{SE} .$$

Indicate exactly how you do this and what variance estimate(s) you use and why. Interpret this interval – roughly what can be said about the chance that such an interval covers the true mean?
(c) Estimate the difference between the two controls by setting up a linear combination of treatment means. What is the estimated SE of this difference?
(d) Estimate linear trend among the levels of TDZ, including the saline water control. [Hint: there are at least two ways to do this, using a regression idea or orthogonal polynomials.]
Be sure in your write-up that you say explicitly how things are done, and verify where appropriate that results are what they should be. For instance show that either the control mean differences, or the corresponding differences in model parameters, are the same as the control contrast specified.

11.3 Growth: Recall that the treatment structure had three levels of BA (0, 0.44, 4.4) and four levels of TDZ (0, 0.2, 2, 20), but the combination of the highest levels was missing. (The added control (`trt` = 20) is dropped here.) The aim here is to investigate the effect of these two chemicals on the number of adventitious plants (advplt). Be sure to show your work. Write down models and assumptions explicitly. Explain your choices. Interpret the several types of sums of squares and justify your choice (it is not necessary to delve into abstract algebra – you may use numerical examples to make your points as appropriate). Again, the suggested transformation to stabilize variance is log10(1+advplt).

(a) Look at the balanced treatment structure obtained by deleting the high level of BA. Analyze and interpret the main effects and interactions. Be sure to indicate why sums of squares of Types I and III are or are not different.

(b) Consider the balanced treatment structure with the high level of TDZ deleted. Repeat the analysis as in (a).

(c) Now examine the unbalanced treatment structure with all 11 combinations. Interpret the four types of sums of squares. Analyze results (as well as you can given the SAS output) and indicate what more you would do or need to complete analysis.

(d) Conduct an analysis of the 11 combinations assuming an additive model. How would you interpret this analysis? Comment briefly on the relative merits of the additive and full models for this problem.

(e) Compare your results with Problem 11.1. Summarize your findings (briefly) from both. That is, what would you tell the scientist?

11.4 Product: This problem arose while consulting with a private company which is trying to design a new product with the right balance of taste and odor. They created 16 combinations of sux factors in a fractional factorial design. Each of the 16 units was evaluated and given scores for taste and odor on a 0–100 scale by ten judges. Only the average scores are available. Your task is to advise the company about which factors are most important. In particular, should there be any follow-up experiments, which three factors would you recommend be studied further?

(a) First consider an additive model with all six factors: carefully write down the model and consider what happens to solutions of the normal equations. What are the estimable functions? What do you learn from analyzing the additive model for the two traits?

(b) Determine the aliasing pattern of two-factor interactions. That is, which interactions are totally confounded with each other? How many can be estimated? It might help to recode the factors as +1 and −1 for this exercise.

(c) Simplify the models. That is, analyze the traits in the presence of possible interactions and arrive at parsimonious models for each.

(d) Present appropriate graphics to support your findings. These might include something about an investigation of assumptions as well as summaries of best estimates of effects.

(e) Interpret your results for the company. That is, address the questions asked above.

CHAPTER 12

Linear Models Inference

This chapter is meant as a quick review of linear models estimation. Connections between least squares, weighted least squares and maximum likelihood are briefly developed without proof. Restricted maximum likelihood is introduced. While reference is made to likelihoods, there is no attempt to develop formal theory of inference. A more complete treatment can be found in Searle (1987) or Seber (1977), along with a review of the necessary matrix algebra.

The material in this chapter is used explicitly in only a few isolated portions of the remainder of this text, notably Chapters 18, 20, 21 and 26. It could be skipped by readers less interested in the general formulation of linear models theory.

Section 12.1 reviews some matrix preliminaries. Sections 12.2 and 12.3 present ordinary and weighted least squares in matrix form. The assumption of normality allows consideration of maximum likelihood and restricted maximum likelihood methods in Sections 12.4 and 12.5. Formal inference about model parameters is developed in Section 12.6, followed in Section 12.7 by recasting of special cases from preceeding and following chapters of this text.

12.1 Matrix preliminaries

In this chapter, notation follows conventions used in linear models, with the observed response vector $\mathbf{y}$ being of length n and a vector of p unknown parameters $\boldsymbol{\beta}$ associated with an $n \times p$ design matrix $\mathbf{X}$. Bold lower-case letters $\mathbf{a}$ signify vectors and bold capitals $\mathbf{A}$ are used for matrices. Elements of a matrix $\mathbf{A}$ are denoted by a_{ij} in this text. A matrix or vector transpose is indicated as $\mathbf{A}^{\mathrm{T}}$ or $\mathbf{a}^{\mathrm{T}}$, respectively. A **generalized inverse** $\mathbf{A}^{-}$ of a matrix $\mathbf{A}$ satisfies

$$\mathbf{A}\mathbf{A}^{-}\mathbf{A} = \mathbf{A}.$$

If $\mathbf{A}$ is square and of full rank, $\mathbf{A}^{-} = \mathbf{A}^{-1}$ is the unique inverse satisfying $\mathbf{A}\mathbf{A}^{-1} = \mathbf{I}$. However, there may be an infinite number of generalized inverses.

The trace of a square matrix is the sum of the diagonal elements $\mathrm{tr}(\mathbf{A}) =$

$\sum_i a_{ii}$. Clearly, the trace of a scalar c is that value. The trace of a sum $\text{tr}(\mathbf{A} + \mathbf{B})$ is the sum of the traces $\text{tr}(\mathbf{A}) + \text{tr}(\mathbf{B})$. Further, the trace is unaffected by cycling, $\text{tr}(\mathbf{ABC}) = \text{tr}(\mathbf{BCA})$. A square matrix is **idempotent** if $\mathbf{A} = \mathbf{A}^\text{T} = \mathbf{A}^2$. For an idempotent matrix, $\text{rank}(\mathbf{A}) = \text{tr}(\mathbf{A})$. The determinant of a square matrix is defined recursively as the alternating sum of products of a row's elements a_{ij} and the corresponding cofactors

$$\det(\mathbf{A}) = \sum_j (-1)^{i+j} a_{ij} \times \det(\mathbf{A}_{ij})$$

in which $\mathbf{A}_{ij}$ is the submatrix of $\mathbf{A}$ with row i and column j removed.

In general, if $\mathbf{y} \sim N(\boldsymbol{\mu}, \mathbf{V})$ and $\mathbf{AV}$ is idempotent, then the **quadratic form** $\mathbf{y}^\text{T}\mathbf{Ay}$ has a $\chi^2_{r;\delta^2}$ distribution with $r = \text{rank}(\mathbf{A})$ and non-centrality parameter $\delta^2 = \boldsymbol{\mu}^\text{T}\mathbf{A}\boldsymbol{\mu}$. The distribution is central ($\delta^2 = 0$) if in addition $\mathbf{A}\boldsymbol{\mu} = \mathbf{0}$. Two quadratic forms $\mathbf{y}^\text{T}\mathbf{Ay}$ and $\mathbf{y}^\text{T}\mathbf{By}$ are independent if $\mathbf{AVB} = \mathbf{0}$.

More detailed development of matrix algebra can be found in chapters of Searle (1987) or Seber (1977). Smaller books focused on matrices such as Healy (1986) can provide a quick review.

12.2 Ordinary least squares

Fixed effect models considered up to this point include factorial arrangements in a completely randomized design. They can be summarized as

$$E(\mathbf{y}) = \mathbf{X}\boldsymbol{\beta} \text{ and } V(\mathbf{y}) = \sigma^2\mathbf{I} ,$$

with vectors $\mathbf{y}$ and $\boldsymbol{\epsilon}$ of length equal to the sample size n, while the design matrix $\mathbf{X}$ has size $n \times p$ and the parameter vector $\boldsymbol{\beta}$ is of length p, with $n > p$. The criteria for 'best' estimators involves minimizing the sum of squares,

$$(\mathbf{y} - \mathbf{X}\boldsymbol{\beta})^\text{T}(\mathbf{y} - \mathbf{X}\boldsymbol{\beta})$$

which leads to the normal equations,

$$\mathbf{X}^\text{T}\mathbf{X}\boldsymbol{\beta} = \mathbf{X}^\text{T}\mathbf{y} ,$$

having a solution $\hat{\boldsymbol{\beta}}$ given by

$$\hat{\boldsymbol{\beta}} = (\mathbf{X}^\text{T}\mathbf{X})^{-}\mathbf{X}^\text{T}\mathbf{y} .$$

This solution is in general neither unique nor an estimator of $\boldsymbol{\beta}$. However, if the model is not over-specified and hence $\mathbf{X}$ is of full rank $r = p$, the solution is the unique minimum variance estimator of $\boldsymbol{\beta}$,

$$\hat{\boldsymbol{\beta}} = (\mathbf{X}^\text{T}\mathbf{X})^{-1}\mathbf{X}^\text{T}\mathbf{y} .$$

If instead $\text{rank}(\mathbf{X}) = r < p$, the only **estimable (linear) functions** are linear combinations of the expected responses, $\mathbf{b}^\text{T} E(\mathbf{y})$ of the form $\mathbf{b}^\text{T}\mathbf{X}\boldsymbol{\beta}$.

That is, $\mathbf{a}^T\boldsymbol{\beta}$ is an estimable function if it has a linear unbiased estimate $\mathbf{b}^T\mathbf{y}$, with $\mathbf{a} = \mathbf{X}^T\mathbf{b}$. The estimators $\mathbf{a}^T\hat{\boldsymbol{\beta}} = \mathbf{b}^T\mathbf{y}$ are **best linear unbiased (BLUE)**, since they have the smallest variance among all unbiased estimators which are linear combinations of $\mathbf{y}$. In particular, $\mathbf{X}\hat{\boldsymbol{\beta}}$ are BLUE for the expected response vector $E(\mathbf{y}) = \mathbf{X}\boldsymbol{\beta}$.

The residuals $\hat{\boldsymbol{\epsilon}} = \mathbf{y} - \mathbf{X}\hat{\boldsymbol{\beta}}$ can be combined to provide an unbiased estimator of variance,

$$\hat{\sigma}^2 = \hat{\boldsymbol{\epsilon}}^T\hat{\boldsymbol{\epsilon}}/(n-r) ,$$

reflecting the error degrees of freedom Under certain conditions, e.g. when the kurtosis is $E[y_i - E(y_i)]^4 = 3\sigma^4$ as is the case for normal data, $\hat{\sigma}^2$ is uniformly minimum variance unbiased among all estimators of quadratic form in $\mathbf{y}$.

Geometrically, the data are projected onto two orthogonal spaces, model and error. That is, define the projection matrix

$$\mathbf{P} = \mathbf{I} - \mathbf{X}(\mathbf{X}^T\mathbf{X})^{-}\mathbf{X}^T .$$

A projection is **idempotent**, $\mathbf{P}^2 = \mathbf{P}$. Further, $\mathbf{I} - \mathbf{P}$ is a projection and $\mathbf{P}(\mathbf{I} - \mathbf{P}) = \mathbf{0}$. Notice that $\mathbf{P}\mathbf{y} = \hat{\boldsymbol{\epsilon}}$ and $(\mathbf{I} - \mathbf{P})\mathbf{y} = \mathbf{X}\hat{\boldsymbol{\beta}}$. That is,

$$\mathbf{y} = \mathbf{X}\hat{\boldsymbol{\beta}} + \hat{\boldsymbol{\epsilon}} = (\mathbf{I} - \mathbf{P})\mathbf{y} + \mathbf{P}\mathbf{y} .$$

12.3 Weighted least squares

Ordinary least squares estimators are inefficient if the variances are not all equal. That is, there are other estimators with smaller variances which take account of the variance-covariance matrix. Weighted least squares, sometimes known as **generalized least squares**, are appropriate for the situation when

$$E(\mathbf{y}) = \mathbf{X}\boldsymbol{\beta} \text{ and } V(\mathbf{y}) = \mathbf{V} ,$$

in which $\mathbf{V}$ is non-singular, having an inverse. The normal equations become

$$\mathbf{X}^T\mathbf{V}^{-1}\mathbf{X}\boldsymbol{\beta} = \mathbf{X}^T\mathbf{V}^{-1}\mathbf{y} ,$$

with solution $\hat{\boldsymbol{\beta}}$ given by

$$\hat{\boldsymbol{\beta}} = (\mathbf{X}^T\mathbf{V}^{-1}\mathbf{X})^{-}\mathbf{X}^T\mathbf{V}^{-1}\mathbf{y} .$$

Similar statements can be made about estimability as done above for ordinary least squares. Estimators of estimable functions are BLUE provided that the matrix $\mathbf{V}$ is known.

If the observations $\mathbf{y}$ are independent with possibly different variances, the matrix $\mathbf{V}$ is diagonal, $\mathbf{V} = \mathrm{diag}(\sigma_1^2, \cdots, \sigma_n^2)$. Notice that the rescaled responses $y_i^* = y_i/\sigma_i$ have means $\sum_j X_{ij}\beta_j/\sigma_i$ and common variance 1.

Letting $X^*_{ij} = X_{ij}/\sigma_i$, the rescaled model has ordinary least squares solution
$$\hat{\beta} = (\mathbf{X}^{*\text{T}}\mathbf{X}^*)^{-}\mathbf{X}^{*\text{T}}\mathbf{y}^* ,$$
which is the same as the weighted least squares solution above.

It is difficult to say much about estimators of the variance-covariance matrix, since it potentially has $n(n+1)/2$, parameters, far more than the number of responses. The sample variance-covariance matrix
$$\hat{\mathbf{V}} = \hat{\epsilon}\hat{\epsilon}^{\text{T}} ,$$
is used in multivariate analysis of variance for intermediate computations.

Later parts of this text consider models which structure $\mathbf{V}$ in natural ways connected to design considerations such as blocking and sub-sampling. These random effects introduce covariance among responses $\mathbf{y}$, but this can be exploited to simplify computations. Some multivariate data analysis methods, involving multiple responses or repeated measures, introduce further assumptions about the parameters of $\mathbf{V}$ which can substantially improve inference and interpretability, provided those assumptions are appropriate.

12.4 Maximum likelihood

Assuming normality of data allows model information to be condensed as
$$\mathbf{y} = \mathbf{X}\beta + \epsilon , \; \epsilon \sim N(\mathbf{0}, \mathbf{V}) ,$$
or simply $\mathbf{y} \sim N(\mathbf{X}\beta, \mathbf{V})$. Estimation involves maximizing the likelihood, or equivalently minimizing the deviance, or minus twice the log likelihood,
$$n\log(2\pi) + \log|\mathbf{V}| + (\mathbf{y} - \mathbf{X}\beta)^{\text{T}}\mathbf{V}^{-1}(\mathbf{y} - \mathbf{X}\beta) .$$

Maximum likelihood (ML), or minimum deviance, estimates agree with generalized least squares estimates, up to the choice of variances.

The ML estimator of variance when $\mathbf{V} = \sigma^2\mathbf{I}$ can be shown to be
$$\hat{\sigma}^2_{\text{ML}} = \hat{\epsilon}^{\text{T}}\hat{\epsilon}/n$$
which is biased. In general, the ML estimators of variance can be viewed as projecting the response onto the error space, accounting for the geometry induced by the variance-covariance structure. That is, consider
$$\mathbf{P}_V\mathbf{y} = \mathbf{P}_V\epsilon \sim N(\mathbf{0}, \mathbf{PVP}) ,$$
using the weighted projection matrix $\mathbf{P}_V = \mathbf{I} - \mathbf{X}(\mathbf{X}^{\text{T}}\mathbf{V}^{-1}\mathbf{X})^{-}\mathbf{X}^{\text{T}}\mathbf{V}^{-1}$. Note that the projection matrix is idempotent.

Given the variance-covariance matrix $\mathbf{V}$, the ML estimates of predicted values are
$$\hat{\mathbf{y}} = \mathbf{X}\hat{\beta} = (\mathbf{I} - \mathbf{P}_V)\mathbf{y} = \mathbf{X}(\mathbf{X}^{\text{T}}\mathbf{V}^{-1}\mathbf{X})^{-}\mathbf{X}^{\text{T}}\mathbf{V}^{-1}\mathbf{y} .$$

Unfortunately, the projection matrix $\mathbf{P}_V$ depends on the unknown parameters in the variance-covariance matrix. Typically, solution of ML involves repeated iteration between $\mathbf{X}\boldsymbol{\beta}$ and $\mathbf{V}$. There are serious numerical challenges that arise for complicated models, which are beyond the scope of this text.

12.5 Restricted maximum likelihood

The bias of the maximum likelihood estimator of variance is somewhat annoying. The situation gets more complicated for general $\mathbf{V}$ with several variance components, leading to consideration of **restricted maximum likelihood (REML)** in which the data are first projected onto the error space using the ordinary projection matrix $\mathbf{P}$. REML estimates of variance components have a natural interpretation in **random and mixed effects models**, which are the subject of the final two parts of this text. Estimation of the model parameters is done conditioning on REML estimates of variance components in $\mathbf{V}$.

In the completely randomized design, the unbiased variance estimator $\hat{\sigma}^2$ can be shown to be the REML estimator. That is, consider the projection of the response onto the error space,

$$\mathbf{Py} = \mathbf{P}\boldsymbol{\epsilon} \sim N(\mathbf{0}, \sigma^2 \mathbf{P}) \ .$$

Since $\text{rank}(\mathbf{P}) = n - r$, maximizing the restricted likelihood yields the unbiased variance estimator

$$\hat{\sigma}^2_{\text{REML}} = \mathbf{y}^T \mathbf{Py}/(n-r) \ ,$$

which coincides exactly with $\hat{\sigma}^2$. Estimates of $\boldsymbol{\beta}$ are the natural ones considered earlier.

In general, the ordinary projection yields

$$\mathbf{Py} = \mathbf{P}\boldsymbol{\epsilon} \sim N(\mathbf{0}, \mathbf{PVP}) \ .$$

Maximizing the restricted likelihood, or minimizing the corresponding deviance

$$(n-r)\log(2\pi) + \log|\mathbf{PVP}| + \mathbf{y}^T \mathbf{PV}^{-1}\mathbf{Py} \ ,$$

yields the estimator $\hat{\mathbf{V}}_{\text{REML}}$. Maximizing the REML deviance to obtain estimates of $\mathbf{V}$ involves repeated iterations, again beyond the scope of this book.

In practice, scientists compute a solution to the normal equations $\hat{\boldsymbol{\beta}}$,

$$\hat{\boldsymbol{\beta}} = (\mathbf{X}^T \mathbf{V}^{-1} \mathbf{X})^- \mathbf{X}^T \mathbf{V}^{-1} \mathbf{y} \ ,$$

by substituting the REML estimate for $\mathbf{V}$. Notice that the fixed parameters $\boldsymbol{\beta}$ do not appear in the REML deviance. In fact, REML has no formal procedure for finding solutions for $\boldsymbol{\beta}$.

12.6 Inference for fixed effect models

Consider an experiment with factorial arrangements in a completely randomized design, which could be summarized, with appropriate assumptions, as $\mathbf{y} \sim N(\mathbf{X}\boldsymbol{\beta}, \sigma^2 \mathbf{I})$. The total sum of squares $\mathbf{y}^T\mathbf{y}$ can be partitioned into **quadratic forms** involving explained and unexplained variation,

$$SS_{\text{TOTAL}} = SS_{\text{MODEL}} + SS_{\text{ERROR}}$$
$$\mathbf{y}^T\mathbf{y} = \mathbf{y}^T(\mathbf{I} - \mathbf{P})\mathbf{y} + \mathbf{y}^T\mathbf{P}\mathbf{y}$$

with the projection matrix $\mathbf{P} = \mathbf{I} - \mathbf{X}(\mathbf{X}^T\mathbf{X})^-\mathbf{X}$ as defined earlier. The quadratic forms SS_{MODEL} and SS_{ERROR} are uncorrelated since $\mathbf{P}(\mathbf{I} - \mathbf{P}) = \mathbf{0}$. The error sum of squares has a distribution proportional to a central chi-square with $n - r$ degrees of freedom, $SS_{\text{ERROR}} \times (n - r)/\sigma^2 \sim \chi^2_{n-r}$. In other words, $\hat{\sigma}^2 = SS_{\text{ERROR}}/(n - r)$ is the usual unbiased estimator of variance. The model sum of squares, on the other hand, has a possibly non-central distribution, $SS_{\text{MODEL}} \times r/\sigma^2 \sim \chi^2_{r;\delta^2}$, with

$$\delta^2 = \boldsymbol{\beta}^T\mathbf{X}^T(\mathbf{I} - \mathbf{P})\mathbf{X}\boldsymbol{\beta} = \boldsymbol{\beta}^T\mathbf{X}^T\mathbf{X}\boldsymbol{\beta} ,$$

the non-centrality parameter vanishing under the assumption that $\boldsymbol{\beta} = \mathbf{0}$. Therefore, the statistic

$$F = \frac{SS_{\text{MODEL}}/r}{\hat{\sigma}^2}$$

has a (possibly non-central) F distribution with r and $n - r$ degrees of freedom.

Usually, questions about model parameters concern only some of the elements of $\boldsymbol{\beta}$. The simplest case concerns whether all mean values are the same. That is, suppose $\mathbf{X}$ has a column $\mathbf{1}$ of all 1s and a corresponding parameter μ. Consider isolating the grand mean from the other parameters, say $\mathbf{X}\boldsymbol{\beta} = \mathbf{1}\mu + \mathbf{X}_2\boldsymbol{\beta}_2$. Partition the model sum of squares as

$$SS_{\text{MODEL}} = R(\boldsymbol{\beta}) = R(\mu) + R(\boldsymbol{\beta}_2|\mu) ,$$

with $R(\mu) = \mathbf{y}^T\mathbf{1}(\mathbf{1}^T\mathbf{1})^{-1}\mathbf{1}^T\mathbf{y} = n\bar{y}_.^2$ and $\bar{y}_. = \mathbf{1}^T\mathbf{y}/n$. The conditional sum of squares due to $\boldsymbol{\beta}_2$ in the presence of μ can be found by first subtracting the mean $\bar{y}_.$ from $\mathbf{y}$. This is the same as multiplying by the matrix $(\mathbf{I} - \mathbf{J}/n)$ with $\mathbf{J}$ an $n \times n$ matrix of all 1s. That is, the model becomes

$$(\mathbf{I} - \mathbf{J}/n)\mathbf{y} = (\mathbf{I} - \mathbf{J}/n)\mathbf{X}_2\boldsymbol{\beta}_2 + (\mathbf{I} - \mathbf{J}/n)\boldsymbol{\epsilon} .$$

Solving the normal equations yields weighted least solutions for $\boldsymbol{\beta}_2$ given μ. The sum of squares is thus

$$R(\boldsymbol{\beta}_2|\mu) = (\mathbf{y} - \mathbf{1}\bar{y}_.)^T\mathbf{X}_2[\mathbf{X}_2^T(\mathbf{I} - \mathbf{J}/n)\mathbf{X}_2]^-\mathbf{X}_2^T(\mathbf{y} - \mathbf{1}\bar{y}_.)$$

with non-centrality parameter

$$\delta^2 = \boldsymbol{\beta}^T\mathbf{X}^T(\mathbf{I} - \mathbf{J}/n)\mathbf{X}\boldsymbol{\beta} = \boldsymbol{\beta}_2^T\mathbf{X}_2^T(\mathbf{I} - \mathbf{J}/n)\mathbf{X}_2\boldsymbol{\beta}_2$$

which vanishes if $\beta_2 = 0$ or equivalently $E(y) = X\beta = 1\mu$. The test statistic is thus

$$F = \frac{R(\beta_2|\mu)/(r-1)}{\hat{\sigma}^2} \sim F_{r-1,n-r;\delta^2}$$

for the hypothesis that $E(y_i) = \mu$ for all responses.

In general, suppose β is partitioned into two vectors of length p_1 and p_2, with a similar partition of the design matrix X,

$$X\beta = [X_1 : X_2]\begin{bmatrix} \beta_1 \\ \beta_2 \end{bmatrix} = X_1\beta_1 + X_2\beta_2 \ .$$

The matrices X_1 and X_2 have respective ranks $r_1 \leq p_1$ and $r_2 \leq p_2$. The model sum of squares can be partitioned by sequential fitting. Let $P_1 = I - X_1(X_1^T X_1)^- X_1^T$ be the projection matrix for β_1 by analogy with P above. Partition the model sum of squares as

$$R(\beta) = R(\beta_1) + R(\beta_2|\beta_1) \ ,$$

with $R(\beta_1) = y^T(I - P_1)y$ the sum of squares involving β_1 on its own. However, the distribution of $R(\beta_1)$ depends on both β_1 and β_2 in general. For instance,

$$E[R(\beta_1)] = r_1\sigma^2 + (X_1\beta_1 + X_2\beta_2)^T(I - P_1)(X_1\beta_1 + X_2\beta_2)$$

does not vanish if $\beta_1 = 0$.

The additional model sum of squares due to β_2 in the presence of β_1 is

$$R(\beta_2|\beta_1) = y^T P_1 X_2 (X_2^T P_1 X_2)^- X_2^T P_1 y \ .$$

That is, after fitting β_1, the data reduce to $P_1 y$ with expectation $P_1 X_2 \beta_2$ not depending on β_1. The quadratic form $R(\beta_2|\beta_1)$ has a non-central chi-square distribution with non-centrality parameter $\delta^2 = (X_2\beta_2)^T P_1 X_2 \beta_2$ which is zero under the assumption that $\beta_2 = 0$. $R(\beta_2|\beta_1)$ and SS_E are independent since $P_1 P = 0$. Hence,

$$F = \frac{R(\beta_2|\beta_1)/r_2}{MS_E} \sim F_{r_2,n-r;\delta^2}$$

which has a central F distribution if $\beta_2 = 0$.

Put another way, $R(\beta_1) - R(\mu)$ is the Type I sum of squares while $R(\beta_2|\beta_1)$ is the Type III sum of squares. The subspaces spanned by $X_1\beta_1$ and $X_2\beta_2$ are *not* orthogonal to one another in general (that is, for unbalanced designs). Instead, it is appropriate to compare the full model $X\beta$ with a reduced model $X_1\beta_1$ using the conditional SS for the Type III hypothesis

$$R(\beta_2|\beta_1) = ||X\hat{\beta} - X_1\hat{\beta}_1||^2 \ .$$

This measures the Euclidean distance, the length of the vector in response n-space, between the projection for the full model fit and that for the reduced model fit.

The **general linear hypothesis** can be thought of as placing $q \leq r$ linear constraints on a model. That is, there is a $q \times p$ matrix $\mathbf{C}$ such that the hypothesis can be written as

$$H_0 : \mathbf{C}\boldsymbol{\beta} = \mathbf{0}.$$

While it is possible to consider $\mathbf{C}\boldsymbol{\beta} = \mathbf{c}$, most situations of interest concern contrasts. The vector $\mathbf{C}\boldsymbol{\beta}$ must be estimable for the hypothesis to make sense. That is, $\mathbf{C} = \mathbf{A}\mathbf{X}$ for some $q \times n$ matrix $\mathbf{A}$, although it is not necessary to find $\mathbf{A}$ in practice. The best linear estimator of $\mathbf{C}\boldsymbol{\beta}$ is

$$\mathbf{C}\hat{\boldsymbol{\beta}} = \mathbf{C}(\mathbf{X}^\mathrm{T}\mathbf{X})^-\mathbf{X}^\mathrm{T}\mathbf{y} = \mathbf{A}(\mathbf{I} - \mathbf{P})\mathbf{y}.$$

This estimator has distribution $\mathbf{C}\hat{\boldsymbol{\beta}} \sim N(\mathbf{C}\boldsymbol{\beta}, \sigma^2 \mathbf{C}(\mathbf{X}^\mathrm{T}\mathbf{X})^-\mathbf{C}^\mathrm{T})$. The variation explained by the hypothesis is in general

$$\begin{aligned}R(\mathbf{C}\boldsymbol{\beta}) &= (\mathbf{C}\hat{\boldsymbol{\beta}})^\mathrm{T}[\mathbf{C}(\mathbf{X}^\mathrm{T}\mathbf{X})^-\mathbf{C}^\mathrm{T}]^{-1}\mathbf{C}\hat{\boldsymbol{\beta}} \\ &= \mathbf{y}^\mathrm{T}(\mathbf{I}-\mathbf{P})\mathbf{A}^\mathrm{T}[\mathbf{A}(\mathbf{I}-\mathbf{P})\mathbf{A}^\mathrm{T}]^{-1}\mathbf{A}(\mathbf{I}-\mathbf{P})\mathbf{y}\end{aligned}$$

and the corresponding test statistic is

$$F = \frac{R(\mathbf{C}\boldsymbol{\beta})/q}{\hat{\sigma}^2} \sim F_{q,n-r;\delta^2}$$

with non-centrality parameter $\delta^2 = (\mathbf{C}\boldsymbol{\beta})^\mathrm{T}[\mathbf{C}(\mathbf{X}^\mathrm{T}\mathbf{X})^-\mathbf{C}^\mathrm{T}]^{-1}\mathbf{C}\boldsymbol{\beta}$ vanishing when the null hypothesis H_0 is true.

It is more economical in terms of computing effort to partition the model as developed above. However, sometimes it may not be possible to do this, requiring a more general approach. In addition, the general linear hypothesis approach is compact and mathematically elegant. It is particularly useful when considering multivariate hypotheses, as is done for multiple responses and repeated measures in later parts of this text.

12.7 Anova and regression models

This section briefly recasts the one-factor and two-factor models into the general linear model framework. In addition, regression and analysis of covariance are briefly developed, anticipating the next part of this book.

The **one-factor model** for a groups introduced earlier had the form

$$y_{ij} = \mu_i + e_{ij} = \mu + \alpha_i + e_{ij}, \quad e_{ij} \sim N(0, \sigma^2)$$

$i = 1, \cdots, a$, $j = 1, \cdots, n_i$, with independent errors and suitable side conditions on the parameters. Let $\mathbf{y}$ be a vector of length $n. = \sum_i n_i$ with entries y_{ij} increasing most rapidly by the second index (j). The $p = 1 + a$ parameter vector $\boldsymbol{\beta}$ consists of $\beta_1 = \mu$ and $\beta_{i+1} = \alpha_i$, $i = 1, \cdots, a$. The $n. \times (a+1)$ design matrix $\mathbf{X}$ has a first column of all 1s (**1**) and a columns with the ith containing n_i 1s opposite the responses for that group and all other elements being 0,

$$E(\mathbf{y}) = \mathbf{X}\boldsymbol{\beta} = [\mathbf{1} : \mathbf{X}_2] \begin{bmatrix} \mu \\ \boldsymbol{\alpha} \end{bmatrix} = \mathbf{1}\mu + \mathbf{X}_2\boldsymbol{\alpha} .$$

Placing a linear constraint on the parameters α_i corresponds to reducing the design matrix to one of full rank a. Examples can be found in Chapter 7, Part C.

The ML deviance is

$$n.\log(2\pi) + n.\log(\sigma^2) + \sum_{ij}(y_{ij} - \mu_i)^2/\sigma^2 .$$

The ML estimators of the group means are the OLS means $\hat{\mu}_i = \bar{y}_{i.}$. The ML estimator of error variance is

$$\hat{\sigma}^2_{\mathrm{ML}} = \sum_{ij}(y_{ij} - \mu_i)^2/n.$$

which is a biased estimator of σ^2.

The REML deviance is

$$(n.-a)\log(2\pi) + (n.-a)\log(\sigma^2) + \sum_{ij}(y_{ij} - \bar{y}_{i.})^2/\sigma^2 .$$

To see this, note that $\mathbf{X}^{\mathrm{T}}\mathbf{X}$ is an $a \times a$ diagonal matrix with sample sizes n_i as the diagonal elements. The matrix $\mathbf{I} - \mathbf{P}$ is block diagonal with the ith diagonal block being an $n_i \times n_i$ matrix with all elements being $1/n_i$. The projection matrix $\mathbf{P}$ thus has rank $n.-a$ and determinant $|\mathbf{P}| = n.-a$. The projection of the responses $\mathbf{Py}$ centers them as $y_{ij} - \bar{y}_{i.}$ for the jth unit in the ith group. Therefore the REML estimator of variance coincides with the usual unbiased estimator

$$\hat{\sigma}^2 = \sum_{ij}(y_{ij} - \mu_i)^2/(n.-a) .$$

However, there is little difference between $\hat{\sigma}^2_{\mathrm{ML}}$ and $\hat{\sigma}^2$ if the sample size $n.$ is large.

Relaxing the assumption of equal variance, the model

$$y_{ij} = \mu + \alpha_i + e_{ij} , \ e_{ij} \sim N(0, \sigma_i^2)$$

with independent errors leads to the ML deviance

$$n.\log(2\pi) + \sum_i n_i \log(\sigma_i^2) + \sum_{ij}(y_{ij} - \mu_i)^2/\sigma_i^2 .$$

The estimators of group means are the same for OLS, WLS, ML and REML. This is not surprising, since the variances σ_i^2 and weights $1/\sigma_i^2$ are the same within any group. The (biased) ML estimators of group variances are

$$\hat{\sigma}^2_{i,\mathrm{ML}} = \sum_j(y_{ij} - \mu_i)^2/n_i ,$$

dividing by n_i rather than $n_i - 1$. The REML deviance is by analogy

$$(n_{..} - a)\log(2\pi) + \sum_i (n_i - 1)\log(\sigma_i^2) + \sum_{ij}(y_{ij} - \bar{y}_{i.})^2/\sigma_i^2,$$

leading to the the unbiased estimators of variance

$$\hat{\sigma}_i^2 = \sum_j (y_{ij} - \mu_i)^2/(n_i - 1).$$

The **two-factor model** was written as

$$y_{ijk} = \mu_{ij} + e_{ijk} = \mu + \alpha_i + \beta_j + \gamma_{ij} + e_{ijk}, \quad e_{ijk} \sim N(0, \sigma^2)$$

$i = 1, \cdots, a$, $j = 1, \cdots, b$, $k = 1, \cdots, n_{ij}$ and independent errors. As in the one-factor model, the responses can be strung out as a vector **y** of length $n_{..} = \sum_{ij} n_{ij}$, the model parameters catenated as a vector β of length $1 + a + b + ab$ (provided $n_{ij} > 0$) and the design matrix **X** a matrix of 0s and 1s. In other words, with vectors for the parameters, including here $\beta = (\beta_1, \cdots, \beta_b)$,

$$E(\mathbf{y}) = [\mathbf{1} : \mathbf{X}_2 : \mathbf{X}_3 : \mathbf{X}_4] \begin{bmatrix} \mu \\ \alpha \\ \beta \\ \gamma \end{bmatrix} = \mathbf{1}\mu + \mathbf{X}_2\alpha + \mathbf{X}_3\beta + \mathbf{X}_2\gamma.$$

This is made explicit in Examples 7.10 and 7.11 using two different sets of linear constraints.

The ML deviance is

$$n_{..}\log(2\pi) + n_{..}\log(\sigma^2) + \sum_{ijk}(y_{ijk} - \mu_{ij})^2/\sigma^2.$$

ML and REML estimators of estimable functions agree with OLS estimators developed previously. The development of partition of quadratic forms in the previous section shows that the Type III sums of squares are most appropriate. That is, the distribution of the Type III sum of squares $R(\alpha|\mu, \beta, \gamma)$ depends on σ^2 and α but not on other model parameters provided there are no empty cells ($n_{ij} > 0$).

Regression and **analysis of covariance** are considered in the next part. Simple linear regression can be written as

$$y_i = \beta_1 + \beta_2 x_i + e_i, \quad e_i \sim N(0, \sigma^2)$$

with independent errors. This clearly fits within the framework of the linear model developed in this chapter, with design matrix **X** having first column **1** and second column filled by x_is. Analysis of covariance combines the flavor of factorial models and regression.

12.8 Problems

12.1 Infer: Consider a general linear hypothesis $C\beta$ for a vector β of length three.
(a) Show that the column vectors $c_1 = (1, -1, 0)^T$ and $c_2 = (1, 1, -2)^T$ are orthogonal in a three-dimensional space. Write down the corresponding contrasts of parameters.
(b) Find another set of two column vectors which are orthogonal.
(c) Show that any contrast among β can be written in terms of the column vectors in (a) or (b).

12.2 Infer: (a) Show that the examples in Chapter 7, Part C, can be written in compact matrix form.
(b) Write down the model and normal equations in matrix form for Example 7.2. Use matrix symbols, but in addition determine exactly $X^T X$.
(c) Show explicitly the normal equations for Examples 7.3 and 7.4 (sum-to-zero and set-to-zero constraints, respectively). Again, multiply out $X^T X$ to show its values explicitly.
(d) Write down the unique matrix solution for $\hat{\beta}$ for a full rank X. Solve the normal equations in (c) using either constraint for the one-factor design.

PART E

Questioning Assumptions

Data analysis requires assumptions. While one may examine data freely, interpretation relies heavily on the assumed nature of the experimental design and the distribution of response measurements. This is especially true for formal statistical inference, but is equally important for informal investigation. Chapter 13 looks at the major assumptions that the model (and hence the design) is correctly determined and the observations are independent, as well as the more minor assumptions that responses have equal variance and a normal distribution.

Chapter 14 addresses the suspicion that variance may not be constant across all levels of the factors. Tests for unequal variance may formally uncover a problem, or may be too weak to be effective. Some methods have been developed for adjusting analysis for unequal variance, but these have certain tradeoffs. Many situations naturally suggest variance-stabilizing transformations, although even this does not rectify all problems.

Sometimes it is possible to transform data to satisfy assumptions. On the other hand, various methods have been developed which do not make many assumptions about the data. Some introductory remarks on such nonparametric, or distribution-free, methods are developed in Chapter 15. The main ideas here concern replacing observed values by their ranks, or using permutations to evaluate the strength of evidence.

CHAPTER 13

Residual Plots

This chapter considers the nature of the assumptions commonly made for analysis of designed experiments. The basic philosophy for data analysis involves building a full model that incorporates beliefs about all possible sources of variability. Evidence from data may suggest ways to simplify, or reduce, the model. However, there is no formal way within this structure to question the full model itself. While it is possible to consider fuller models, with fewer assumptions, it is more feasible in practice to investigate departures from model assumptions by examining plots of residuals, adapting many of the diagnostic ideas from regression. An excellent source for practical advice 'beyond anova' is Miller (1997).

There are four basic assumptions commonly employed in the comparison of group means, or more generally in the analysis of variance:

- the model is correct
- responses are uncorrelated
- variances are equal
- responses have a normal distribution.

It may be best in some situations simply to ignore potential difficulties with assumptions. Some studies indicate that certain violations of assumptions have negligible effects on estimates and tests. Alternatively, evidence from residual plots may suggest removing outliers, employing suitable transformations, using weighted analysis, or redefining the model relationship between factors and response.

With small samples, these four assumptions cannot be investigated in depth. Most formal tests for violations of an assumption have low power; some are very sensitive to other assumptions. Further, they raise the dilemma of multiple stage testing on the same data set. Miller (1997) provides an excellent overview of results to date, with particular reference to Scheffé (1959, ch. 10).

Judgement calls based on hints gleaned from residual plots may be all that is certain, beyond prior information about related research problems. The basic residual plot is a scatter plot of the predicted values against the residuals (response minus predicted), using plot symbols to identify factor levels, or group assignments. Many improvements have been made on this plot over the years, particularly in the area of regression diagnostics (*cf.* Belsley, Kuh and Welsch 1980). Several recent books focusing on graphical methods have innovative ideas worth considering (Hoaglin, Mosteller and Tukey 1991; Cleveland 1993).

Section 13.1 explores ideas about plotting residuals to detect departures from these assumptions with some attention to outliers and zeros. Subsequent sections briefly consider consequences of having the incorrect model, correlated responses, unequal variances or non-normal data. The latter two assumptions are examined more deeply in Chapters 14 and 15. Various aspects of correlated responses permeate the remaining parts of this book. While the discussion is presented in terms of the single-factor models, the ideas are quite general and concern all the experiments considered in this text.

13.1 Departures from assumptions

Many problems with the basic assumptions can be examined in the regression setting by plotting residuals against predicted values or against other factors. This idea carries over with slight modification to the present situation. Factors usually have fewer levels than regressors, leading to fewer distinct predicted values. However, it can be very valuable to augment residual plots with identifying symbols for factor levels. Sometimes several copies of the same plot with different identifiers can stress different relationships. The predicted values in the one-factor analysis of variances are the group means. Thus plot $\bar{y}_{i\cdot}$ against $y_{ij} - \bar{y}_{i\cdot}$ using different plotting symbols to identify each group. It is often helpful to **jitter** the means slightly. (Jittering involves adding some small uniform random variate to a value. Cleveland (1985) recommended that 'small' be about 4% of the range of the value.) For some purposes, it is useful to plot the group mean $\bar{y}_{i\cdot}$ against individual responses y_{ij} although it is vital to remember that these are naturally correlated.

It is certainly possible to plot factor level (group identifier $-1,2,3,\cdots$) against response. However, it is important to recognize that identifiers may impose an arbitrary ordering of groups which has no relationship to average response or to any structure among factor levels. A remedy for this would be to order groups by mean response when plotting, which is somewhat like plotting predicted against actual response.

Example 13.1 Cloning: The residual plot for the cloning experiment in Figure 13.1(a) shows a 'bulge' in the middle. This might suggest a pattern similar to binomial variation. However, in this case, `clones` 3 and 7 come from commercial cultivars, and `clones` 2 and 10 are also unusual for other reasons. Figure 13.1(b) drops these four, suggesting equal variance across

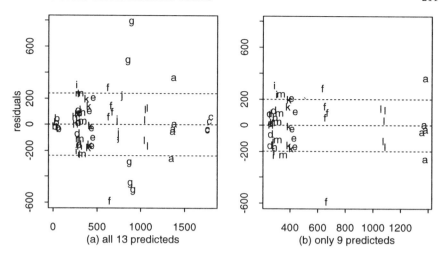

Figure 13.1. *Cloning residual plots: residual against predicted group means (jittered) for (a) all 13* clones *and (b) the balanced subset of nine* clones. *Horizontal lines mark one standard deviation.*

the remaining clones. However, there appears to be an outlying observation for clone 6. The predicted values have been jittered (Cleveland 1985) in Figure 13.1 to make it easier to see patterns.   ◇

With enough data per group, a histogram or normal plot of residuals can suggest violations of assumptions. In large data sets it may be feasible to plot separate histograms for each factor level. Box-plots or stem-and-leaf plots can often capture the key features of data and be more convenient, depending on the number of groups and the sample sizes per group.

More complicated models involving factorial arrangements suggest other ways to plot data to examine model assumptions. For instance, plot residuals against the levels (or mean response per level) of other factors not considered in the model. Covariates (regressors) which are not considered in a model can be examined by scatter plot in a similar fashion to the regression setting. A **scatter plot** is a graphical representation of two columns of numbers plotted so that the projection onto each axis hits the scale at its value. Again, it is usually a good idea to increase information about relationships by identifying points with plot characters or symbols for different factor levels. It may be useful to examine the same scatter plot with a variety of identifiers, highlighting different factors which may or may not be in the model as it stands.

Outliers, data which lie outside the ordinary range of response, can arise for a variety of reasons, including transcription errors and unusual circumstances for a particular experimental unit. Residual plots, particularly with identification of factor levels by symbols, can usually uncover outliers. The various diagnostics developed for regression can be adapted for analysis of variance in a straightforward manner.

Some experiments yield responses which are generally positive, such as yield, weight, length, or counts. **Zeros** may represent a very low response (e.g. pathogen below a threshold), no response (e.g. pathogen not present at all), problems with the experiment (e.g. plants dried out), or missing data which were incorrectly recorded. A few zeros may not affect assumptions for analysis, but a large proportion (say 10%) can upset several assumptions at once. For instance, if one group has all zeros, its variance is zero. The distribution of residuals when there are many zeros is likely to be affected, leading either to a large number of zero residuals, or to a symptomatic diagonal banding in plots of predicted values against residuals.

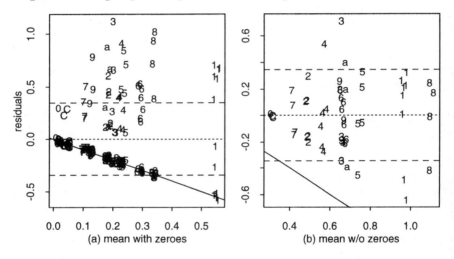

Figure 13.2. *Growth residual plots with zeros. Solid line at* advplt $= 0$ *identifies zeros in (a). Plot symbols are* 0-9, a *for the 11 chemical combinations (*BA, TDZ $= 0, 0; 0, 0.2; \cdots 4.4, 2)$ *and* C *for added BHTA control. Dotted line at zero residual; dashed lines at one SD.*

Example 13.2 Growth: Many samples had no adventitious plants (advplt $= 0$). Residual plots against the 12 group means (11 chemical combinations and added control) for fits of log(1+advplt) both (a) with and (b) without zeros are shown in Figure 13.2. The predicted and residual values have been jittered to make it easier to see the patterns. Notice the strong diagonal banding (the solid lines indicate advplt $= 0$). ◇

It is difficult to ignore a large proportion of zeros in residual plots. They appear along a line with slope –1. It may make sense to analyze the experiment with and without the zeros, overlooking problem with assumptions if results are the same regardless. Another approach is to replace zeros by a normal variate using an estimate of variance from the remaining data; however, this adds variation to the data, which is generally not acceptable to a scientist. However, it raises the question of what those zeros really mean, which is important.

It may also be useful to tabulate the pattern of zeros with respect to the factorial arrangement, perhaps revealing an unexpected effect of factors. Questions about such tables of zeros can be formalized with chi-square tests, although that is beyond the scope of this text.

13.2 Incorrect model

The first and most important, assumption is that the model has been correctly expressed in terms of the expectation of the response, up to some unknown but fixed set of parameters. Any mistake in **model specification** would lead to a **bias** in the estimation of the central tendency of present or future response measurements. For instance, consider two groups with equal variance and unequal means,

$$E(y_{ij}) = \mu_i , \ V(y_{ij}) = \sigma^2 , \ i = 1, 2, \ j = 1, \cdots, n_i ,$$

that were mistakenly combined as one group, stating (incorrectly) that

$$E(y_{ij}) = E(\bar{y}..) = \mu , \ i = 1, 2, \ j = 1, \cdots, n_i ,$$

with $\mu = \tilde{\mu}.. = (n_1\mu_1 + n_2\mu_2)/(n_1 + n_2)$. The expected **mean square error (MSE)**, or variation around the hypothesized center μ, is

$$E[(y_{ij} - \tilde{\mu}..)^2] = (\mu_i - \tilde{\mu}..)^2 + \sigma^2 , \ i = 1, 2 ,$$

which has a bias component and a variance component,

$$E(\text{MSE}) = (\text{bias})^2 + \text{variance}.$$

The MSE would be different for the two groups – equal means but unequal variances – unless the sample sizes were the same ($n_1 = n_2 = n$). A combined estimate of variance for the incorrect model would have expectation under equal sample sizes of

$$E\left[\sum_i\sum_j(y_{ij} - \bar{y}..)^2/(2n-2)\right] = (\mu_1 - \mu_2)^2/4 + \sigma^2 .$$

Thus the variance estimate would be inflated, or biased upward. The bias in mean and variance estimates would not diminish by increasing sample sizes.

Problems in model fit can be investigated with histograms by group, and/or plots of group means against residuals. **Interaction plots** (see Part C) of mean response for factor combinations can reveal deficiencies of a model. Again, it helps to identify groups by separate symbols when plotting all data together. Some packages may allow this, or it may be simpler to annotate figures by hand with reference to the raw data or a print of predicted and residual values.

Example 13.3 Cloning: Consider again the cloning experiment. A histogram of the raw data (Figure 13.3(a)) shows some severe skew to the right. The model appears to be wrong. That is, these data are from 13 clones which may have different means. A histogram of residuals, the raw data centered by group means, looks more 'bell-shaped'. However, even this may hide some evidence of unequal variance which appears in the table of means and SDs of Example 4.5 on page 56. ◊

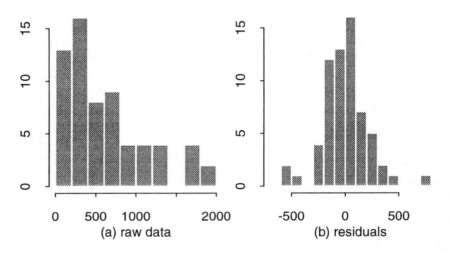

Figure 13.3. *Cloning histograms: (a)* titer *concentration for all 64 measurements; (b) all 64 residuals (deviations from group means). The raw data appear skewed due to a few groups with large means, while the residuals appear symmetric with a few outliers.*

Margin plots of one margin against cell means provide an alternative to interaction plots. They can suggest particular alternative forms of interaction which can be particularly useful if there are few degrees of freedom available, as examined in Chapter 9, Part B. Other plots that can be helpful in sorting out interactions are effect plots and half-normal plots.

Formal tests are possible by considering the full model to be more complicated than was first supposed. With small sample sizes, the reduced (original) model may be 'accepted' due to a lack of statistical power. This does not mean this model is correct, but only that there is insufficient evidence to reject it. The same issues arise for testing lack of model fit as those found in multiple (e.g. stepwise) regression. The guiding principle of parsimony suggests opting for a simpler model unless there is overwhelming evidence that a more complicated model leads to better understanding. However, it may be wise to be skeptical about too simple an interpretation, noting informally any indications of discrepancies or trends for future studies.

13.3 Correlated responses

If responses are uncorrelated, then group variances summarize the most relevant information about variability in the data. This assumption is central to least squares estimation and to a ready geometric interpretation of the interrelation of full and reduced models. It simplifies the computation of projections from full to reduced sample spaces. Dependence among responses typically enters through either blocking into larger units or correlation across a sequence, such as time or space.

Correlated responses demand attention to modelling the variation among responses as well as the mean. This increases the complexity of models, adds further parameters to estimate and reduces the power of comparisons if it is, in fact, not needed. However, if it is ignored, tests and estimates can be severely biased. Tests which ignore positive correlations among responses, which is the typical situation, may have unrealistically small p-values, because the error variance tends to be underestimated. Thus, correlation can seriously inflate the Type I error rate, rejecting the null hypothesis too often.

Correlation induced by blocking can be addressed in a straightforward manner. The next two parts of this book examine ways to include independent random effects in addition to experimental unit error. In fact, the key concepts of blocking and sub-sampling are central to understanding the value and consequences of nested experimental designs.

Serial correlation, or correlation of measurements over time or space, is not uncommon in experimental settings. For instance, measurements taken over a morning may be affected by the rising ambient air temperature. A scientist may be scoring leaf damage and unconsciously getting more critical in evaluation (or learning how to do the task better) over the course of the study. Plotting residuals against run order or run time may reveal suspicious patterns of dependence. This may show a time trend, or a hint of periodic (sine) wave. This signal could by a systematic change in mean value, or could reflect autocorrelation among responses, with high values leading to further high values and so on. Spatial patterns of dependence could be examined in a similar manner by plotting the residuals on the same places occupied by the original responses. A contour plot of residuals might reveal evidence of systematic trends or spatial autocorrelation.

Autocorrelation in long sequences suggests using time series methods, which are beyond the scope of this text (see Box, Jenkins and Reinsel

1994). However, many experiments have only short sequences due to the heavy cost of maintaining subject material. Methods involving such *repeated measures*, in the last part of this book, are natural extensions of nested experiments.

13.4 Unequal variance

Equal variance and uncorrelated responses lead to elegant geometric interpretations and simplified tests. Unequal variances do not cause appreciable problems in comparing means if all groups have the same sample sizes. Tests of equality of variance and tests of equality of means when variances are unequal are considered in Chapter 14, while transformations to stabilize variances are discussed in Chapter 15. However, with small to modest data sets, plots and a skeptical attitude may be the most valuable tools for investigation of unequal variance.

Residual plots of group mean against response (Figure 4.3, page 54) or of group mean against residual (Figure 13.1) can show evidence of unequal variance. Box-plots by group can be very effective as well when sample sizes per group are moderate to large. Systematic spreads may suggest transformations or problems in model specification. Again, using different plot characters highlights factors of interest. A plot of group means against absolute residuals $|y_{ij} - \bar{y}_{i\cdot}|$ focuses on the magnitude of deviation. Further, these absolute residuals are the basic ingredients for Levene's test of unequal variance, discussed in the next chapter, which does not depend on the assumption of normality.

13.5 Non-normal data

Normality is the least important assumption. The key features of the shape of the distribution are the degree of symmetry about the center of mass (skewness) and the concentration about the center relative to the tails of the distribution (kurtosis). Provided there are enough data, say 25 experimental units per model degree of freedom, the **central limit theorem** states that estimates of model parameters are roughly normally distributed regardless of the original distribution of the responses. Severe nonnormality might encourage the use of transformations or nonparametric methods, which are discussed in Chapter 15.

Normal plots can sometimes reveal outliers or hints of skewness or kurtosis. The normal, or quantile-quantile (Q-Q), plot consists of plotting the sorted residuals against corresponding quantiles of the normal distribution. That is, if there are n residuals, the ith largest residual is plotted against the $q_i = (i - s)/(n + 1 - 2s)$ quantile of the standard normal, with offset $s = 3/8$ if $n < 10$ and $s = 1/2$ otherwise. If the data are normal, the points should lie along a straight line with slope equal to the standard deviation.

NON-NORMAL DATA

Typically, a line is added to such a plot connecting the lower and upper quartiles (25th and 75th percentile).

In addition, histograms (or box-plots or stem-and-leaf plots) can provide useful graphical summaries for uncovering problems with a distribution. However, normal plots and histograms can be misleading. Small samples of normal data can look far from normal. On the other hand, since least squares model fitting tends to yield nearly normal residuals, violations in distribution can sometimes be hidden by a fit to the 'wrong' model expectation.

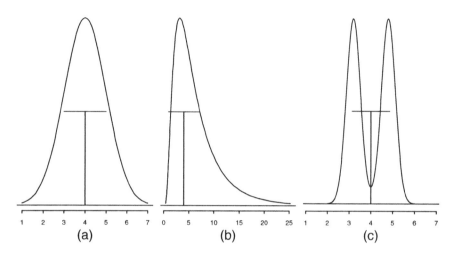

Figure 13.4. *Distribution shape and numerical summaries. Vertical bar marks mean and horizontal bar spans one SD on either side for (a) symmetric, (b) skewed and (c) bimodal distributions. These summary statistics are generally inadequate for histograms similar to (b) or (c).*

Example 13.4 Infer: Ideally the sample means $\bar{y}_{i.}$ estimate the central tendency of the data. They work very well for symmetric distributions of data, but can perform poorly if the data are bimodal or heavily skewed as shown in Figure 13.4. More appropriate methods for non-normal data are discussed in Chapter 15. ◊

Care is needed when checking the distribution of data across groups, or for that matter with any set of data that may not be a simple random sample from a population. Examine the **residuals after model fit** rather than the raw responses. Correct any problems in model specification before questioning normality among the residuals. It may be better to examine a scatter plot of predicted against residual or to examine histograms for subsets of the data rather than aggregating all data together in one histogram.

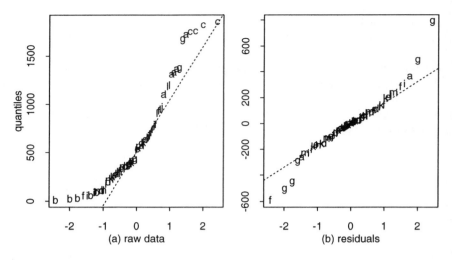

Figure 13.5. *Cloning normal probability plots: (a)* `titer` *concentration for all 64 measurements; (b) all 64 residuals (deviations from group means). Plot symbol is group identifier. Line passes through upper and lower quartiles of data. Note skew in (a) and heavy tails in (b) as found in Figure 13.3.*

Example 13.5 Cloning: Figure 13.5 shows the normal probability plot of (a) the raw clone data and (b) the residual deviations from clone means. Note the asymmetry in the former and the heavy tails in the latter, reflecting the skewed histogram of Figure 13.3. The problem data correspond to clones 6 and 7, which were identified in the residual plots (Figure 13.1). The diagonal lines pass through the upper and lower quartiles as suggested by Cleveland (1993). ◊

If data are not normal, then lack of correlation does not imply independence. In particular, estimates of mean and variance are no longer independent. This means that any test statistic, such as t or F, is not strictly distributed as theory would suggest. However, hopefully sample sizes are sufficiently large that mean and variance estimates are 'almost' uncorrelated and the test statistic is 'almost' distributed as theory would suggest.

Lack of normality is not a serious problem for tests of group means. Skewness, as measured by the third moment,

$$\gamma_1 = E[(y_{ij} - \mu_i)^3]/\sigma^3 \ (= 0 \text{ for normal}) \ ,$$

is somewhat important for comparing means. Group histograms for data which are skew have one tail heavier than the other. One-sided tests will be invalid for such data, being too conservative one way and too liberal the other. This tends to cancel out for two-sided tests. However, it leads to confidence intervals that have roughly correct coverage probability but are symmetric when they should be skewed to reflect the underlying distribution. Comparisons of means from distributions with similar skewness tend to behave adequately, as the skewness cancels out.

Departure from normality can have a great effect on tests of group variances. The behavior of inferential tools for variances depends heavily upon **kurtosis**, as measured by the fourth moment,

$$\gamma_2 = E[(y_{ij} - \mu_i)^4]/\sigma^4 - 3 \ (= 0 \text{ for normal}) \ .$$

Detailed analysis of these effects can be found in Scheffé (1959, ch. 10) along with summaries of earlier simulations results. The variance of the estimate of variance is

$$V(\hat{\sigma}^2) = \sigma^4[2/(n-1) + \gamma_2/n] \approx \frac{2\sigma^4}{n-1}\left(1 + \frac{\gamma_2}{2}\right) \ .$$

Thus if kurtosis is positive, the distribution has heavier tails and is more peaked in the center than normal, leading to highly variable estimates of variance. Negative kurtosis arises with light tails and a flat central part of the distribution; this leads to less variable estimates of variance than for normal data. Thus inference on variances may be either too liberal or too conservative, depending on the shape of the distribution. This can be adjusted, approximately, by appealing once again to Satterthwaite's argument. That is

$$\hat{d}(n-1)\hat{\sigma}^2/\sigma^2 \approx \chi^2_{\hat{d}(n-1)}$$

with approximate degrees of freedom $\hat{d} = 1/(1 + \gamma_2/2)$, where the kurtosis is estimated from the data by the method of moments.

13.6 Problems

13.1 Infer: Consider the residual plot for a response with five levels in a two-factor analysis of variance. Why are there five diagonal lines of points on the residual plot? What is the common slope of these lines? Figure this out by considering the mathematical relationship between the residual and predicted values.

13.2 Growth: Consider the distribution of residuals for the two-factor full model with interactions from Problem 11.3.
(a) Examine a plot of predicted against residual values using plot symbols (e.g. `code`). You may want to jitter values slightly to get a better view of all the data.
(b) Why are there long diagonal strings of points?
(c) What assumptions appear to be violated?

13.3 Growth: Delete zero counts of advplt and redo the analysis and residual plots. Compare these results with those found earlier. How do the conclusions differ and why? That is, briefly interpret results for the scientist.

CHAPTER 14

Comparisons with Unequal Variance

Comparisons of groups can be performed without the assumption of equal variance across groups. However, careful thought about interpretation may question the sense of comparing means if the variances are unequal. Weighted analysis of variance can improve the efficiency of inference. In some situations it may be important to test whether variances are equal. It may help to consider the big picture, that the real interest is in comparing groups. Means may provide a useful summary for some problems but may be inadequate for others. In some situations, a variance-stabilizing transformation, such as log relative data or square root for counts, may change the comparison to one of means with (near-equal) variances.

Inference about means in the presence of unequal variances is commonly known as the **Behrens–Fisher problem**. It is intimately tied to tests for equal variance. Miller (1997) provides a considered review of literature in this area. Weerahandi (1995) develops exact inference when variances may be unequal by introducing generalized tests, p-values and confidence intervals.

Section 14.1 considers the consequences of testing for mean differences while ignoring differences in variances. The next sections examine what happens if unequal variances are seriously considered. Weighted analysis of variance (Section 14.2) is possible when variances depend on some independent measurement, or covariate, in an obvious way. However, this raises the need for approximate methods for weighted sums of χ^2, the simplest being that of Satterthwaite examined in Section 14.3. Exact, or generalized, inference in the presence of unequal variance is briefly introduced in Section 14.4. Finally, Section 14.5 briefly examines formal tests for unequal variance. Transformations to stabilize variances are deferred to the next chapter.

14.1 Comparing means when variances are unequal

Consider the comparison of two means when variances are unequal. The problem of comparing a group means is analogous but more complicated (see Scheffé 1959, ch. 10). The pivot statistic for comparing two means is

$$T = \frac{\bar{y}_{1\cdot} - \bar{y}_{2\cdot} - \mu_1 + \mu_2}{\hat{\sigma}\sqrt{1/n_1 + 1/n_2}},$$

with weighted estimate of variance

$$\hat{\sigma}^2 = [(n_1 - 1)\hat{\sigma}_1^2 + (n_2 - 1)\hat{\sigma}_2^2]/(n_\cdot - 2).$$

For large sample sizes, the pivot is approximately normally distributed,

$$t \approx N\left(0, \frac{n_2\sigma_1^2 + n_1\sigma_2^2}{n_1\sigma_1^2 + n_2\sigma_2^2}\right).$$

With equal sample sizes, this is standard normal. In fact, the t statistic seems to work reasonably well for $n_1 = n_2$ even for moderate sample sizes. Power calculations near the null hypothesis are not greatly affected by unequal variance when sample sizes are equal.

Example 14.1 Power: If sample sizes are unequal, then the variance of t is either larger (if the group with the smaller sample size has the larger variance) or smaller (in the reverse situation) than 1. The problem gets worse the farther the ratio of sample sizes is from 1. That is, any comparison which allows the possibility of widely different variances among groups is likely to have very low power for detecting differences among means (unless sample sizes are equal). The size of such a test (probability of Type I error) could be anywhere between 0 and 1, depending on the ratio of true variances. This is shown in Figure 14.1 with $R = n_1/n_2$ and the variance ratio being σ_1^2/σ_2^2. ◊

Unequal variances imply that the distribution of $\hat{\sigma}^2$ is no longer proportional to a χ^2 variate. Instead, it has the distribution of a weighted sum of χ^2 variates. The arguments above (and in Scheffé 1959 or Miller 1997) are based on the first two moments of the distribution of $\hat{\sigma}^2$, comparing them to the first two moments of a χ^2. This idea is used later in a variety of settings to suggest approximate inferential tools.

14.2 Weighted analysis of variance

Comparison of means while allowing for unequal variance is quite straightforward if the variances are known up to a constant, by considering a weighted analysis. However, if the sample variances provide the main information about the unequal variance, this introduces uncertainty in their measurement which must be addressed for inferential questions.

Weighted analysis of variance ideally involves choosing weights proportional to the inverse of the variance, as this essentially reduces to the equal variance case. That is, if $V(y_{ij}) = \sigma_i^2$, define $z_{ij} = y_{ij}/\sigma_i$. These z_{ij}

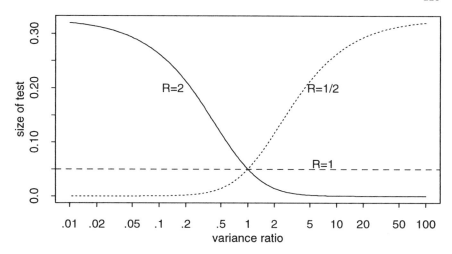

Figure 14.1. *Power under null hypothesis when variances differ. The ratio of variances and ratio of sample sizes ($R = n_1/n_2$) can profoundly affect the power under the null hypothesis, or size, of tests. Tests are conservative when the larger sample has the larger variance. This can occur for high variance ratio (solid) or low variance ratio (dotted). Variance ratio does not affect test size when samples sizes are equal (dashed).*

have equal variance (=1) and means $E(z_{ij}) = E(y_{ij}/\sigma_i) = \mu_i/\sigma_i$. Analysis of the z_{ij} can proceed with the assumption of equal variance, with appropriate modification of hypotheses.

Consider weights $w_i = \sigma_i^{-2}$. The partition of the total sum of squares for $z_{ij} = y_{ij}\sqrt{w_i}$ is a weighted sum of squares for y_{ij},

$$\sum_i \sum_j (z_{ij} - \bar{z}_{..})^2 = \sum_i n_i (\bar{z}_{i.} - \bar{z}_{..})^2 + \sum_{ij} (z_{ij} - \bar{z}_{i.})^2$$
$$\sum_{ij} w_i (y_{ij} - \tilde{y}_{..})^2 = \sum_i n_i w_i (\bar{y}_{i.} - \tilde{y}_{..})^2 + \sum_{ij} w_i (y_{ij} - \bar{y}_{i.})^2 ,$$

with weighted grand mean $\tilde{y}_{..} = \sum_i n_i w_i \bar{y}_{i.}/w_{..}$.

In practice, the group variances are unknown. Estimates of **weights which depend on response**, such as $w_i = \hat{\sigma}_i^{-2}$, or $w_i = 1/\bar{y}_{i.}$ if variance is proportional to mean, should be used with caution if at all. These weighted sums of squares are not quadratic forms in the responses. Consider, for instance, the comparison of $a = 2$ means. Using the inverse of group sample variances as weights, the sum of squares for the hypothesis $H_0 : \mu_1 = \mu_2$ is

$$\sum_i n_i w_i (\bar{y}_{i.} - \tilde{y}_{..})^2 = n_1 w_1 n_2 w_2 (\bar{y}_{1.} - \bar{y}_{2.})^2 / (n_1 w_1 + n_2 w_2)$$
$$= (\bar{y}_{1.} - \bar{y}_{2.})^2 / (\hat{\sigma}_1^2/n_1 + \hat{\sigma}_2^2/n_2)$$

with residual sum of squares

$$\sum_i \sum_j w_i (y_{ij} - \bar{y}_{i\cdot})^2 = \sum_i \hat{\sigma}_i^{-2}(n_i - 1)\hat{\sigma}_i^2 = n_{\cdot} - 2 \, .$$

Thus the approximate F statistic would be

$$F = \frac{(\bar{y}_{1\cdot} - \bar{y}_{2\cdot})^2}{\hat{\sigma}_1^2/n_1 + \hat{\sigma}_2^2/n_2} \, .$$

This statistic does not have an F distribution unless variances are equal. Even though numerator and denominator are independent (assuming normality), the denominator is not proportional to a χ^2 variate. If sample sizes were large enough, then F would be proportional to a χ_1^2 variate. For moderate sample sizes, the denominator is proportional to a weighted sum of χ^2 variates.

Note that the approximate F statistic above for comparing means is the square of an approximate Welch's t test,

$$T = (\bar{y}_{1\cdot} - \bar{y}_{2\cdot})/\sqrt{\hat{\sigma}_1^2/n_1 + \hat{\sigma}_2^2/n_2} \, .$$

14.3 Satterthwaite approximation

The weighted sum of independent χ^2 variates is rather messy. In practice, a rough approximation can be obtained using a method commonly attributed to Satterthwaite. This is developed below for Welch's t test, but the idea can be used quite generally. It is quick and easy to do. Unfortunately, its performance can be quite poor in practice.

The distribution of the estimate of the variance of the difference of group means

$$S = \hat{\sigma}_1^2/n_1 + \hat{\sigma}_2^2/n_2 \, ,$$

is approximated by a χ^2 variate, matching the first two moments. The **Satterthwaite method** involves finding a constant of proportionality a and approximate degrees of freedom r such that S and a variate X agree to the first two moments with $X/a \sim \chi_r^2$. Thus the relations

$$E(S) = ar \text{ and } V(S) = 2a^2 r$$

can be solved for r and a to yield

$$r = 2[E(S)]^2/V(S) \text{ and } a = E(S)/r = V(S)/2E(S) \, .$$

Typically r is not an integer, suggesting interpolation on distribution tables. However, since S does not have a χ^2 distribution anyway, this is a minor problem. It is probably best to round r down to the nearest integer to be somewhat conservative. The approximate degrees of freedom for the

GENERALIZED INFERENCE 225

variance estimator S are
$$r = \frac{(\sigma_1^2/n_1 + \sigma_2^2/n_2)^2}{\sigma_1^4/n_1^2(n_1-1) + \sigma_2^4/n_2^2(n_2-1)}.$$

In general, $r = n_1 + n_2 - 2$ only if $n_1 = n_2$ and $\sigma_1 = \sigma_2$. The approximate degrees of freedom may be greater or less than $n_. - 2$ depending on the ratio of variances and on the ratio of sample sizes. In practice, r is estimated using the sample variances.

This χ^2 approximation can be used to examine the test statistics under the null hypothesis of no mean difference,

$$F \approx F_{1,r} \text{ or } T \approx t_r.$$

Inference on linear combinations of means can be done in an analogous way. The estimate of variance of a contrast $\sum c_i \hat{\mu}_i$ using separate estimates of group variances, $\sum c_i^2 \hat{\sigma}_i^2/n_i$, leads to the pivot statistic

$$T = \left(\sum c_i \hat{\mu}_i - \sum c_i \mu_i\right) / \sqrt{\sum c_i^2 \hat{\sigma}_i^2/n_i}.$$

This has approximately a standard normal distribution, provided sample sizes for all groups with non-zero coefficients c_i are 'large', by appealing to the central limit theorem.

For small sample sizes, the statistic is not distributed as t even if the data are normal. Arguing as above, the approximate degrees of freedom r can be found by matching the first two moments of the variance estimator,

$$\begin{aligned} r &= 2[E(\sum c_i^2 \hat{\sigma}_i^2/n_i)]^2 / V(\sum c_i^2 \hat{\sigma}_i^2/n_i) \\ &= [\sum c_i^2 \sigma_i^2/n_i]^2 / [\sum c_i^2 \sigma_i^4/n_i^2(n_i-1)]. \end{aligned}$$

In practice, once again, r is approximated by replacing the group variances σ_i^2 by their estimates $\hat{\sigma}_i^2$.

14.4 Generalized inference

It is possible to develop inferential tools to solve the Behrens–Fisher problem exactly. A concise review, with development of generalized tests, p-vales and confidence intervals, can be found in Weerahandi (1995). The brief exposition below for two samples is extended to one-factor and two-factor experiments in the reference.

In the following development, it is helpful to distinguish between the realized estimators $(\bar{y}_{i\cdot}, \hat{\sigma}_i^2)$ and random variables with the same distribution $(\bar{Y}_{i\cdot}, S_i^2)$. If the variances were known, the standard normal pivot statistic,

$$Z(\bar{Y}_{1\cdot}, \bar{Y}_{2\cdot}) = \frac{(\bar{Y}_{1\cdot} - \bar{Y}_{2\cdot}) - (\mu_1 - \mu_2)}{\sqrt{\sigma_1^2/n_1 + \sigma_2^2/n_2}},$$

would be all that need be considered. The sample variances have independent χ^2 distributions,

$$X_i^2(S_i^2) = \frac{(n_i - 1)S_i^2}{\sigma_i^2} \sim \chi^2_{n_i-1} , \ i = 1, 2.$$

Their sum is also χ^2 and independent of Z. Therefore, the ratio

$$T = T(\bar{Y}_{1.}, \bar{Y}_{2.}, S_1, S_2) = \frac{Z\sqrt{n. - 2}}{\sqrt{X_1^2 + X_1^2}}$$

has a $t_{n.-2}$ distribution. Unfortunately, its form depends on the unknown variances. This problem can be remedied by a little trick. Define the beta distributed variate

$$B = B(S_1, S_2) = \frac{X_1^2}{X_1^2 + X_1^2} \sim \text{Beta}\left(\frac{n_1 - 1}{2}, \frac{n_2 - 1}{2}\right) ,$$

and consider the following **generalized pivot statistic**,

$$R(\bar{Y}_{1.}, \bar{Y}_{2.}, S_1, S_2) = T\left(\frac{\hat{\sigma}_1^2(n_1 - 1)}{n_1(n. - 2)B} + \frac{\hat{\sigma}_2^2(n_1 - 1)}{n_2(n. - 2)(1 - B)}\right)^{1/2} ,$$

or $R = TC$. At the observed data, this pivot reduces to

$$r = R(\bar{y}_{1.}, \bar{y}_{2.}, \hat{\sigma}_1^2, \hat{\sigma}_2^2) = (\bar{y}_{1.} - \bar{y}_{2.}) - (\mu_1 - \mu_2) ,$$

which does not depend on the variances. The distribution of R is related to that of T in the following way,

$$\text{Prob}\{R \leq r\} = \text{Prob}\{T \leq r/C\} .$$

This distribution can be found by convolving the $t_{n.-2}$ distribution with the beta distribution. That is,

$$\text{Prob}\{T \leq r/C\} = E[G_{n.-2}(r/C)] ,$$

in which $G_{n.-2}(t) = \text{Prob}\{T \leq t\}$ is the cumulative distribution function of $t_{n.-2}$ and the expectation is with respect to the random variable B. Details can be found in Weerahandi (1995, sec. 7.3). These calculations are currently available in the XPro package but have not yet been incorporated into larger packages such as SAS.

Modestly misbehaved examples in Weerahandi (1995) indicate that this problem can be serious. However, these results are largely built around the normal assumption. If a scientist is comfortable with normality, and is very concerned to get exact results, this appears to be the appropriate avenue. In fact, some federal agencies in the USA are now requiring this approach.

14.5 Testing for unequal variances

Evidence of unequal variance is often best detected in residual plots. Sometimes it may not be apparent whether variances are really different or whether observed unequal variances really estimate a common (true) variance. In addition, some scientists just want to have that formal confirmation that variances are or are not equal. Unfortunately, formal tests for differences among variances are disappointing. Further, the process of conducting variance tests changes the significance and interpretation of subsequent formal tests in ways that are poorly understood.

Most tests of variances are very sensitive to violations of normality and unequal sample sizes. In addition, formal tests for variances are not that powerful. In practice, a three-fold range in variances cannot be distinguished from chance, while a ten-fold difference is likely to upset the performance of means tests. In practice, it is better to examine patterns in residual plots than to rely on formal tests. Miller (1997, ch. 7) compares a wide variety of tests for variance. Milliken and Johnson (1992, sec 2.3) have a nice discussion as well. The following is a brief synopsis without explicit references.

The commonly used tests based on normal theory are named for their authors: Bartlett, Cochran and Hartley. Bartlett's test is a slight variant on the likelihood ratio test. Hartley's test considers the ratio of the largest to the smallest variance estimators. It is quick and easy and may give enough information for a judgement call. These three tests rely heavily upon the assumption of normality and do not perform well if sample sizes differ markedly.

There have been many ideas for nonparametric assessment of unequal variance. The most commonly used and accepted approach is Levene's test. In a one-factor experiment, analyze the absolute deviations $|y_{ij} - \bar{y}_{i.}|$ as the responses using anova. This seems to perform well enough in practice and is robust to violations of normality. It is generally recommended for testing unequal variances. In addition, it is easy to perform, and can be adapted to factor combinations and to more than one random effect.

Box suggested analyzing log transforms of variances as responses. This is possible if there are many experimental units which can be subgrouped appropriately. He suggests another way to study to experiments in which there is sub-sampling (see Chapter 22, Part H), or multiple measurements on experimental units, by examining their log variance as a measure of quality.

14.6 Problems

14.1 Growth: Another scientist suggested using a square root transformation rather than log on these data. Conduct formal analysis using this on the growth data without zeros. You may restrict attention to a one-factor approach for the 12 treatments. Include in your analysis an investigation of residuals. For instance, does Levene's test reveal any lingering evidence of unequal variance? It there obvious trend in the residuals?

14.2 Variance: The data set for this problem comes from Weerahandi (1995). There are four groups and a total of 31 observations. As you will see, the standard deviations range from 1.5 to 5.7. The casual observer might choose to ignore this variability, but read on!

(a) Write down the usual assumptions and report the results of the usual analysis of variance.

(b) By hand, make a 'dot-plot' of the data, using letter symbols for group. Comment.

Now drop the assumption that variances are equal.

(c) Note the lack of obvious relationship between mean and variance by group. (Plot them!)

(d) Instead, use the inverse of group variance as weight. Briefly justify this choice (why might this be reasonable?). Now briefly critique it (why is it silly in this problem?). How does the use of estimates of group variances affect the p-value? [Hint: examine the new SDs.]

(e) The exact test differs from all of these. It is based on the randomization principle. That is, if there are no group differences, then all assignments of group labels to the data are equally likely. That is, one could (in theory) examine every permutation of the 31 responses (with nine As, seven Bs, eight Cs and seven Ds) and compute the F statistic for each one. The p-value is then the proportion of F values that are as extreme as or larger than the one observed. The 'right' p-value, based on exact generalized inference, for the raw data is 0.030, or for the ranks (exact Kruskal–Wallis) is 0.06.

(f) Comment on the disparity among p-values (you now have four different ones!). What is your conclusion about differences among groups?

14.3 Variance: Now think about inference on the variances themselves.

(a) Calculate the ratio of the largest to smallest variances. This is Hartley's F-max test. ('Liberal' (using sample size of nine in four groups) critical values for 5% and 1% are, respectively, 7.18 and 11.7. See Milliken and Johnson (1992, Table A.1).) This is a very easy test to perform. Unfortunately, it relies heavily on the normal assumption, and is best for balanced data. Interpret results with caution for this data set.

(b) Conduct Levene's test for unequal variance. This test is not sensitive to departures from normality, and can be used for small samples. Interpret results.

(c) Comment briefly on the dilemma of testing for equal variance before conducting analysis of variance. How is this problem lessened (or greatened) by increasing sample size?

CHAPTER 15

Getting Free from Assumptions

This chapter concerns methods that rely on fewer assumptions than standard analysis. These approaches are sometimes called 'nonparametric' or 'distribution-free' because they tend to make very broad assumptions about the shape of the error distributions. The main ideas presented here concern replacing the actual observed values by ranks, or other derived scores, which have nice properties. In addition, we investigate the relationship between randomizing as a justification for inference and permutation tests based on randomization.

There are still some assumptions, however. Independence of observations remains the most important. While the distribution of errors need not be normal, it is usually assumed that differences between groups can be summarized by a 'shift of center'. These may be expressed in terms of medians instead of means, but the nonparametric methods typically work better if the 'spread' is similar across groups. Therefore, ideas developed in the previous chapter may still be relevant.

Section 15.1 develops the replacement of observations by their ranks or other scores. Section 15.2 examines inference using permutations of observations. The last section introduces Monte Carlo simulation methods which can be useful in this and other broader contexts.

15.1 Transforming data

Transformations can be used to get free of assumptions by attempting to conform to those very assumptions. Sometimes it is possible to transform data to correct problems with model specification, unequal variance or non-normality. However, care is needed since alteration of one of these may affect the other two! In addition, data-based transformations may affect later significance tests in non-obvious ways. A modern, comprehensive treatment of these issues can be found in Carroll and Ruppert (1988).

It is assumed through most of this text that data have been transformed, if needed, to support the belief in equal variances and normal distribution of residual errors. Analysis of variance methods are fairly robust to violations of assumptions of equal variance and normality. Thus it is not necessary to remove all suggestions of problems.

Further, any transformation may require sufficient justification to scientific peers. Theoretical arguments based on prior knowledge are much easier to defend than arcane transformations fine-tuned to the present data. Reminding colleagues of commonly used transformations such as the pH scale in chemistry and the so-called Richter scale in seismology can help open the way for transformations. It is important to be consistent across related problems arising from a larger experiment. Does it make sense to analyze log(biomass) in one place but untransformed biomass in another? Better to transform both or neither.

Transformations can improve the behavior of inferential statistics for testing and interval estimation. However, it is often most appropriate to present data in plots and tables in the original units in order to communicate meaningful information to colleagues. Confidence intervals can be developed on a transformed scale and back-transformed for plotting. While it may take a little adjustment to accept skewed intervals, they more properly reflect the nature of variation in the data. Another approach is to use a transformed scale for plotting but label it using the original units. This is done in many plots throughout this book.

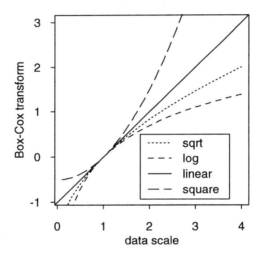

Figure 15.1. *Transformations in common use. The square root (dotted) and log (dashed) transformations shrink large values; the latter also spreads out values near zero. Square (long dashed) and higher powers spread out larger values. Be cautious using transformations with many zeros or negative values.*

Example 15.1 Infer: A common way to transform data is to consider logarithms (natural $\log(y)$ or base 10 $\log_{10}(y)$) or square roots ($\sqrt{y}$). These are examples of a general class of monotone, or order-preserving, functions known as the **Box–Cox transformations**

$$y(\lambda) = (y^\lambda - 1)/\lambda$$

with the convention that $y(0) = \log(y)$. Some common choices are shown in Figure 15.1. The preferred choice of λ minimizes

$$(\lambda - 1) \sum \log(y) - \frac{n}{2} \log(MSE_\lambda)$$

with $(\lambda - 1) \sum \log(y)$ the derivative of the Box–Cox transform. The second term is the normal log likelihood, while the first is a first-order correction term. Note that in practice, this method could be employed to suggest a transform, which would then be rounded off to a convenient value such as $\lambda = 1/2$. ◇

Cautions have been raised about empirical choice of λ followed by transformation of the same data. Carroll and Ruppert (1988) suggest transforming both the data and its expectation if the main aim is to alter the shape of the errors while maintaining the same model specification.

It is well known that a pure interaction two-factor model can often be made into an additive model by a log transform of the response. Hoaglin, Mosteller and Tukey (1991) suggested a plot for a generalization of this concept. That is, consider the main effects and residuals adjusted for mean square as introduced in Chapter 13:

$$\begin{aligned} r_i(A) &= (\bar{y}_{i..} - \bar{y}_{...})\sqrt{an_{i.}/(a-1)} \\ r_j(B) &= (\bar{y}_{.j.} - \bar{y}_{...})\sqrt{bn_{.j}/(b-1)} \\ r_{ijk}(AB) &= (y_{ijk} - \bar{y}_{i..} - \bar{y}_{.j.} + \bar{y}_{...})\sqrt{n_{..}/(n_{..} - ab)} \ . \end{aligned}$$

Plot $r_{ijk}(AB)$ against $r_i(A)r_j(B)/\bar{y}_{...}$ and find the slope m of the regression line. They suggest that the power transformation using $\lambda = 1 - m$ will roughly remove the interaction. For instance, a slope of 1 suggests using the log transform, which annihilates interactions of the form $\gamma_{ij} = \alpha_i \beta_j$. No appreciable slope implies no need for transformation.

With enough data per group it is possible to plot group mean against group variance (or group standard deviation as in Figure 4.4, page 57). A linear relationship would suggest a square root (or log) transformation to stabilize the variance (see Snedecor and Cochran 1989 or Seber 1977). However, such plots can be confusing with small sample sizes per group. Further, some data simply do not exhibit nice relationships! When possible, it is best to appeal to theoretical grounds for transformations of responses to achieve (approximately) equal variance among groups. Such **variance stabilizing transformations** assume the variance is a smooth function of the mean and rely on a Taylor expansion. The ideas presented below can be found in Scheffé (1959, ch. 10) or Snedecor and Cochran (1989, ch. 14).

Suppose that the variance of a response differs from group to group, depending on the group mean,

$$V(y_{ij}) = \sigma_i^2 = g^2(\mu_i) \text{ , with } E(y_{ij}) = \mu_i$$

and $g(\cdot)$ some known smooth function. For instance, concentrations and weights often have the standard deviation proportional to the mean, or $g(\mu) = \mu$.

The idea is to find another function $f(\cdot)$ such that $f(y_{ij})$ has roughly constant variance. One would then analyze the transformed response $f(y_{ij})$ instead of the raw response. Remember that the key questions likely concern detecting differences in a broad sense rather than necessarily focusing on the means. Questions about means of transformed responses still concern comparing groups, but they satisfy one more assumption made for inference with tools such as pivot statistics based on t or F distributions.

Suppose the function f is smooth and can be approximated locally by a straight line,

$$f(y_{ij}) \approx f(\mu_i) + (y_{ij} - \mu_i)f'(\mu_i) .$$

The variance of $f(y_{ij})$ is roughly constant across groups provided

$$V[f(y_{ij})] \approx V(y_{ij})[f'(\mu_i)]^2 = [g(\mu_i)f'(\mu_i)]^2 = \text{ constant.}$$

In other words, set $f'(\mu) = c/g(\mu)$ for an arbitrary constant c. Solving for f by integrating, yields

$$f(y) = c \int \frac{dy}{g(y)} .$$

For problems with a constant coefficient of variation (standard deviation proportional to mean), $g(\mu) = a\mu$ and

$$f(y) = c \int \frac{dy}{y} = log(y) .$$

In practice, it may be necessary to consider $log(y + c)$ for some small constant c if there are zero response values.

Counts, such as radiation or fluorescence, often exhibit Poisson variation. That is, counts usually have the variance proportional to the mean, or $g(\mu) = a\sqrt{\mu}$, with μ the expected count. Thus a square-root transformation is appropriate for counts, by appealing to the Taylor approximation developed above:

$$f(y) = c \int \frac{dy}{\sqrt{y}} = \sqrt{y} .$$

Example 15.2 Infer: Proportions, either as frequencies of successes or as fractions of a control, often have standard deviation which is largest around 0.5 and diminishes near 0 and 1, or $g(\mu) = a\sqrt{\mu(1-\mu)}$ with μ the

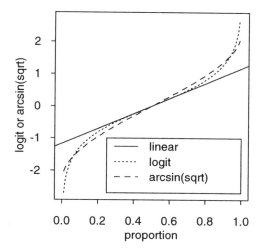

Figure 15.2. *Transformations of proportions. The logit (dotted) and arcsin square root (dashed) curves have been standardized to mean 0 and variance 1. The solid line is the tangent to the logit at the median. Both transformations spread out values near 0 and near 1 in slightly different ways. The logit makes proportions more normally distributed, while arcsin square root stabilizes the variance across a wide range of proportions.*

expected proportion. Again, the Taylor approximation leads to

$$f(y) = c \int \frac{dy}{\sqrt{y(1-y)}} = \arcsin(\sqrt{y}) ,$$

the **arcsin** (inverse of sine trigonometric function) of the square root of the proportion. Another common transformation for proportions is the **logit**

$$logit(y) = \log(y) - \log(1-y) ,$$

which draws proportions closer to a normal distribution. The logit is used in conjuntion with weighted least squares for categorical data analysis, or generalized linear models (McCullagh and Nelder 1989). Figure 15.2 shows both the logit and arcsin square root transforms standardized to have variance 1 and mean 0. The straight line is tangent to the logit transform to emphasize the differences. ◊

While normality is the least important assumption, it is still helpful at times to consider transformations to improve normality. The logit for proportions mentioned above is an obvious example, stretching out proportions near 0 and 1. Transformations toward symmetry tend to improve estimates of means. Skewed data sometimes may appear to arise from the log-normal distribution, suggesting the log as a natural transformation. Carroll and

Ruppert (1988) discussed a modification of the Box–Cox transform for symmetric data to reduce kurtosis. Suppose the data y are symmetric with median 0. The transformation

$$y(\lambda) = [|y|^\lambda sign(y) - 1]/\lambda$$

appears to correct moderate kurtosis problems. Choices of $0 < \lambda < 1$ decrease sharp peaks while $\lambda > 1$ heightens the peak. Again, it is best to guide transformations for skew or kurtosis by theoretical arguments. In the absence of theory or strong empirical evidence for a particular transformation, many researchers turn to rank-based methods, a nonparametric monotone transformation of the data which is briefly discussed in the next section.

15.2 Comparisons using ranks

Many problems with assumptions can be overcome by replacing observations by their ranks. Lehmann's (1975) book on nonparametrics employs this idea extensively. A full treatment of these methods is beyond the scope of this text. However, we can indicate how they can be used in some settings. Ranks are most often used in testing hypotheses about differences in the center of distributions. However, they can also be used for some plots and for estimation in some settings. Unfortunately, to date the tools available do not have the flexibility of least squares methods. In particular, they do not easily lend themselves to the experiments discussed in the latter parts of this book.

The basic idea is to order all the responses y_{ij}, regardless of group, and replace each response by its rank r_{ij}, or position in the order. Tied responses would all be assigned the tie of their ranks. Inference is then carried out on the ranks as if they were the responses. Tests based on ranks of this fashion are sometimes called Wilcoxon-type tests, after one of their pioneers.

Tests comparing groups based on ranks are simple to perform for single-factor experiments. Calculations can often be done by hand. Expected values and variances reduce to simple formulas, unless there are ties. Some efforts have been made at two-factor experiments, but hand calculations soon get tedious. Alternatively, statistical packages can be used to replace responses by ranks and then use standard analysis of variance methods.

Several alternatives to ranks have been developed, but seem to attract little general use in practice. A natural choice replaces each ranked observation by its **normal score**, the expected value if the data were truly a sample from the normal distribution. A close approximation due originally to van der Waerden (see Lehmann 1975, sec. 2.7B) uses the $i/(n+1)$ normal quantile. This has been improved slightly, as noted for normal Q-Q plots : approximate the ith normal score by the $q_i = (i-s)/(n+1-2s)$ quantile of the standard normal, with offset $s = 3/8$ if $n < 10$ and $s = 1/2$ otherwise. Notice that a normal plot of these normal scores produces a straight line! In practice, inference on ranks tends to be much easier to justify to scientific colleagues than replacing the observed responses by some score. Therefore this more complicated approach is seldom used.

15.3 Randomization

Sir Ronald Fisher noticed that the exact distribution of the t statistic (and similarly the F statistic) under the null hypothesis of no group differences can be derived by randomly permuting the assignment of experimental units to groups. The histogram of t values computed for all permutations is very close to the t distribution with the appropriate (error) degrees of freedom (and similarly for the F) if the assumptions more or less hold. Fisher used this as a justification for the t test, and more generally for the analysis of variance which he introduced. For a historical account see Fisher (1935; 1990) and the biography by Box (1978). In addition, it is well worth reading Cox (1958, ch. 5).

Permutation tests form the basis for justifying randomization in assignment of experimental units to groups. Under the assumption of no difference among groups, any permutation of assignments of experimental units would not change the observed response. This leads to unbiased estimates of group means and variances. Further, it ensures that there is a small chance of large error due to some systematic pattern among experimental units that was not known in advance.

Permutations must preserve the randomization restrictions imposed by the design structure, such as blocking and sub-sampling. In a sense, this is a justification for keeping design structure in a model even if some aspects are not significant.

Permutation tests can be derived by permuting assignment *after* completing the experiment. The test consists in determining how extreme the observed responses with their actual groups are with respect to all other possible permutations. The basic idea consists of calculating a pivot statistic for the data at hand. Under the null hypothesis of no difference, this should have the same distribution as the pivot statistic calculated for any possible permutation of the data. The p-value for the observed data can be calculated as the fraction of permuted pivot statistics which are as extreme or more extreme than the one observed.

Note however that if assumptions hold reasonably well, it may not be worth the extra effort to perform such tests. Fisher (1935), in fact, preferred to use the approximate t and F tests unless there were compelling reasons to abandon them. This may have been due in part to the tremendous effort needed to perform permutation tests before the age of computers.

This idea can be applied in any situation where there is a hypothesis that different group assignments would yield the same results. It is a simple idea to convey, but can be very powerful. For instance, the experimental units need not be independent. The basic requirement is that randomization be used in the assignment of factor levels. A permutation test can be constructed based on the nature of randomization.

15.4 Monte Carlo methods

Small-sample problems adapt well to permutation or rank tests. The possibilities can sometimes be enumerated by hand, or found in existing tables or through computer programs. However, larger problems can be much less tractable. For instance, with 30 samples divided evenly into two groups, there are over 155 million possible permutations of assignments. It would be better to use knowledge about sampling from *this* population. That is, use a Monte Carlo simulation method to estimate the p-value.

Example 15.3 Permute: Consider the problem of having a t value for comparing two groups of 15 each, when the assumption of normality is in question. The scientist would like to know the p-value to within 10% (that is 0.01 ± 0.001 or 0.05 ± 0.005). Assign a sequence number $1, \cdots, 30$ to each of the 30 observed responses. For $j = 1, \cdots, m$ construct Monte Carlo trials and pivot statistics as follows: (a) draw 30 uniform (0,1) random numbers; (b) set the responses for sequence numbers corresponding to the lowest 15 random numbers to group 1 and the other 15 to group 2; (c) calculate the pivot statistic, calling it t_j. Sort the m pivot statistics, determine the fraction that are more extreme (larger in absolute value) than the observed t, and report that as the estimated p-value.

The number of Monte Carlo trials, m, should be chosen large enough to ensure the desired precision. In this case, under the null hypothesis, the estimated p-value has binomial distribution, with variance $p(1-p)/m$. Therefore it may be interpreted that the goal is to have m large enough that two standard errors (roughly 95% confidence interval for true p) is about $0.1p$ or $m \approx 10/\sqrt{p}$. Thus the sample size depends on the actual significance probability, which is not surprising. In practice, the scientist would chose some sufficiently small value, or use the rough guideline from classical approaches. These days it is not a problem to double the simulation size for such an effort. ◊

Monte Carlo methods can be employed in a large variety of other situations besides permutations. The bootstrap (Efron 1982; Efron and Tibshirani 1994) and other resampling methods sample from the data in a variety of ways. More recently, Markov chain Monte Carlo methods construct a Markov chain which samples from the data likelihood (Gelman *et al.* 1995).

Diggle (1983) resamples from spatial point patterns. Rather than just considering a single value as the pivot statistic, he examines the cumulative distribution of measurements of interest. He notes how they fall relative to samples from 'random' distributions of points.

15.5 Problems

15.1 Variance: Replace the observations by their ranks and rerun the usual analysis of variance. [This test is an approximation of the Kruskal-Wallis test.] Comment on whether this a reasonable approach for this problem. [Hint: what assumption(s) are helped by using ranks?] Compare this to the other four results in Problem 14.2.

15.2 Variance: A nonparametric version of Levene's test for unequal variances can be constructed by replacing each observation by the rank of its absolute deviation from the group's median. Perform such a test for these data, and compare results with Problem 14.3.

15.3 Growth: Reconsider the reduced data set without zeros in Problem 13.3. Replace the non-zero observations by their ranks and redo the plots and analysis. Do the conclusions change at all? Would you recommend this procedure to the scientist?

15.4 Running: Theory surrounding Problem 8.2 suggests that log(vo2) may be more appropriate than vo2 in terms of inference assumptions.
(a) Why does this seem reasonable?
(b) Reanalyze the data using log(vo2).
(c) Interpret the results for the scientist. How do conclusions change? Is it merely a matter of which main effects and interactions are significant, or is there a qualitative difference in the interpretation of factor effects on response?

PART F

Regressing with Factors

This part considers the role of regression in the relationship between responses and factors. In some cases it makes sense to regress ordered factor levels on a response. In other situations, a regressor, known as a covariate, can be used to adjust response means by group to reduce bias and variance. Further, many experiments have multiple responses which may be affected by a factor of interest.

Chapter 16 concerns factors that have ordered levels in which the mean response may increase or decrease with level. The connections between a single-factor analysis of variance and simple linear regression are explored. This provides an opportunity to introduce path coefficients as a convenient summary of relationships. The concept of errors in variables is explored in two related contexts.

Chapter 17 develops the idea of adjusting the group mean by a covariate which may be different for every experimental unit. Least squares estimates for the adjusted means are developed, along with pivot statistics which suggest more formal inference. At the same time, the relationship between the response and the covariate, adjusted for factor level, is investigated. The covariate may be a nuisance that must be removed in order to focus attention on factors.

In many studies, however, the covariate is simply another type of response. In some situations, one response may 'drive' a second response, based on scientific knowledge of the system under study, suggesting that the apparent effect of the factor on the second response can be explained at least in part by the first. However, multiple responses may have no clear 'cause and effect' ordering, requiring a more symmetric approach. For instance, how do two responses covary, adjusting for the fact that they were measured across a range of experimental conditions? Chapter 18 explores in particular how to assess group differences when there may be several responses of interest.

CHAPTER 16

Ordered Groups

Some groups can be ordered, at least in a qualitative fashion. This ordering may be based on the level of a factor determining groups or on another measurement which characterizes those groups. The scientist may wish to investigate whether there is an association between this ordering and the measured responses. If the association is linear, the problem reduces to simple linear regression.

Judges may rate contestants on a scale from poor to excellent with five levels. Perhaps the mean response could be associated with these ordered levels, say increasing linearly from 1=poor through 5=excellent. In other situations, quantitative measurements may be unreliable, suggesting instead scoring on a less precise but more repeatable scale. The scientist may want to establish differences in response between these groups.

Sometimes the scientist attempts to set factor levels but can only do this within a small range, or margin of error. Is it appropriate in this situation to ignore the **error in variables** during analysis? Rather, what is an appropriate way to investigate this in order to determine how sensitive the response is to the exact settings?

Section 16.1 considers groups whose mean response may lie along a line with respect to the group levels. Section 16.2 provides formal testing of linear association between ordered groups and mean group responses. Section 16.3 introduces path diagrams as a graphical aid for thinking about questions when regressing with factors. Sections 16.4 and 16.5 investigate two approaches to the study of error in variables.

16.1 Groups in a line

Suppose there are a groups under study and that they are ordered based on some characteristic x. For instance, the group levels may represent increasing doses of some treatment. What is the relationship between group order and the measured response? The simplest relationship is linear,

$$\mu_i = \beta_0 + \beta_1 x_i \ , \ i = 1, \cdots, a \ .$$

That is, the mean response rises (or falls if β_1 is negative) with characteristic value x_i. The single-factor model can be written as

$$y_{ij} = \mu_i + e_{ij} = \beta_0 + \beta_1 x_i + e_{ij},$$

for $j = 1, \cdots, n_i$. This is almost the model for simple linear regression – except that all responses in group i have the same characteristic value x_i. Thus simple linear regression can be viewed as a special case of the single-factor analysis of variance model.

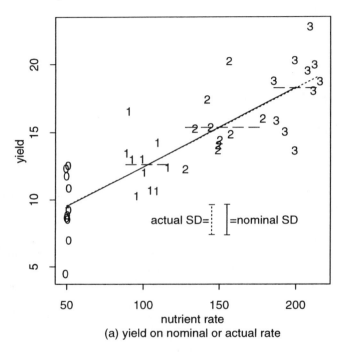

Figure 16.1. *Nutrient yield on nominal and actual rates. The linear fits for the* **nominal** *(solid line) and* **actual** *(dotted line) rates are nearly the same and have similar estimates for the standard deviation of* **yield** *about the line.*

Example 16.1 Nutrient: A scientist studied the effect on yield of nutrient-enriched fertilizers by adding one of three levels of nutrient (50, 100, 150) above a baseline of about 50 units. However, there was considerable variability in application. The actual nutrient levels applied were 10–20 units off (rate SD = 10). Can this variability be ignored? Figure 16.1 shows linear fit of **yield** on the **nominal** (solid line) and **actual** (dotted line) rates. It appears that little is lost in terms of the mean value (regression line). The standard deviation around the regression line for the **nominal** rate is

TESTING FOR LINEARITY

(2.30) while the SD using the `actual` rates is 2.26, about 2% smaller. The `actual` rate explains slightly more variation (69%) than does the `nominal` rate (68%). ◇

The linear relationship above could be extended to quadratic or higher-order polynomials in a natural way,

$$\mu_i = \beta_0 + \beta_1 x_i + \beta_2 x_i^2 \ .$$

Interpretation of higher-order polynomials can be difficult. Further, polynomial curve fitting is very sensitive to variation in the data. Another alternative is to consider nonlinear relationships between group level and mean response, such as the exponential,

$$\mu_i = \beta_0 + \beta_1 e^{\beta_3 x_i} \ .$$

However, if the levels correspond to a logarithmic scale, then the relationship between mean response and level could be linear. Consider the pH scale, which is logarithmic already. For instance, an increase in pH from 5.5 to 6.5 might correspond to a linear decrease in bacterial viability.

Put another way, the relationship between ordered group levels i and mean responses μ_i can be viewed as a curve. The simplest form of this curve is a straight line. A flat line (slope $\beta_1 = 0$) corresponds to no association. In general, the association could be quite complicated and nonlinear. That is, it is possible to use analysis of variance to investigate nonlinear relationships. However, if the number of groups is large, it may be better to consider nonlinear (Bates and Watts 1988) or nonparametric regression (Green and Silverman 1994) approaches.

16.2 Testing for linearity

A linear relationship between group level and mean response may be adequate in many experiments. This can be tested formally by comparing the two models

$$y_{ij} = \mu_i + e_{ij}$$
$$y_{ij} = \beta_0 + \beta_1 x_i + e_{ij} \ .$$

In other words, it is possible to test the null hypothesis $H_0 : \mu_i = \beta_0 + \beta_1 x_i$ against the alternative that the group means μ_i are unrestricted. This can be achieved by the general F test. First partition the response into

$$y_{ij} = (\hat{y}_{ij} - \bar{y}_{..}) + (\bar{y}_{i.} - \hat{y}_{ij}) + (y_{ij} - \bar{y}_{i.}) + \bar{y}_{..} \ ,$$

with $\hat{y}_{ij} = \hat{\beta}_0 + \hat{\beta}_1 x_i$ being the regression estimate. Recall that the slope and intercept regression estimates are

$$\hat{\beta}_1 = \sum_{i=1}^{a} n_i (\bar{y}_{i.} - \bar{y}_{..})(x_i - \tilde{x}.) / \sum_{i=1}^{a} n_i (x_i - \tilde{x}.)^2$$
$$\hat{\beta}_0 = \bar{y}_{..} - \hat{\beta}_1 (x_i - \tilde{x}.)$$

source	df	SS	E(MS)
linear	1	$\sum_{ij}(\hat{y}_{ij} - \bar{y}_{..})^2$	$\sigma^2 + \beta_1^2 \sum_i n_i(x_i - \tilde{x}_.)^2$
nonlinear	$a - 2$	$\sum_{ij}(\bar{y}_{i\cdot} - \hat{y}_{ij})^2$	$\sigma^2(1 + \delta_{\text{NONLINEAR}}^2/(a-2))$
error	$n_. - a$	$\sum_{ij}(y_{ij} - \bar{y}_{i\cdot})^2$	σ^2
total	$n_. - 1$	$\sum_{ij}(y_{ij} - \bar{y}_{..})^2$	---

Table 16.1. *One-factor anova for testing linearity*

with $\tilde{x}_. = \sum_{i=1}^a n_i x_i / n_.$. The total sum of squares around the mean $\bar{y}_{..}$ can be partitioned as shown in Table 16.1. The sum of the linear and nonlinear SS is the variation explained by groups, presented in earlier chapters. The sum of the linear and error SS is the unexplained variation from simple linear regression. Thus the nonlinear SS represents the difference in explained variation between the group mean model and the regression model. That is, the appropriate test of whether the group means are linear, $H_0 : \mu_i = \beta_0 + \beta_1 x_i$, is

$$F = \frac{SS_{\text{NONLINEAR}}/(a-2)}{MS_{\text{ERROR}}}$$

which has an $F_{a-2, n_.-a}$ distribution under the null hypothesis. If the group means μ_i do not lie along a line $\beta_0 + \beta_1 x_i$, then F has a non-central F distribution, with non-centrality parameter

$$\delta_{\text{NONLINEAR}}^2 = \sum_i n_i[(\mu_i - \tilde{\mu}_.)^2 - \beta_1^2(x_i - \tilde{x}_.)^2]/\sigma^2$$

with $\tilde{\mu}_. = \sum_i n_i \mu_i / n_.$. This measures any nonlinear pattern in group means after removing any linear trend.

Example 16.2 Nutrient: Formal analysis of the effect of nutrient rate (**nominal** or **actual**) confirms the findings presented in Figure 16.1. A test for nonlinearity comparing the SS_{MODEL} between the anova and regression fits for the **nominal** rates is not significant:

$$F = ((428.1 - 427.5)/(3-1))/5.31 = 0.056 \ (p > 0.5) \ .$$

Table 16.2 contains the model sums of squares, mean squares for model and error, and F values for an anova on the four rate groups and for regression on the **nominal** and **actual** rates. The 4% difference in MS_{ERROR} between **nominal** and **actual** translates into a 2% difference in standard deviations. In this case, the formal tests are highly significant either way. ◇

source	df	SSmodel	MSmodel	MSerror	F
anova	3	428.1	142.7	5.59	25.5
nominal	1	427.5	427.5	5.31	80.5
actual	1	435.5	435.5	5.10	85.4

Table 16.2. *Nutrient anova for nominal and actual rates*

16.3 Path analysis diagrams

Path diagrams can provide a very convenient visual representation of relationships among factors and multiple responses. Path analysis diagrams were invented by Sewall Wright in 1918 (see Wright 1934). For a modern treatment in the setting of latent variables, see Bollen (1989). This approach has attracted some controversy as it involves assigning cause and effect, which is outside the purview of statistical inference. However, if the 'direction' of relationships is based on insights beyond the experiment under study, along the lines presented earlier in this chapter, path diagrams can provide concise graphical summaries to stimulate understanding.

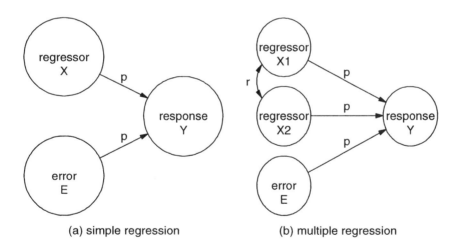

(a) simple regression (b) multiple regression

Figure 16.2. *Path diagrams for regression. Circles represent model components. Unidirectional arrows point in direction of (prior belief about) cause and effect; bidirectional arrows only indicate association. Path coefficients (p) and correlations (r) show strength of paths.*

The basic idea is to use paths between the response or responses and factors, covariates and errors. Weights are attached to the connecting paths.

Consider simple linear regression,

$$y = \beta_0 + \beta_1 x + e$$

and standardize both the regressor and response. That is, replace y by $Y = (y - \mu_y)/\sigma_y$ and x by $X = (x - \mu_x)/\sigma_x$, with μ_y and σ_x the population mean and standard deviation for y, and similarly for x. The new model is

$$Y = p_1 X + p_e E$$

with $p_1 = \beta_1 \sigma_x/\sigma_y$ and $p_e = \sigma/\sigma_y$. Notice that p_e^2 is the proportion of variation of the ys that is not explained by the xs. In addition, $p_1^2 + p_e^2 = 1$. That is, the path coefficients partition the total variation in a natural way, which can be depicted graphically as in Figure 16.2(a).

Similarly, regression with two covariates can be represented as

$$Y = p_1 X_1 + p_2 X_2 + p_e E \ .$$

The proportion of variation in Y explained by X_1 after adjusting for X_2 is p_1^2. If X_1 and X_2 are uncorrelated, then $p_1^2 + p_2^2 + p_e^2 = 1$. More generally, the explained variation is partitioned as

$$1 - p_e^2 = p_1^2 + p_2^2 + 2 p_1 r_{12} p_2$$

with r_{12} the correlation between X_1 and X_2. This is depicted schematically in Figure 16.2(b) with path coefficients on directed paths and correlations on undirected paths. The values of path coefficients can be estimated using the usual least squares estimators.

Wright (1934) showed that the correlation between a response and a covariate can be found by summing the product of path coefficients and correlations over all other variables,

$$\mathrm{corr}(X_i, Y) = r_{iy} = \sum_j p_j r_{ji} \ .$$

This relationship is in fact generally applicable to quite complicated path diagrams. Since the correlation of Y with itself is $1 = \sum p_i r_{iy}$, substituting the previous relation for r_{iy} gives

$$1 = \sum_i p_i^2 + 2 \sum_{i<j} p_i r_{ij} p_j \ .$$

In particular, this demonstrates the relation for a response influenced by two covariates and independent errors shown above.

Since analysis of variance can be recast as multiple regression, it makes sense to consider path diagrams for anova models. For instance, if an experiment compares three groups, two linear contrasts of group means can be constructed, say X_1 and X_2, to address the two degrees of freedom among these groups. Thus analysis of variance models can be represented with path diagrams, as shown in Figure 16.3(a). If the contrasts are not

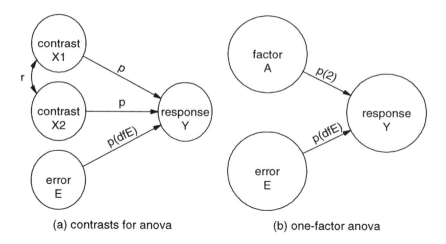

Figure 16.3. *Path diagrams for analysis of variance. Consider factor* A *to have three levels and hence two degrees of freedom. Path diagram (a) shows contrasts in anova recast as multiple regressors while (b) formally represents factor* A. *Degrees of freedom per path are in parentheses, or 1 if missing. See Figure 16.2 for further details.*

orthogonal, then the contrast variables have some correlation r_{12} as shown in the figure. For convenience, Figure 16.3(b) presents a simplified path diagram for a one-factor analysis of variance. To make this precise, the path coefficient p_a is defined by

$$p_a^2 = p_1^2 + p_2^2 + 2p_1 r_{12} p_2 \ .$$

For instance, if A is set to the mean for observations in each group, suitably centered and rescaled to have variance 1, then the sign of p_a is positive. In practice, it may be important to consider specific one-degree-of-freedom contrasts, but the formal reduction of a group to one node on a path diagram is handy for the following developments.

Example 16.3 Nutrient: The path diagram for the analysis of variance for yield on the nominal rate in Figure 16.4 shows (a) the explained variation (68% = 0.825^2) for the anova fit and (b) the test of linearity using values from Table 16.2. The squared path coefficients add up to 1.0 within roundoff error. The first bubble in (b) corresponds to the one-degree-of-freedom contrast to examine linear trend. The second bubble collapses the other two degrees of freedom as measuring any evidence of nonlinearity. The degrees of freedom are included in parentheses so that formal tests can be reconstructed from the figure. ◇

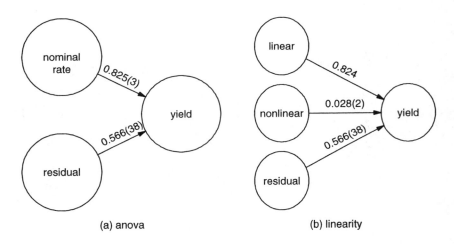

Figure 16.4. *Nutrient path diagrams for linearity. (a) The analysis of variance of nominal nutrient* rate *on* yield *explains* $0.825^2 = 68\%$ *of the variation. (b) The evidence strongly suggests that there is a linear relation, as only* $0.028^2 = 0.08\%$ *of the variation can be attributed to nonlinearity.*

16.4 Regression calibration

Most models only capture a part of the real problem. Coding a factor as low, medium or high may be a reasonable approximation, even though these levels are not exactly set (and may not be precisely known). Covariates may be measured with error but may be treated as known and fixed. Further, factors or covariates may be omitted because their effects are believed to be negligible, or because of limited time or other resources. These complications can introduce bias into a model and can inflate the variance. These ideas are only partly developed here and in the next section. The subject of **error in variables** has been addressed in the linear regression setting by Fuller (1987). See also Carroll, Ruppert and Stefanski (1995) for nonlinear models.

This section considers **regression calibration models** in which the experimenter attempts to calibrate a factor at nominal levels. However, the actual level administered to the experimental unit may differ from this nominal level due to situations beyond the control of the scientist. For instance, a certain level of fertilizer may be measured, but only a portion can actually be absorbed by a plant based on the local soil characteristics. Alternatively, all the fertilizer may be usable, but the actual amount applied may be variable due to a rough measuring device. Consider a single-factor model

$$y_{ij} = \mu_i + e_{ij},$$

REGRESSION CALIBRATION

with $e_{ij} \sim N(0, \sigma^2)$. Suppose that the true model is

$$y_{ij} = \mu_i + \gamma z_{ij} + e_{ij},$$

for some level displacement $z_{ij} \sim N(0, \sigma_z^2)$. The natural questions to ask are: (a) if we knew z_{ij}, could estimation be improved significantly; and (b) how much is lost by ignoring z_{ij}? The first part concerns inference about γ, while the second part concerns the effect of the unknown γ on inference for μ_i.

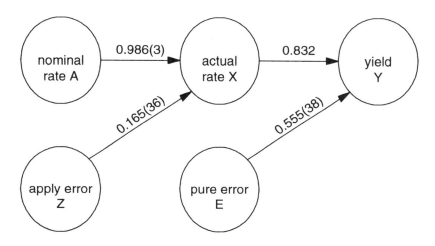

Figure 16.5. *Nutrient regression calibration: nominal rate is applied with error, leading to actual rate of nutrient which directly affects yield.*

Example 16.4 Nutrient: The experiment involved setting the nominal rates for the nutrient and measuring yield. However, there was some error in application (Figure 16.5). Thus the actual rate has a higher path coefficent (0.832) than the nominal rate (0.824 in Figure 16.4). ◇

If z_{ij} is measured, its effect after explaining group effects can be calculated. This problem is deferred to the chapter on analysis of covariance in the next part of the book.

Ignoring the error in variables does not introduce bias into the estimates of group means, since

$$E(y_{ij}) = \mu_i + \gamma E(z_{ij}) = \mu_i.$$

However, it does inflate the variance,

$$V(y_{ij}) = \sigma^2 + \gamma^2 \sigma_z^2.$$

The greater the effect γ of the displacement, the greater the inflation of variance. This makes it more difficult to detect significant differences among groups.

The above formulation assumes the random displacement z_{ij} affects response in the same manner for all factor levels. This may be reasonable if the μ_i are linear, e.g. $\mu_i = \beta_0 + \beta_1 x_i$. In general, the slope may vary with factor level,

$$y_{ij} = \mu_i + \gamma_i z_{ij} + e_{ij} ,$$

In this case, there is still no bias but the responses from different groups have different variances,

$$V(y_{ij}) = \sigma^2 + \gamma_i^2 \sigma_z^2 .$$

Therefore, if the γ_i are suspected of varying widely among the groups, methods appropriate to unequal variances should be used.

Bias in estimation of group means can arise if the displacements are not centered about zero. That is, if $E(z_{ij}) = \eta_i$ are not all equal, then the expected difference of sample means is not the corresponding difference in population means,

$$E(\bar{y}_{1.} - \bar{y}_{2.}) = (\mu_1 - \mu_2) + \gamma(\eta_1 - \eta_2) .$$

This is further exacerbated if the slopes are not equal.

16.5 Classical error in variables

The classical approach to error in variables, or error calibration, is a bit different from the regression calibration presented in the previous section. Here the factor of interest cannot be measured directly. Rather, a 'surrogate' can be measured instead. It may not even be clear whether the factor of interest has discrete levels or has a continuous distribution. Again, see the references cited in the previous section for more details beyond this introduction. For convenience, only simple regression is presented.

Example 16.5 Nutrient: The scientist managed to measure the nutrient indirectly. Figure 16.6(a) plots the observed rate against the yield. Regression of yield on observed rate (solid line) is attenuated. That is, the observed slope (0.054) is lower than the slope (0.058) for the actual rate (dotted line). A comparison of observed and actual rates in Figure 16.6(b) shows how far these deviate from the nominal rate (dashed identity line). ◊

Example 16.6 Nutrient: The attenuated relationship between yield and the observed rate relative to the actual rate is evident in the anova information in Table 16.3. This has been summarized graphically in the path diagram of Figure 16.7. The correlation between observed rate and yield

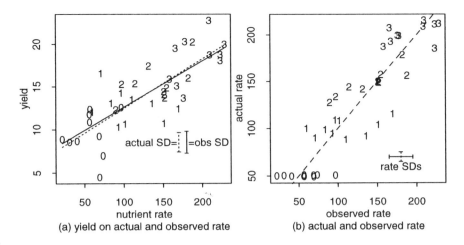

Figure 16.6. *Nutrient yield on actual and observed rates. (a) The linear fit of yield on observed rates (solid line) is attenuated relative to that on actual rates (dotted line). In addition, the SD for yield on observed rate is larger. (b) The observed rates have considerable more spread than the actual rates.*

source	df	MSmodel	MSerror	F
actual	1	435.5	5.10	85.4
observed	1	381.6	6.52	58.5

Table 16.3. *Nutrient anova for classical error in variables*

(0.779) is somewhat stronger than might be supposed from the path coefficients for the **actual** rate ($0.903 \times 0.832 = 0.751$) due to some correlation between measurement error and **yield**, which is not represented in this diagram. ◇

The classic error in variables model has a true model

$$y_i = \beta_0 + \beta_1 x_i + e_i$$

but the regressor x_i cannot be observed directly. Instead, the scientist can measure

$$w_i = x_i + u_i$$

with u_i some independent zero-mean noise. If the observable w_i is regressed on y_i, the slope is attenuated. That is,

$$y_i = \beta_0 + \lambda \beta_1 w_i + e_i$$

with attenuating factor, sometimes called reliability ratio, $\lambda = \sigma_x^2 / \sigma_w^2$,

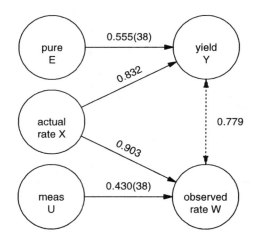

Figure 16.7. *Nutrient classical error in variables. Actual rate is observed with measurement error. Regression of yield on observed rate is attenuated, with lower correlation than for actual rate.*

which is at most one. The relationship is illustrated in Figure 16.6, which differs in important ways from the regression calibration model shown in Figure 16.5.

Thus the slope using the surrogate measurement w_i is biased, even if the measurement error u_i has mean zero. In addition, the variance is inflated,

$$V(y_i|w_i) = \sigma^2 + \lambda \beta_1^2 \sigma_u^2 .$$

Corrections for this attenuation have been proposed, in particular if there is extra knowledge about the measurement variance σ_u^2. See Fuller (1987) or Carroll, Ruppert and Stefanski (1995) for further details. In addition, there is some treatment of this model in Seber (1977, sec. 6.4), Snedecor and Cochran (1989, sec. 18.7) and Miller (1997, sec. 5.5).

16.6 Problems

16.1 Iron: Gray cast iron is a common commercial iron-based casting alloy. The two principal alloying elements are carbon and silicon. This problem examines the relationship between carbon level (C) and a particular melting property (resp), specifically called the 'primary austenite liquidus arrest temperature of the cooling curve'. Three levels of added carbon were used, low (1), medium (2) and high (3). However, it is not possible to set these levels exactly, for a variety of practical reasons. Instead, the scientist can measure the actual carbon afterward (Cactual). The questions for this problem concern how to relate carbon level to the response. Is it enough just to consider the three discrete levels (coded as `Clevel` = 1, 2, 3) or is there further information about the response in the actual carbon? Use plots, models and tables of means and/or analyses of variance to present your arguments.

(a) Run a one-factor analysis of variance comparing the effect of the three levels of carbon on response. Include multiple comparison of means. In particular, consider whether the linear contrast of means $(1\mu_3 + 0\mu_2 - 1\mu_1)$ is zero.

(b) Regress the discrete carbon levels on response. Why does the variation explained differ from the contrast considered in (a)?

(c) Test lack of linear fit. That is, is there evidence that the relationship between carbon level and response could be nonlinear?

(d) Regress the actual carbon on response. Why is the fit worse than in part (b), even though a more 'accurate' measure of carbon is used? [You may want to use plots to show this.]

(e) Fit a model with both discrete level and actual carbon to examine the question of error in variables. How strong is the evidence?

16.2 Design: A new drug is being considered to treat blood pressure but the investigators are unsure of the proper dose level. They also must compare it to the current drug treatment. Suppose that four levels of the new drug, increasing in powers of 2, are to be examined. The scientist wants to know ahead of time how to analyze the data in order to complete a grant application.

(a) Start with a one-factor model with five levels. Construct appropriate orthogonal contrasts (assuming equal sample sizes) to compare the new drug with the existing therapy and to assess the log-linear trend for the new drug. Ascertain their standard errors in terms of the model SD.

(b) Construct 'dummy variables' to reproduce the contrasts in (a). That is, define new variables in terms of the factor levels.

(c) Write down the anova table using the dummy variables or the contrasts, showing how the model sum of squares and degrees of freedom are partitioned. Include expected mean squares.

(d) Indicate how data from this experiment could be analyzed with a standard statistical package. That is, write out appropriate statements for the said package.

16.3 Forage: Again consider only the mature cows. The goal is to draw inference about possible linear or quadratic trends with respect to the amount of alfalfa. That is, consider linear and quadratic contrasts in terms of the rank of the `diets` (1–5).

(a) Construct 'orthogonal' polynomials using the SAS `contrast` phrase as in Example 5.4, page 75.

(b) Use 'regressors' to examine `linear` and `quadratic` trends as in Example 5.5.

(c) Compare results between (a) and (b). Why are they different? How can you demonstrate this using mathematical notation?

(d) Test whether there is evidence of lack of fit of a straight line. That is, is there evidence of higher-order polynomial trend, or more generally of some nonlinear trend with the amount of alfalfa in the `diet`?

(e) Interpret your results in light of graphical and/or tabular summaries.

16.4 Forage: Consider a simple linear regression of `dmi` the `diet` rank.

(a) Perform ordinary regression.

(b) State assumptions clearly.

(c) Why is the regression slope estimate different from the estimate of the linear contrast found earlier?

(d) Perform weighted regression using the inverse of the sample sizes by `diet` group.

(e) Why does the slope estimate agree with the linear contrast now?

CHAPTER 17

Parallel Lines

The analysis of variance developed so far assumes that the mean responses at different factor levels are constant but possibly different. Many studies have further complications, with the mean response within any group depending on other variables as well as the factor levels identifying the group.

This other measurement, or **covariate**, may be a different response to different factor levels, or some factor that could not be precisely controlled in the experimental design, but that may affect the response. In a sense, **analysis of covariance**, or **ancova**, as this problem is sometimes called, compares factor levels while adjusting for the association between response and covariate. In other words, the response and this other measurement may 'covary', or vary together, regardless of factor levels.

The principal aim is to remove a known source of variation so that factors can be more properly assessed. This removal can increase precision and, in some cases, overcome the confounding effect of a covariate. For instance, if the covariate happens to be larger at one factor level than another in the sample, this could indicate possible bias in the response to the factor. Analysis of covariance can be very important in removing some bias for **observational studies**, where comparisons can involve data collected for rather different purposes. The analysis of covariance, adjusting cell means by a covariate, can be readily extended to several factors. It is further possible to adjust for several covariates at the same time.

Quite often plots can help clarify relationships, suggesting more formal models and tests for confirmation. Consider plotting the response against the covariate, using different plotting symbols for each level of the experimental factor (or each cell for factor combinations). If there are enough factor levels, or enough observations per factor, separate plots by factor level or subsets of factor levels could be helpful.

The simplest relationship which involves both covariate and factor has parallel regression lines, with possibly different intercepts for different factor levels (Section 17.1). Section 17.2 introduces adjusted estimates. Section 17.3 reemphasizes the importance of diagnostics for developing models and checking assumptions. Sequential approaches to testing in Sections 17.4

and 17.5 first consider the factor or the covariate, respectively. The recommended approach is the adjusted, or Type III, developed in Section 17.6. More complicated relationships might involve crossing (non-parallel) lines, or curvilinear relationships.

17.1 Parallel lines model

The classical situation in analysis of covariance has the experimental factor and the covariate affecting response in ways that are readily separable. That is, their effects are assumed to be additive. Consider measured responses y_{ij} and covariates x_{ij} for experimental units $j = 1, \cdots, n_i$, assigned at random to factor levels $i = 1, \cdots, a$. The **additive model** is

$$\text{response} = \text{factor} + \text{covariate} + \text{error} ,$$

or in symbols,

$$y_{ij} = \mu_i + \beta(x_{ij} - \bar{x}_{..}) + e_{ij} ,$$

with $i = 1, \cdots, a$, $j = 1, \cdots, n_i$, and random errors $e_{ij} \sim N(0, \sigma^2)$. The regression coefficient (or slope) β is the change in response associated with a unit change in the covariate x_{ij}. The factor is assumed to affect response only through the group mean μ_i. Sometimes the model is written in terms of factor effects, $\alpha_i = \mu_i - \mu$, as

$$y_{ij} = \mu + \alpha_i + \beta(x_{ij} - \bar{x}_{..}) + e_{ij} ,$$

with some side conditions such as $\sum_i n_i \alpha_i = 0$. Some authors leave off the $\bar{x}_{..}$, but it actually makes interpretation of factor effects much simpler below.

Example 17.1 Feed: A scientist in food microbiology and toxicology (Chin *et al.* 1994) examined the effect of additives on weight gain in animals. He considered two levels (0.25% and 0.5%) of an additive CLA, a control and another additive (0.5% LA), which were labelled, respectively, as 1, 2, 3, 4. The initial concern was the tremendous differences in spread of weight gain for the four treatments. Plots of weight gain against feed intake (Figure 17.1) showed that this could be readily explained by feed intake. The feed intake slope is significant ($p = 0.0088$), as is the difference between LA and CLA ($p = 0.0082$) and the linear effect of %CLA ($p = 0.013$). The p-values are all computed using a Type III approach. Further, the effect of additive level seems to be linear and uncorrelated with the feed intake. ◊

The expected response for factor level i with covariate value x_{ij} is

$$E(y_{ij}) = \mu_i + \beta(x_{ij} - \bar{x}_{..}) .$$

The expected mean response for factor level i is

$$E(\bar{y}_{i\cdot}) = \mu_i + \beta(\bar{x}_{i\cdot} - \bar{x}_{..}) ,$$

PARALLEL LINES MODEL

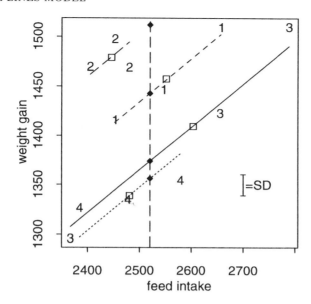

Figure 17.1. *Feed treatment adjusted for intake. The additive* LA *(4 dotted) reduces* feed intake *but does not appear to influence* weight gain *directly relative to the* control *(3 solid). However, the additive* CLA *(1 = 0.25, 2 = 0.5, dashed) reduces* feed intake *while increasing* weight gain *linearly with the level of* CLA. *Solid diamonds are LS means for residuals; open squares are marginal means.*

which depends not only on factor level but also on the average covariate value across experimental units at that level. If different factor groups have different mean covariate values, then a straight comparison of sample means could be very misleading. Instead, adjust sample means by the covariate, as discussed in the next section.

The sample grand mean has expectation

$$E(\bar{y}..) = \tilde{\mu}. = \sum_i n_i \mu_i / n. \; ,$$

which does not depend on the covariate. However, if there are unequal sample sizes, then $E(\bar{y}..) \neq \bar{\mu}.$. This problem was encountered earlier with unbalanced designs.

It may be helpful to notice some similarities with regression and analysis of variance. If there is no treatment effect ($\mu_i = \mu$), the model reduces to **simple regression**,

$$y_{ij} = \mu + \beta(x_{ij} - \bar{x}..) + e_{ij} \; ,$$

with slope β and intercept $\mu - \beta\bar{x}..$. Letting $\beta = 0$ instead reduces the analysis of covariance model to a single-factor analysis of variance. Here is another perspective which shows a connection to the two-way additive model.

Example 17.2 Size: A quality improvement team wants to compare production across several of the manufacturing units in its company under three different management strategies. The units are scored based on their size on a 1–4 scale and three units of each size are chosen (12 in all). It is assumed, based on prior studies, that larger units have higher productivity regardless of management practice. The experiment is performed and productivity is measured.

One possible model for this process is a **two-factor additive model**

$$y_{ij} = \mu + \alpha_i + \beta_j + e_{ij}$$

with α_i the effect of management strategy and β_j the effect of unit size. However, this ignores the believed strong association between size and productivity. Perhaps it would be more fruitful to consider a **linear relationship**, $\beta_j = \beta_0 + \beta j$. Letting $x_{ij} = j$, the model could be written as

$$y_{ij} = \mu + \alpha_i + \beta(x_{ij} - \bar{x}_{..}) + e_{ij} .$$

This reduces the number of model parameters, as there is now only one degree of freedom for size as opposed to $b - 1 = 3$. However, a visual check of the assumption of linearity is always a good idea, using a plot of residuals against predicted by plot symbols for strategy and/or unit size. ◇

This type of model, in a simpler form, was considered in Chapter 16. These and variants of the analysis of covariance model considered in the next chapter can be placed easily in the linear model framework. Thus all the results on linear models carry over directly to the present situation.

17.2 Adjusted estimates

This section finds least squares estimates for factor levels adjusted by covariate, and covariate slope adjusted by factor levels. Variances of these estimates are determined, with the aid of some notation for partitioning variation.

It is helpful to adopt some notation (see Searle 1977) for **partitioning sums of squares** in this and subsequent sections. The Total variation of response can be partitioned as Between-group plus Within-group variation,

$$\sum_{ij}(y_{ij} - \bar{y}_{..})^2 = \sum_i n_i(\bar{y}_{i\cdot} - \bar{y}_{..})^2 + \sum_{ij}(y_{ij} - \bar{y}_{i\cdot})^2$$

$$T_{yy} = B_{yy} + W_{yy} .$$

Similar partitions can be made of the total sum of squares for the covariate, $T_{xx} = B_{xx} + W_{xx}$, and the total sum of cross-products of covariate and

ADJUSTED ESTIMATES

response, $T_{xy} = B_{xy} + W_{xy}$. In an analogous manner, consider the between-group covariation of the true mean response to factor and the covariate,

$$B_{x\mu} = \sum_i n_i(\mu_i - \tilde{\mu}.)(\bar{x}_{i\cdot} - \bar{x}_{..}) \ .$$

Since $W_{x\mu} = 0$, the total covariation is $T_{x\mu} = B_{x\mu}$.

Consider the **unadjusted estimates**, which arise naturally from analysis of variance and regression. They are, using inverted hats, the group means,

$$\check{\mu}_i = \bar{y}_{i\cdot}$$

and the simple regression slope estimator,

$$\check{\beta} = T_{xy}/T_{xx} = \sum_{ij}(x_{ij} - \bar{x}_{..})(y_{ij} - \bar{y}_{..})/\sum_{ij}(x_{ij} - \bar{x}_{..})^2 \ .$$

As it turns out, these coincide with adjusted estimates developed below if the mean covariate values per factor level are the same ($\bar{x}_{i\cdot} = \bar{x}_{..}$). This is analogous to orthogonality in analysis of variance (and in multiple regression), where coefficient estimates are unaffected by the presence of other factors (or regressors) which are uncorrelated with the factor (or regressor) of concern. However, in general, estimates of factor means and covariate slope are more complicated.

The 'best' (in the sense of least squares) estimates arise by minimizing the residual sum of squares,

$$SS_E = \sum_{ij}[y_{ij} - \mu_i - \beta(x_{ij} - \bar{x}_{..})]^2 \ .$$

That is, solving the normal equations,

$\mu_i:$ $\sum_j y_{ij} = n_i\mu_i + \beta\sum_j(x_{ij} - \bar{x}_{..})$

$\beta:$ $\sum_{ij} y_{ij}(x_{ij} - \bar{x}_{..}) = \sum_{ij}\mu_i(x_{ij} - \bar{x}_{..}) + \beta\sum_{ij}(x_{ij} - \bar{x}_{..})^2 \ ,$

leads to the least squares (LS) estimates

$\hat{\mu}_i = \bar{y}_{i\cdot} - \hat{\beta}(\bar{x}_{i\cdot} - \bar{x}_{..})$

$\hat{\beta} = W_{xy}/W_{xx} = \sum_{ij}(x_{ij} - \bar{x}_{i\cdot})(y_{ij} - \bar{y}_{i\cdot})/\sum_{ij}(x_{ij} - \bar{x}_{i\cdot})^2 \ .$

The **adjusted slope** $\hat{\beta}$ is the regression of the within-group residuals on the covariate. The estimates $\hat{\mu}_i$ are known as the **adjusted factor means**, corrected for the deviation of the covariate mean value $\bar{x}_{i\cdot}$ from the covariate sample grand mean $\bar{x}_{..}$. The estimate of population grand mean for the response is $\hat{\mu} = \bar{y}_{..}$ which does not depend on the covariate. Note that $E(\hat{\mu})$ is not the simple average of expected group means $\bar{\mu}.$ unless the sample sizes are equal.

The adjusted group means are modified by the covariate, and the adjusted slope is modified by the group structure, unless there is a particular type of balance. That is, if in all treatment groups the covariates average to the same value ($\bar{x}_{i\cdot} = \bar{x}_{\cdot\cdot}$), then the adjusted group means agree with the unadjusted means ($\hat{\mu}_i = \check{\mu}_i = \bar{y}_{i\cdot}$) and the adjusted slope agrees with the regression slope ($\hat{\beta} = \check{\beta}$).

Since the estimates of μ_i and β are linear combinations of the response, their properties are inherited from those assumed for the errors. The adjusted slope has variance

$$V(\hat{\beta}) = \sigma^2/W_{xx}$$

and $\hat{\beta}$ is uncorrelated with the raw sample means ($\bar{y}_{i\cdot}$). Therefore, the variances of the adjusted factor means are

$$V(\hat{\mu}_i) = \sigma^2[1/n_i + (\bar{x}_{i\cdot} - \bar{x}_{\cdot\cdot})^2/W_{xx}] \ ,$$

which looks very much like the variance of the estimate of the regression line at $\bar{x}_{i\cdot}$. The adjusted factor means are correlated with the slope estimate, with covariance

$$\text{cov}(\hat{\mu}_i, \hat{\beta}) = \sigma^2(\bar{x}_{i\cdot} - \bar{x}_{\cdot\cdot})/W_{xx} \ .$$

Further, adjusted factor means are correlated,

$$\text{cov}(\hat{\mu}_i, \hat{\mu}_k) = \sigma^2(\bar{x}_{i\cdot} - \bar{x}_{\cdot\cdot})(\bar{x}_{k\cdot} - \bar{x}_{\cdot\cdot})/W_{xx} \ ,$$

since they are each adjusted by this slope. Great care is needed when constructing comparisons of adjusted means.

Pivot statistics can be constructed using these variances, along with an unbiased estimate of σ^2,

$$\hat{\sigma}^2 = \sum_{ij}[y_{ij} - \hat{\mu}_i - \hat{\beta}(x_{ij} - \bar{x}_{\cdot\cdot})]^2/(n_{\cdot} - a - 1) \ ,$$

leading to simple hypothesis tests and/or confidence intervals in the usual fashion. However, it is usually advisable to conduct overall tests for factor effects adjusted by the covariate before intensive investigation of means comparisons.

Example 17.3 Diet: The cows introduced in Problem 4.2, page 68, varied considerably in terms of their capacity. One way to account for this is to include a measure of their initial capacity as a covariate. In this case, it is natural to consider the dry matter intake at three weeks (covar). Figure 17.2 shows the initial capacity against the average dmi over the study period, suggesting that the covariate appears to be important. ◇

PLOTS WITH SYMBOLS

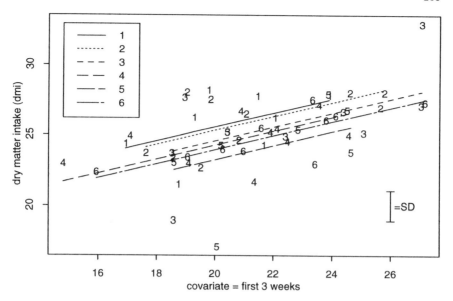

Figure 17.2. *Diet adjusted for initial capacity: initial capacity (*`covar`*) against dry matter intake (*`dmi`*) over the study by* `diet`*. Standard deviation (SD) is based on analysis of covariance with parallel lines.*

17.3 Plots with symbols

Interaction plots, introduced earlier for factorial arrangements, can suggest the form of relationship between covariate and response for each level of the factor. These are easily employed if the covariate has only a few discrete values; otherwise, covariate values can be grouped into intervals for graphical purposes. Other **diagnostic plots** suggest themselves. Residuals from an analysis of variance with the factor alone can be plotted against the covariate (again using plotting symbols for factor levels). Alternatively, regress y on x and examine a plot of residuals against predicted values (with plot symbols for the factor levels).

Inference requires the usual assumptions that the model is correct and that errors are independent, having equal variance and roughly normal distribution. As always, it is important to check the residuals

$$r_{ij} = y_{ij} - \hat{\mu}_i - \hat{\beta}(x_{ij} - \bar{x}_{..})$$
$$= y_{ij} - \bar{y}_{i.} - \hat{\beta}(x_{ij} - \bar{x}_{i.}) \ .$$

It is often very revealing to plot residuals against the predicted values (or against other covariates) to question assumptions about the model. In particular, using different plot symbols for each factor level can uncover other relationships which may merit examination.

Example 17.4 Feed: Ancova residuals in Figure 17.3 might suggest that there are different slopes by treatment group. For instance, replicates in group 4 (control) show a negative relation, indicating that the common slope is too steep. This is apparent in retrospect by examining Figure 17.1, but might be missed at first. ◊

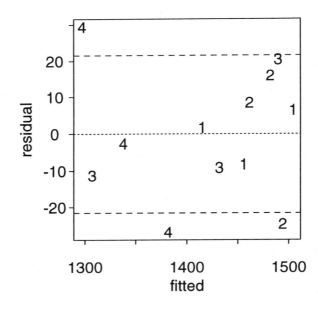

Figure 17.3. *Feed ancova residuals with plot symbols. Treatment group codes (1 = 0.25% CLA, 2 = 0.5% CLA, 3 = 0.5% LA, 4 = control) reveal patterns not apparent without symbols. Dashed lines are at one SD = 21.5.*

In a balanced experiment ($\bar{x}_{i\cdot} = \bar{x}_{\cdot\cdot}$), there is no difficulty separating effects of factor and covariate. However, in most situations, these effects are partially confounded. That is, the covariate affects the adjusted group means, and the factor affects the adjusted slope. If these adjustments are large, conclusions based on adjusted or unadjusted estimates could be rather different. If adjusting by the covariate removes any difference between mean response across factor levels, does that mean the factor has no effect? If the adjusted slope is near zero, does that mean the covariate has no relationship with response?

If the adjusted means are roughly the same and the adjusted slope is nearly zero, do neither have an effect? This latter situation can arise if $\bar{x}_{i\cdot}$ and the $\bar{y}_{i\cdot}$ are nearly collinear. The differences in group means could be explained equally well by factor or covariate. Judgements must rely on other information, both theoretical and empirical, concerning this and similar experiments.

17.4 Sequential tests with multiple responses

Suppose that X and Y are both responses which may be affected by the experimental factor A. It may be important to ask first if there are any factor effects, and then if there is any association between X and Y after adjusting for the factor. This approach assumes that differences in response Y among factor groups can be attributed to the factor A or to chance. This assumption is shown below to be untenable, but the development is useful for approaches considered in the next two sections.

This approach first fits a single-factor analysis of variance of Y on factor A, then regresses the residuals on the covariate X. The single-factor anova table for treatment has been previously shown to be

source	df	SS
A = factor	$a - 1$	$SS_A = B_{yy}$
error	$n. - a$	$SS_E = W_{yy}$
total	$n. - 1$	$SS_T = T_{yy}$

with $SS_A = \sum_{ij}(\bar{y}_{i.} - \bar{y}_{..})^2$. The variation explained by the covariate after adjusting for the factor is

$$SS_{X|A} = \sum_{ij}(\hat{y}_{ij} - \bar{y}_{i.})^2 = \sum_{ij}[\hat{\beta}(\bar{x}_{i.} - \bar{x}_{..})]^2 = \hat{\beta}^2 W_{xx}^2 = \hat{\beta} W_{xy} .$$

This suggests partitioning the error sum of squares from the single-factor model (W_{yy}) into parts explained by the covariate and by the factor. In table form,

source	df	Type I SS
A	$a - 1$	$SS_A = B_{yy}$
$X|A$	1	$SS_{X|A} = \hat{\beta} W_{xy}$
error	$n. - a - 1$	$SS_E = W_{yy} - \hat{\beta} W_{xy}$
total	$n. - 1$	$SS_T = T_{yy}$

The variation explained by factor and covariate together is

$$SS_{A,X} = B_{yy} + \hat{\beta} W_{xy} .$$

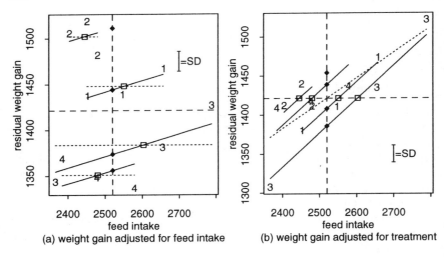

Figure 17.4. *Feed residuals and confounding. Residuals with added grand mean are from (a) simple regression of weight gain on feed intake or (b) anova of weight gain on treatment. Solid lines indicate proper further adjustments; dotted lines are naïve fits. Dashed cross highlights mean of feed intake and weight gain. Solid diamonds are LS means for residuals; open squares are marginal means.*

Example 17.5 Feed: It is sometimes useful to examine graphically the effect of simply removing a factor or covariate. Figure 17.4 removes (a) the covariate of feed intake or (b) the factor of treatment level from the response weight gain. Figure 17.4(a) was constructed by using the residuals from the simple regression of weight gain on feed intake with the (estimate of population) grand mean added back to keep values more or less in the original range. Since the means of feed intake per treatment group are not all the same, there is some residual non-zero slope. The naive anova of residual weight gain on treatment groups (dotted lines) is incorrect. For one thing, the marginal means for these residuals (open squares) do not agree with either the marginal means or the LS means presented in Figure 17.1. Further, the sums of squares and test in Table 17.1 are not correct. An adjusted (Type III) approach includes both feed intake and treatment in the fit (see Table 17.1 and solid lines on Figure 17.4(a)). The LS means (solid diamonds) agree exactly with the LS means from Figure 17.1, and the adjusted sums of squares and tests (Table 17.1) agree with results shown in Example 17.6.

The plot in Figure 17.4(b) arises from first removing the effect of treatment groups, again adding back the grand mean. Notice that the marginal group means (open squares) are now all identical, but the slope of the simple regression (dotted line) is different from the adjusted slope from

source	df	SS	MS	F	p-value
naïve anova					
treatment	3	40656	13552	20.6	0.0004
error	8	5262	658		
total	11	45918			
adjusted (Type III) ancova					
feed intake	1	2021	2021	4.4	0.075
treatment	3	42677	14226	30.7	0.0002
error	7	3242	463		

Table 17.1. *Feed anova table after removing covariate*

source	df	SS	MS	F	p-value
naïve simple regression					
feed intake	1	19096	19096	19.9	0.0012
error	8	9592	658		
total	11	28688			
adjusted (Type III) approach					
feed intake	1	25446	25446	54.9	0.0001
treatment	3	42677	14226	30.7	0.0002
error	7	3242	463		

Table 17.2. *Feed anova table after removing factor*

ancova (solid lines). Table 17.2 presents the naïve and adjusted approaches for sums of squares and tests. Notice that the adjusted mean square error agrees with that in Tables 17.4 and 17.1. ◊

Expected mean squares for terms can be found by considering expected values of within-group and total sums of squares. It can be readily shown that

$$E(B_{yy}) = \sigma^2(a-1) + \sum_i n_i[\mu_i - \tilde{\mu}_\cdot + \beta(\bar{x}_{i\cdot} - \bar{x}_{\cdot\cdot})]^2$$

$$E(W_{yy}) = \sigma^2(n_\cdot - a) + \beta^2 W_{xx}$$

$$E(\hat{\beta} W_{xy}) = \sigma^2 + \beta^2 W_{xx} .$$

To derive these, recall that $E(X^2) = V(X) + [E(X)]^2$ and carefully substitute the full model into each of the quadratic forms. This information leads to the following expected mean squares for the sequential fit of the factor followed by covariate:

source	df	$E(MS)$
A	$a-1$	$\sigma^2 + \sum_i n_i[\mu_i - \tilde{\mu}. + \beta(\bar{x}_i. - \bar{x}..)]^2/(a-1)$
$X\|A$	1	$\sigma^2 + \beta^2 W_{xx}$
error	$n. - a - 1$	σ^2
total	$n. - 1$	—

Thus, justification of this approach, in particular to interpret any inference involving the unadjusted group means, involves assuming

$$\beta(\bar{x}_i. - \bar{x}..) = 0 \; .$$

That is either $\beta = 0$, or there is a balance in covariate levels ($\bar{x}_i. = \bar{x}..$), by design (or chance). In general the unadjusted mean square MS_A confounds the effects of the factor and covariate.

However, the adjusted means square $MS_{X|A}$ isolates the slope β, and focuses attention on the within-group covariation of X and Y. That is, after removing any **spurious correlation** between covariate and response which could be attributed to the factor, what remains is pure association.

17.5 Sequential tests with driving covariate

In some situations, the covariate X may drive the response Y, with the latter modified by the factor. Covariates measured before the experimental factor are applied are surely unaffected by the factor. Thus it may be reasonable at times to assume that any linear relationship between X and Y can be attributed to the covariate. Differences in response to factor A are then evaluated after adjusting for the covariate. This Type I approach involves regressing y_{ij} on x_{ij}, and then running an analysis of variance of the residuals $y_{ij} - \bar{y}.. - \check{\beta}(x_{ij} - \bar{x}..)$ on the factor A.

The ancova table for simple regression is

source	df	SS
X=Covariate	1	$SS_X = \check{\beta} T_{xy}$
error	$n. - 2$	$SS_E = T_{yy} - \check{\beta} T_{xy}$
total	$n. - 1$	$SS_T = T_{yy}$

The unadjusted SS for the covariate can be verified by

$$SS_X = \sum_{ij}(\hat{\hat{y}}_{ij} - \bar{y}..)^2 = \sum_{ij}[\check{\beta}(x_{ij} - \bar{x}..)]^2 = \check{\beta}^2 T_{xx}^2 = \check{\beta}T_{xy} .$$

The previous section derived the variation explained by A and x together, regardless of order. Consider another partition, explaining the response first by covariate, and then by factor adjusted for covariate:

source	df	Type I SS	
X	1	SS_X	$=\check{\beta}T_{xy}$
$A\|X$	$a-1$	$SS_{A\|X}$	$=B_{yy} + \hat{\beta}W_{xy} - \check{\beta}T_{xy}$
error	$n. - a - 1$	SS_E	$=W_{yy} - \hat{\beta}W_{xy}$
total	$n. - 1$	SS_T	$=T_{yy}$

Expected mean squares can be derived using calculations from the previous section, with the following additional term,

$$E(\check{\beta}T_{xy}) = \sigma^2 + T_{xx}[\beta + T_{x\mu}/T_{xx}]^2 ,$$

leading to the following table:

source	df	$E(MS)$
X	1	$\sigma^2 + T_{xx}[\beta + T_{x\mu}/T_{xx}]^2$
$A\|X$	$a-1$	$\sigma^2 + \sum_i n_i[\mu_i - \tilde{\mu}. + T_{x\mu}/T_{xx}]^2/(a-1)$
error	$n. - a - 1$	σ^2
total	$n. - 1$	—

Suppose all the linear relationship between response and covariate is attributed to the effect of the covariate. This adds an **assumption** to the model, namely that

$$T_{x\mu} = B_{x\mu} = \sum_i n_i(\mu_i - \tilde{\mu}.)(\bar{x}_{i.} - \bar{x}..) = 0 .$$

In other words, the covariate and the factor effects are assumed to be **uncorrelated**. If valid, this justifies a Type I approach, with expected mean squares as follows:

source	df	Type I SS	$E(MS)$
X	1	$\check{\beta} T_{xy}$	$\sigma^2 + \beta^2 T_{xx}$
$A\|X$	$a-1$	$B_{yy} + \hat{\beta} W_{xy} - \check{\beta} T_{xy}$	$\sigma^2 + \sum_i n_i (\mu_i - \tilde{\mu}.)^2/(a-1)$
error	$n. - a - 1$	$W_{yy} - \hat{\beta} W_{xy}$	σ^2
total	$n. - 1$	T_{yy}	—

Thus, first evaluate the effect of covariate, and then examine the additional variation explained by the factor after adjusting for the covariate.

If the covariate is constant within each factor level ($x_{ij} = \bar{x}_{i\cdot}$), the adjusted slope estimate $\hat{\beta}$ is zero. That is, all information about the slope rests in the sample means $\bar{y}_{i\cdot}$. It is not possible to separate the effect of the factor from the covariate since they are completely confounded. Even with partial confounding, the data provide no evidence to separate the linear effect of the factor on the response from the effect of the covariate. The above approach resolves this confounding by making a further assumption that the factor effects have no linear trend with respect to the covariate.

17.6 Adjusted (Type III) tests of hypotheses

The adjusted, or Type III, approach conditions on all other factors being in the model. This section develops pivot statistics and tests of hypotheses for the main inferential questions concerning the effect of treatments adjusted for the covariate and the effect of covariate adjusted for the treatments. It is shown that the proper approach in general involves Type III SS for both questions.

The tables of expected mean squares in the previous two sections show that, in general, the Type I SS for A and X are each confounded with the other factor. However, under certain circumstances the Type I approach for the covariate seems reasonable.

The pivot statistic for $H_0 : (\mu_i = \mu | \beta)$ would therefore be

$$F = MS_{A|X}/MS_E \sim F_{a-1, n.-a-1; \delta_A^2}$$

with the non-centrality parameter

$$\delta_A^2 = \sum_i n_i [\mu_i - \tilde{\mu}. + T_{x\mu}/T_{xx}]^2/(a-1)\sigma^2$$

depending on the covariate but not on β. Note that under the null hypothesis, $T_{x\mu} = 0$ and $\mu_i = \tilde{\mu}. = 0$, which implies $\delta_A^2 = 0$. Thus the pivot statistic does not depend on any unknown parameters under the null hypothesis. Power, on the other hand, depends on the pattern of μ_is and the

source	df	Type III SS	$E(MS)$
$X\|A$=covariate	1	$\hat{\beta}W_{xy}$	$\sigma^2(1+\delta_X^2)$
$A\|X$=treatment	$a-1$	$B_{yy}+\hat{\beta}W_{xy}-\check{\beta}T_{xy}$	$\sigma^2(1+\delta_A^2)$
error	$n.-a-1$	$W_{yy}-\hat{\beta}W_{xy}$	σ^2
total	$n.-1$	T_{yy}	—

Table 17.3. *Type III adjusted sums of squares for ancova*

correlation between them and the $\bar{x}_i$ values.

Inference for the slope concerns the conditional mean square $MS_{X|A}$, leading to the pivot statistic for the null hypothesis $H_0 : (\beta=0|\mu_i)$ of

$$F = MS_{X|A}/MS_E \sim F_{1,n.-a-1;\delta_X^2}$$

with non-centrality parameter

$$\delta_X^2 = \beta^2 W_{xx}/\sigma^2$$

depending on the covariate but not upon μ_i. Clearly, $\delta_X^2 = 0$ if $\beta = 0$. Power depends on β and on the within-group variation of the $(x_{ij}-\bar{x}_{i\cdot})$ values.

This information is summarized in the ancova in Table 17.3. Note once again that the Type III sums of squares do not add up unless there is a particular type of balance, namely $\bar{x}_{i\cdot} = \bar{x}_{\cdot\cdot}$, which implies $B_{xx} = B_{xy} = 0$. Under this condition, Type III SS reduce to Type I SS.

Example 17.6 Feed: The analysis of covariance in Table 17.4 shows three approaches. The first two are sequential, considering the `treatment` or the `feed intake` as driving the `weight gain` response. In general, the adjusted (Type III) approach is most appropriate, as it provides unconfounded tests of `treatment` after adjusting for `feed intake` ($p = 0.0002$) and `feed intake` after adjusting for `treatment` ($p = 0.0001$). ◇

17.7 Different slopes for different groups

Sometimes the plots suggested earlier in this chapter show evidence of non-parallel lines. That is, the (linear) relationship between response y and covariate x may have a different slope for different levels of the factor A. Essentially, this is interaction between covariate and factor, and is handled in an analogous way to interaction in a two-factor model.

Consider the following model, which allows different slopes and intercepts

source	df	SS	MS	F	p-value
sequential (driving factor)					
treatment	3	34721	11574	25.0	0.0004
feed intake	1	25446	25446	55.0	0.0001
error	7	3242	463		
total	11	63409			
sequential (driving covariate)					
feed intake	1	17490	17490	37.8	0.0005
treatment	3	42677	14226	30.7	0.0002
error	7	3242	463		
total	11	63409			
adjusted (Type III)					
feed intake	1	25446	25446	55.0	0.0001
treatment	3	42677	14226	30.7	0.0002

Table 17.4. *Feed analysis of covariance*

for different factor levels,

$$y_{ij} = \mu_i + \beta_i(x_{ij} - \bar{x}_{..}) + e_{ij},$$

with β_i the covariate slope in group i, and μ_i its intercept. For some purposes, it may be useful to partition the slope into a common slope β and a deviation from common slope γ_i,

$$y_{ij} = (\mu + \alpha_i) + (\beta + \gamma_i)(x_{ij} - \bar{x}_{..}) + e_{ij},$$

with appropriate side conditions on both α_i and γ_i.

The (interaction) hypothesis $H_0 : \gamma_i = \bar{\gamma}.$ or $\beta_i = \beta$ addresses the question of **parallel lines**. If this hypothesis is rejected, then there is evidence for different slopes at different factor levels, complicating interpretation of the effect of the factor adjusted for covariate. For instance, the expected mean response at factor level i is

$$E(\bar{y}_{i.}) = \mu_i + \beta_i(\bar{x}_{i.} - \bar{x}_{..}),$$

and the expected value of the grand mean is

$$E(\bar{y}_{..}) = \tilde{\mu}. + \sum_i n_i\beta_i(\bar{x}_{i.} - \bar{x}_{..})/n. = \tilde{\mu}. - \beta\bar{x}_{..} + \sum_i n_i\beta_i\bar{x}_{i.}/n.,$$

which depends on all the slopes β_i. Note that both of these confound the

DIFFERENT SLOPES FOR DIFFERENT GROUPS

effects of factor and covariate unless the design is **balanced** ($\bar{x}_{i\cdot} = \bar{x}_{\cdot\cdot}$). In that case, $\bar{\mu}_{i\cdot} = \mu_i$ and $\bar{\mu}_{\cdot\cdot} = \mu$. However, these only summarize the mean response at the average level of the covariate.

Contrasts between factor levels would typically be done at some covariate level, in order to remove its effect. Denoting the mean response of factor level i at covariate level x by

$$\mu_i(x) = \mu_i + \beta_i(x - \bar{x}_{\cdot\cdot}) ,$$

the contrast of two-factor levels is

$$\mu_1(x) - \mu_2(x) = (\mu_1 - \mu_2) + (\beta_1 - \beta_2)(x - \bar{x}_{\cdot\cdot}) .$$

Thus, unless $\beta_1 = \beta_2$, this contrast depends on covariate level x. In plain language, it is important to consider the effect of factors with respect to particular levels of the covariate. Again, this is interaction!

The LS estimates correspond to the within-group intercepts and slopes for regression,

$$\hat{\mu}_i = \bar{y}_{i\cdot} - \hat{\beta}_i(\bar{x}_{i\cdot} - \bar{x}_{\cdot\cdot}) ,$$
$$\hat{\beta}_i = \sum_j (x_{ij} - \bar{x}_{i\cdot})(y_{ij} - \bar{y}_{i\cdot}) / \sum_j (x_{ij} - \bar{x}_{i\cdot})^2 .$$

Variances of the estimates, and pivot statistics can be derived in the usual fashion. For the mean response of factor level i at a fixed covariate value, the LS estimate and variance are

$$\hat{\mu}_i(x) = \bar{y}_{i\cdot} + \hat{\beta}_i(x - \bar{x}_{i\cdot}) ,$$
$$V(\hat{\mu}_i) = \sigma^2[1/n_i + (x - \bar{x}_{i\cdot})^2 / \sum_j (x_{ij} - \bar{x}_{i\cdot})^2] ,$$

which can be used to develop pivot statistics for contrasts of factor levels at that covariate value.

For **testing**, in general only a Type III approach is appropriate. The following ancova table summarizes the sums of squares. Expected mean squares could be developed in the usual fashion.

source	df	Type III SS
X=covariate	1	$SS(X\|A, X*A)$
A=factor	$a-1$	$SS(A\|X, X*A)$
$X*A$	$a-1$	$SS(X*A\|A, X)$
error	$n_{\cdot\cdot} - 2a$	$SSE = \sum_i \sum_j (y_{ij} - \bar{y}_{i\cdot} - \hat{\beta}_i(x_{ij} - \bar{x}_{i\cdot}))^2$
total	$n_{\cdot\cdot} - 1$	$SS_{\text{TOTAL}} = \sum_i \sum_j (y_{ij} - \bar{y}_{\cdot\cdot})^2$

If the design in balanced ($\bar{x}_{i\cdot} = \bar{x}_{\cdot\cdot}$), then Type III SS agree with Type I SS.

Example 17.7 Feed: Evidence for different slopes in Figure 17.5 suggests conducting a formal test for interaction. Table 17.5 shows that the interaction is not significant. There may in fact be interaction, or these variations may simply be due to small sample sizes.

source	df	SS	MS	F	p-value
model building (Type II)					
feed intake	1	25446	25446	116.2	0.0004
treatment	3	42677	14226	64.9	0.0008
fi*trt	3	2365	788	3.6	0.12
error	4	876	219		
adjusted (Type III)					
feed intake	1	2916	2916	13.3	0.022
treatment	3	2396	799	3.6	0.12
fi*trt	3	2365	788	3.6	0.12
error	4	876	219		

Table 17.5. *Feed ancova of different slopes*

source	df	SS	MS	F	p-value
model building (Type II)					
feed intake	1	25446	25446	116.0	<0.0001
treatment	3	42677	14226	92.8	<0.0001
fi*high	1	2328	2328	15.1	0.0081
error	6	920	153		
adjusted (Type III)					
feed intake	1	27382	27382	178.6	<0.0001
treatment	3	44668	14889	97.1	<0.0001
fi*high	1	2328	2328	15.1	0.0081
error	6	920	153		

Table 17.6. *Feed anova for different high slopes*

Table 17.6 considers a special kind of interaction, different slopes for the two high (0.5%) treatments from the other two. Now there is evidence for interaction ($p = 0.008$). However, it is important to keep in mind that there are only 12 data points, and this investigation was conducted post hoc, after noticing a pattern in the plots. Further experiments should help clarify this issue. ◇

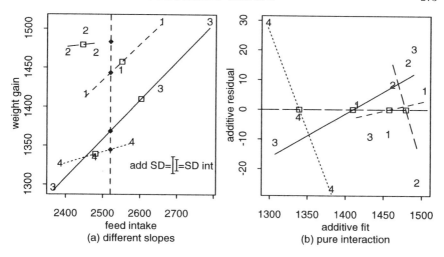

Figure 17.5. *Feed evidence for different slopes. (a) Fits of different slopes by treatment groups suggests that the* 0.5% *treatments (2* = CLA, *4* = LA *may result in slower increases in weight gain with feed intake. (b) Regression lines by group overlaid on a plot of residuals reinforces this pattern. Symbols are the same as Figure 17.1, page 257. Standard deviations (SD) shown for additive (parallel lines) and interaction (different slopes) models.*

Note that if the lines are not parallel, then interpretation of the (main effects) hypothesis $H_0 : \alpha_i = 0$ needs some careful consideration. For some problems, the **pencil of lines** model with common intercept but different slopes has merit. That is, the model

$$y_{ij} = \mu + \beta_i x_{ij} + e_{ij}$$

assumes that the mean response at $x = 0$ is μ, and as x increases, the mean response increases or decreases at a rate depending on the factor level i. This may be appropriate for some growth models or intervention studies where there is a clear interpretation of $x = 0$. Alternatively, this model may be deemed adequate because it allows the mean values to diverge as the covariate increases. However, if the covariate values are far from zero, it may be preferable to consider pencils of lines originating at a point x_0 other than zero by shifting the covariate, such as

$$y_{ij} = \mu + \beta_i(x_{ij} - x_0) + e_{ij} \ .$$

17.8 Problems

17.1 Feed: Conduct the formal analysis for `weight gain` following the example through this chapter. Be sure to specify carefully the model and assumptions, and document your steps in analysis. Interpret the results in your own words for the scientist.

17.2 Diet: Verify the information presented in Figure 17.2. That is, perform an analysis of covariance of `dmi` on `covar` and `diet`. Do the diets differ significantly after adjusting for initial capacity? Is the covariate significant?

17.3 Wasp: Consider the study detailed in Examples 18.3 and 18.5.
(a) Plot gonadium width (e.g. `G1Wa`) against head width (`HW`) using plot symbols for `caste`.
(b) Conduct an analysis of covariance of `G1Wa` on `HW`.
(c) Add the ancova lines to the plot.
(d) Could the lines be parallel? Perform an appropriate formal test.
(e) Could the lines have the same intercept at zero? Perform an appropriate formal test.
(f) Interpret the results for the scientist. In particular, what do (d) and (e) say about possible hypotheses about how shapes of queens and workers developed?

CHAPTER 18

Multiple Responses

Many experiments have several responses. This can arise for a variety of reasons. For instance, the scientist may have a general idea of how experimental conditions might affect the process under study, but may be unsure of the 'best' response to measure. Alternatively, interest may focus on the effect of groups on one or a few responses, which may be further influenced by other variables, which in turn may or may not differ among groups. For convenience, this chapter restricts attention to comparing groups in a one-factor completely randomized design. However, the ideas generalize readily to more complicated designs.

Multivariate analysis of variance, or **manova**, has potentially two advantages over separate 'univariate' anovas for each response. First, it can be a more powerful method to detect differences among groups if some of the responses are correlated. Second, it can enhance interpretation by reducing consideration to one or a few linear combinations of responses. However, neither of these 'features' is guaranteed for any particular experiment. For instance, manova can be less powerful if responses are uncorrelated. In addition, it may not be obvious how to interpret the manova results. In general, it is advisable to use a combination of univariate and multivariate approaches when there are multiple responses of interest. More detailed treatment of multivariate methods presented in an accessible manner can be found in Morrison (1976, ch. 5–6), Bray and Maxwell (1985) and Krzanowski (1990, ch. 11–13).

The general procedure adopted below is similar to the univariate methods presented earlier. Begin with an overall test to ascertain whether evidence supports significant differences among groups. Then analyze those differences in detail and interpret their significance. Here it is important to examine which groups differ as well as which responses contribute to those differences.

There is no one 'right' approach to the analysis of multiple responses for all situations. Instead, there are a variety of methods which form a palette of possibilities. Section 18.1 considers several ways to summarize information about variation and covariation among multiple responses explained by deviations from the null hypothesis, relative to unexplained error variation and covariation. This has led traditionally to four overall tests, with differing strengths and weaknesses. Section 18.2 examines ways to analyze differences and attribute them to responses, or combinations of responses. These include univariate anovas, with or without adjustment for multiple testing and discriminant analysis. Once the important responses have been identified, multiple comparisons can be addressed either on individual responses or on linear combinations suggested by analysis. While data analysis cannot assess cause and effect, outside knowledge of the relationship among responses in an experiment can greatly enhance interpretation, as well as streamline analysis. The connection between causal models and analysis of covariance is examined through step-down analysis in Section 18.3. Section 18.4 explores the use of path analysis to guide methods developed in earlier sections.

18.1 Overall tests for group differences

It is wise at some early stage of analysis to assess whether there is any evidence for overall differences among groups. This may be reflected in only one response. More likely, any significant differences might be attributed to several of the measured responses. This section develops the overall null hypothesis of no group differences and corresponding pivot statistics for multiple responses. However, there appears to be no one way to assess evidence of differences which is 'best' under all circumstances.

The univariate model considered up to now for comparing groups was

$$y_{ij} = \mu_i + e_{ij} \; , \; e_{ij} \sim N(0, \sigma^2) \; ,$$

with $i = 1, \cdots, a$, $j = 1, \cdots, n_i$, and all errors (and hence all responses) independent. The null hypothesis of no group differences was

$$H_0 : \mu_1 = \cdots = \mu_a \text{ or } H_0 : \mu_i = \bar{\mu}_{..} \; .$$

The total variation sum of squares for a univariate response was partitioned into variation between groups (explained by deviations from the null hypothesis H_0) and variation within group (unexplained, or error),

$$SS_T = SS_H + SS_E \; .$$

The general F test compared the variation explained by the hypothesis, SS_H, with the error variation under the full model, SS_E, using the overall pivot statistic

$$F = \frac{SS_H / df_H}{SS_E / df_E} \; ,$$

which is distributed as F_{df_H, df_E} under the null hypothesis H_0. Here $df_H = a - 1$ and $df_E = n_{.} - a$.

Now consider p multiple responses. For each experimental unit (i, j), there is a vector of responses $\{y_{ijk}\}_{k=1}^{p}$ which has mean $\{\mu_{ik}\}_{k=1}^{p}$. The

OVERALL TESTS FOR GROUP DIFFERENCES 277

univariate model for the kth response,

$$y_{ijk} = \mu_{ik} + e_{ijk} , \ e_{ijk} \sim N(0, \sigma_k^2) ,$$

has independent errors across the n_i experimental units. However, multiple responses on the same experimental unit may be correlated.

Example 18.1 Size: Consider an experiment to compare growth in three species (or groups). There may be several correlated measures of 'size' and several other correlated measures of 'shape'. Key questions in the experiment may revolve around finding evidence of group differences in terms of 'size' and 'shape', and finding a linear combination of responses which characterize each of these differences. ◇

The **composite null hypothesis** of no difference among groups for the kth response is

$$H_0 : \mu_{1k} = \cdots = \mu_{ak} \ \text{or} \ H_0 : \mu_{ik} = \bar{\mu}_{.k} .$$

Separate tests could be employed for each of the p multiple responses. That is, partition the sums of squares,

$$SS_T(\mathbf{y}_k) = SS_H(\mathbf{y}_k) + SS_E(\mathbf{y}_k) ,$$

with $\mathbf{y}_k = \{y_{ijk}\}_{i,j=1}^{a,n_i}$ being the vector of measurements for the kth response. The kth pivot statistic is

$$F(\mathbf{y}_k) = \frac{SS_H(\mathbf{y}_k)/df_H}{SS_E(\mathbf{y}_k)/df_E} ,$$

which is again distributed as F_{df_H, df_E} under the null hypothesis H_0. However, these p pivot statistics are not independent if the responses are correlated.

Example 18.2 Path: Consider the path coefficients for two correlated responses as shown in Figure 18.1,

$$\begin{aligned} Y_1 &= p_{a1}A + p_{e1}E_1 \\ Y_2 &= p_{a2}A + p_{e2}E_2 \end{aligned}$$

with explained variation for each response being

$$\begin{aligned} 1 - p_{e1}^2 &= p_{a1}^2 + 2p_{a1}p_{e1}r_{12}p_{a2}p_{e2} \\ 1 - p_{e2}^2 &= p_{a2}^2 + 2p_{a1}p_{e1}r_{12}p_{a2}p_{e2} .\end{aligned}$$

That is, there is a common portion of the explained variation due to the correlation r_{12} between the errors. ◇

It may be valuable to consider some or all of the p multiple responses together. One way to do this is to examine questions about linear com-

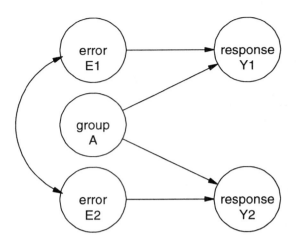

Figure 18.1. *Path diagram for multivariate analysis. One factor may affect multiple responses that are correlated. Correlation is modeled through the error structure. See Figure 16.2, page 245, for further details.*

binations across the responses. Consider a 'new' response $\mathbf{v} = \{v_{ij}\}_{i,j=1}^{a,n_i}$ constructed as $v_{ij} = \sum_k b_k y_{ijk}$, with means and common variance being

$$E(v_{ij}) = \nu_i = \sum_k b_k \mu_{ik} \text{ and } V(v_{ij}) = \sum_k b_k^2 \sigma_k^2 + \sum_{k \neq l} b_k b_l \sigma_{kl} ,$$

in which $\sigma_{kl} = E(e_{ijk} e_{ijl})$ are the covariances between y_{ijk} and y_{ijl}. These new responses are independent since the $n.$ experimental units are independent. Inferential questions might focus on the null hypothesis of no group differences $\nu_i = \bar{\nu}.$, with the pivot statistic depending on the ratio of sums of squares between hypothesis and error for the new variable, $\lambda = SS_H(\mathbf{v})/SS_E(\mathbf{v})$. In fact, it would be possible to analyze one or more such new response vectors $\mathbf{v}$ in place of the original multiple responses $\{\mathbf{y}_k\}_{k=1}^p$.

Example 18.3 Wasp: Scientists studying social insects such as wasps and ants have noted that queen and worker castes can have very different sizes and shapes. For some species, the queens are simply larger, suggesting that they continue to grow with the same basic 'plan' as workers. They are just fed for longer. Jeanne, Graf and Yandell (1995) examined a wasp species which does not follow this pattern. They measured 13 responses on 50 workers and 50 queens. The measurements were coded based on body part – head (H), thorax (T), wing (W) or gonadium (G) – and kind of measurement – width (W), height (H) and length (L). The gonadium (also known as gastral tergite) has two length and three width measurements.

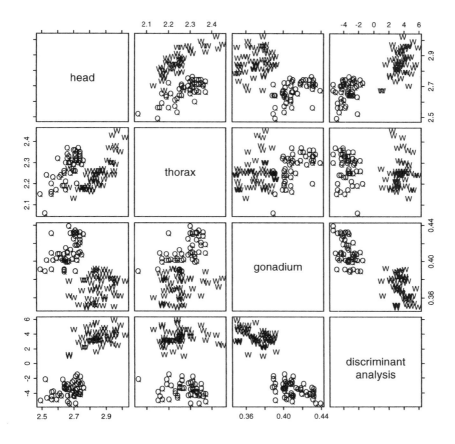

Figure 18.2. *Wasp scatter plots. Widths of head (HW), thorax (TW) and gonadium (G1Wa) are plotted against each other and against the best discriminator (DA). Plot symbols identify queens (Q) and workers (W).*

Figure 18.2 shows width measurements for head (HW), thorax (TW) and gonadium (G1Wa), along with the linear combination of the 13 responses that does the best to discriminate between queens and workers (DA). It would be easy to give the linear combination $\sum_k b_k y_{ijk}$ used to create the best discriminator DA, but the b_ks are not well determined because the responses are highly correlated. Instead, examine the correlations of the 13 measurements with DA. Figure 18.3 shows the correlations plotted against the t statistic for comparing queens and workers. Note that those responses that are most correlated with DA, regardless of sign, have high (absolute) values of t. In addition, the thorax width (TW) has the smallest (absolute) correlation and the smallest t statistic. ◇

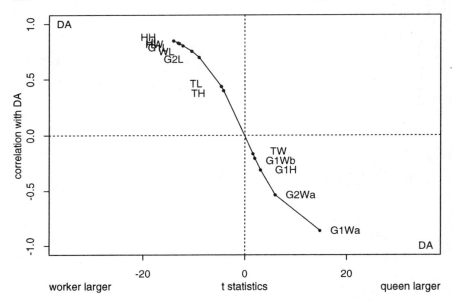

Figure 18.3. *Wasp discriminant analysis: t statistics to compare queens and workers are shown against correlation of each of the 13 measured responses with the best discriminator (DA). The three head measurements (HH, HL, HW) are highly correlated with each other, and hence appear on top of each other. DA is plotted as positive and negative to emphasize the arbitrariness of its sign.*

Example 18.4 Path: Figure 18.4 shows the new discriminant response **v** as a linear combination of multiple responses $\mathbf{y}_1$ and $\mathbf{y}_2$ to explain group differences. Building on the path coefficients for manova, the equation for the new response V can be presented as

$$\begin{aligned} V &= p_{1v}Y_1 + p_{2v}Y_2 \\ &= (p_{a1}p_{1v} + p_{a2}p_{2v})A + p_{e1}p_{1v}E_1 + p_{e2}p_{2v}E_2 \\ &= p_{av}A + p_{ev}E \ . \end{aligned}$$

The path coefficients are constrained such that V has variance 1:

$$p_{kv}^2 = \frac{b_k^2 \sigma_k^2}{b_1^2 \sigma_1^2 + b_2^2 \sigma_2^2 + 2b_1 \sigma_{12} b_2} \ .$$

The goal is to choose these path coefficients such that the explained variation is largest, or equivalently the unexplained variation p_{ev}^2 is as small as possible. In other words, pick p_{1v} and p_{2v} to minimize

$$p_{ev}^2 = (p_{e1}p_{1v})^2 + (p_{e2}p_{2v})^2 + 2p_{e1}p_{1v}r_{12}p_{e2}p_{2v}$$

with $r_{12} = \sigma_{12}/\sigma_1\sigma_2$ the correlation between responses. Notice that the proportion of unexplained variation is related to the ratio of hypothesis to error sums of squares: $p_{ev}^2 = 1/(1+\lambda)$. In addition there is a geometric interpretation. That is, the vector (p_{1v}, p_{2v}) lies along the minor axis of the elliptical contours for the joint distribution of $p_{e1}E_1$ and $p_{e2}E_2$. ◊

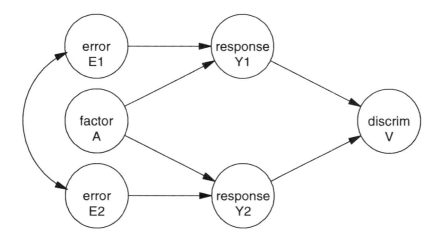

Figure 18.4. *Path diagram for discriminant analysis. The principal effect of factor* **A** *on multiple correlated responses can be summarized in a new discriminant variable* **V**. *Correlation is modeled through the error structure.*

The multivariate approach makes a suitable choice of weights $\{b_{1k}\}_{k=1}^p$ using matrix algebra such that the first canonical variate $\mathbf{v}_1 = \sum_k b_{1k}\mathbf{y}_k$ has the largest possible ratio of sums of squares

$$\lambda_1 = \frac{SS_H(\mathbf{v}_1)}{SS_E(\mathbf{v}_1)}.$$

Put another way, this choice of weights $\{b_{1k}\}_{k=1}^p$ maximizes the F ratio for testing the null hypothesis of no difference in group means. It is further possible with more than two groups ($a > 2$) to find constants $\{b_{2k}\}_{k=1}^p$ such that $\mathbf{v}_2 = \sum_k b_{2k}\mathbf{y}_k$ is uncorrelated with $\mathbf{v}_1$ and the ratio of sums of squares $\lambda_2 = SS_H(\mathbf{v}_2)/SS_E(\mathbf{v}_2)$ is maximized. This can be repeated until there are up to $s = \min(p, a-1)$ new **canonical variates** $\mathbf{v}_1, \cdots,$ $\mathbf{v}_s$, which are uncorrelated, with corresponding ratios of sums of squares in decreasing order of the **eigenvalues** $\lambda_1 \geq \cdots \geq \lambda_s \geq 0$.

The overall composite null hypothesis of no difference among groups considers the null hypotheses for all s multiple responses simultaneously,

$$H_0 : \nu_{il} = \bar{\nu}_{\cdot l} \text{ for } l = 1, \cdots, s,$$

with linear combinations $\nu_{il} = \sum_k b_{lk}\mu_{ik}$ and weights $\{b_{lk}\}_{k=1}^p$ chosen as described in the previous paragraph. This composite hypothesis is multivariate in nature and not readily reduced to a single test statistic. However, the eigenvalues λ_l contain the key information about the compromise between explained and unexplained variation, as argued in the next section. Four methods have been widely used in practice to summarize this information. Each test has its strengths and weaknesses. In practice, it is wise to consider all four in conjunction with univariate inference. As always, take time to examine plots, including pairwise scatter plots of responses with symbols for groups.

Roy's greatest root corresponds to the most significant linear combination,
$$\text{ROY} = \lambda_1 = \frac{SS_H(\mathbf{v}_1)}{SS_E(\mathbf{v}_1)} .$$
It is the most powerful if in fact there is only one dimension to the differences among groups. The other three methods use all the eigenvalues in one way or another. Wilks's likelihood ratio criterion,
$$\text{WILKS} = 1/\prod_{l=1}^s (1+\lambda_l) = \prod_{l=1}^s \frac{SS_E(\mathbf{v}_l)}{SS_T(\mathbf{v}_l)} ,$$
is the product of proportions of unexplained variation across the canonical variates. It can be derived from likelihood principles, and is therefore appealing from a theoretical perspective. However, it is no better or worse than other statistics in practice. The Hotelling–Lawley trace,
$$\text{HOTEL} = \sum_{l=1}^s \lambda_l = \sum_{l=1}^s \frac{SS_H(\mathbf{v}_l)}{SS_E(\mathbf{v}_l)} ,$$
is the sum of the ratios of sums of squares of hypothesis to error across the canonical variates.

The explained variations for each canonical variate are sometimes referred to as canonical correlations,
$$r_l^2 = \lambda_l/(1+\lambda_l) = \frac{SS_H(\mathbf{v}_l)}{SS_T(\mathbf{v}_l)} .$$
The Pillai–Bartlet trace, which is the sum of these proportions,
$$\text{PILLAI} = \sum_{l=1}^s r_l^2 ,$$
has intuitive appeal as a generalization of the sample multiple correlation averaged over multiple responses,
$$R_{\text{MULT}}^2 = \sum_{l=1}^s r_l^2/s = \text{PILLAI}/s .$$

If $p = 1$ or $a = 2$, these four overall statistics agree. However, their properties in general are not well understood. They have been studied through published simulations. All four appear to be robust to violations of assumptions (e.g. normality, equal variance) if sample sizes are equal. However, they all have difficulty as sample sizes become more unequal, with PILLAI being the most robust. No one statistic is 'best' in all situations. If differences can be explained in terms of a single linear combination of responses, then the order of statistics in terms of power is ROY > HOTEL > WILKS > PILLAI. However, this order is reversed if differences are more diluted, requiring consideration of several dimensions. In practice, it is advisable to consider several approaches in concert. That is, examine all four statistics, as well as univariate tests, to sort out whether or not significant differences exist. Contradictions between different approaches may indicate borderline significance, or may highlight lack of power of some statistics for particular forms of group differences.

18.2 Matrix analog to F test

The development in this section gives some indication why inference can be based solely on the eigenvalues λ_l. It can easily be skipped. More detail can be found in Morrison (1976) or Krzanowski (1990) or in SAS (see Littell, Freund and Spector 1991). Simultaneous inference for multiple responses involves considering a **matrix partition of sums of squares and crossproducts**

$$\mathbf{T} = \mathbf{H} + \mathbf{E}.$$

The $p \times p$ matrix $\mathbf{H}$ for variation and covariation explained by the hypothesis has elements

$$h_{kl} = \sum_i \sum_j (\bar{y}_{i \cdot k} - \bar{y}_{\cdot \cdot k})(\bar{y}_{i \cdot l} - \bar{y}_{\cdot \cdot l}),$$

with $h_{kk} = SS_H(\mathbf{y}_k)$. The matrix $\mathbf{E}$ similarly comprises error variation and covariation, and has elements

$$e_{kl} = \sum_i \sum_j (y_{ijk} - \bar{y}_{i \cdot k})(y_{ijl} - \bar{y}_{i \cdot l}),$$

with $e_{kk} = SS_E(\mathbf{y}_k)$. The analog to a ratio of variances in this multivariate situation is matrix multiplication by the inverse. Therefore, overall pivot statistics have traditionally been derived from the matrix

$$\mathbf{HE}^{-1}.$$

The rank of $\mathbf{H}$, and hence rank($\mathbf{HE}^{-1}$), is s. It is not necessary in practice to construct the inverse $\mathbf{E}^{-1}$. Instead, a statistical package solves the **characteristic equation**

$$(\mathbf{H} - \lambda_l \mathbf{E})\mathbf{b}_l = \mathbf{0}, \ l = 1, \cdots, s,$$

in terms of its s **eigenvectors** and corresponding **eigenvalues**. In fact, the lth largest eigenvalue is just the ratio λ_l described earlier, with corresponding eigenvector being the weights $\mathbf{b}_l = \{b_{lk}\}_{k=1}^{p}$. The eigenvalues contain the key information for overall tests of significant differences among groups.

No single statistic can summarize all the information contained in the eigenvalues of $\mathbf{HE}^{-1}$. In practice, the four statistics developed in the previous section are used in conjunction with univariate statistics. Here are the four statistics in terms of matrices (recalling that $\det(\mathbf{A})$ is the determinant and $\text{tr}(\mathbf{A})$ is the trace of a square matrix $\mathbf{A}$):

$$\begin{aligned} \texttt{ROY} &= \lambda_1 \\ \texttt{WILKS} &= \det(\mathbf{HE}^{-1} + \mathbf{I}) \\ \texttt{HOTEL} &= \text{tr}(\mathbf{HE}^{-1}) \\ \texttt{PILLAI} &= \text{tr}(\mathbf{H}(\mathbf{H}+\mathbf{E})^{-1}) \ . \end{aligned}$$

The distributions of these statistics are somewhat complicated. Each can be approximated by an F distribution (see Morrison 1976). The distribution of Roy's greatest root has a conservative upper bound which is exact only if $s = 1$. The exact distribution for Wilks's is known but is complicated if there are many degrees of freedom in hypothesis and error.

18.3 How do groups differ?

Evidence of differences suggests multiple comparisons of group means. It is natural and appropriate to examine individual responses as in the univariate anova. That is, for the kth response, examine contrasts of the form $\sum_i c_i \mu_{ik}$ with $\sum_i c_i = 0$. The same issues apply as found earlier with a single response, except now they are compounded by considering p multiple responses. Some experimenters prefer to control **experiment-wise error rate** per response. However, a BONFERRONI approach could be used across the p multiple responses.

Multiple comparisons can be employed on linear combinations of responses, $\sum_k b_k \sum_i c_i \mu_{ik}$, for pre-selected constants b_k. Strong evidence of significant differences in only one dimension suggests contrasts for the first canonical variate could be used, of the form $\sum_k b_{1k} \sum_i c_i \mu_{ik} = \sum_i c_i \nu_{i1}$. These are relatively robust to violations of assumptions. Other canonical variates could be examined with some caution. In practice, it is a good idea to examine the canonical variates to see if they are readily interpretable in terms of simple combinations of responses. That is, could the weights b_{lk} be rounded off, with some set to zero, to approximate an intuitive contrast? However, be somewhat cautious with data-determined (post-hoc) contrasts. Simulations suggest that contrasts based on the Pillai–Bartlet statistic are not very robust, even though the overall test is.

Discriminant analysis is a method to formalize the process of attributing differences in group means to a few linear combinations of responses. This involves interpreting each significant canonical variate $\{\mathbf{v}_l\}_{l=1}^{s}$ in terms of its ability to discriminate among the a groups and its association with the p multiple responses. Often, researchers prefer to employ a stepwise procedure first to reduce the number of multiple responses, and then interpret the canonical variates associated with this reduced set.

The basic idea in **canonical discriminant analysis** (e.g. proc candisc in SAS) is to use the data to suggest the number of significant canonical variates, or dimensions, required to discriminate among the a groups. Each significant variate $\mathbf{v}_l$ can then be examined to assess how groups differ using, for example, multiple comparisons on contrasts $\sum_i c_i \nu_{il}$. Usually the scientists is interested in determining what responses contribute to each significant canonical variate. The **canonical variate correlation** between each response $\mathbf{y}_k$ and each canonical variate $\mathbf{v}_l$ measures the **factor loading**, or degree of shared variation. These factor loadings indicate the responses that are most 'similar' to each canonical variate, ideally suggesting a 'candidate' response, or combination of responses, which could approximate that canonical variate. A valuable follow-up analysis would use only those candidate responses – how well do they do to discriminate among groups?

Example 18.5 Wasp: The scientist wanted to understand the differences in sizes and shapes between queens and workers without having to resort to fancy multivariate methods such as discriminant analysis. However, single measurements were not enough. The pairs plot in Figure 18.2 suggests considering two traits at a time. Since gonadium width (G1Wa) had the largest t statistic for comparing castes, it seemed natural to plot it against head width (HW), measured at the opposite end of the wasps. Figure 18.5(a) shows both responses along with parallel (solid) and separate (dotted) ancova lines of HW on G1Wa adjusting for caste. Table 18.1 summarizes formal tests, showing no evidence for parallel lines (all tables use adjusted, or Type III, sums of squares). It would appear that the evidence for caste differences after adjusting for G1Wa is modest ($p = 0.085$). However, refitting with parallel lines yields a highly significant adjusted caste effect ($p < 0.0001$). Thus gonadium width can explain some but not all of the differences in head width between queens and workers.

It seemed important somehow to remove size. Notice that thorax width (TW) had the smallest t statistic, indicating that it appears to measure size without regard to caste. Analysis of covariance of gonadium width (G1Wa) on TW adjusting for caste is highly significant. This further suggests adjusting head width (HW) by both G1Wa and TW to remove size and a major part of caste differences. Table 18.2 summarizes both operations. Notice

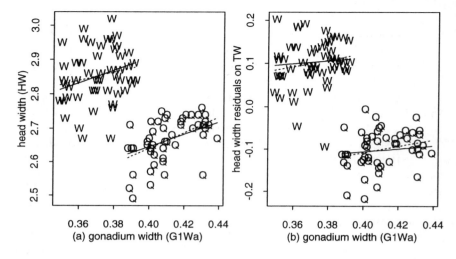

Figure 18.5. *Wasp analysis of covariance. (a) Gonadium width (*G1Wa*) can be partially explained by head width (*HW*) and* caste. *Parallel lines (solid) are adequate, but separate lines (dotted) for each* caste *are shown for comparison. (b)* G1Wa *and* HW *residuals after regressing on thorax width (*TW*) show no association adjusted for* caste. *Ordinary ancova lines (dotted) adjusting only for* G1Wa *and* caste *are visually appealing but incorrect; proper ancova lines adjusting for* caste *and both covariates (solid) of* G1Wa *and* TW *are shown evaluated at the mean value of* TW. *Plot symbols identify queens (*Q*) and workers (*W*).*

source	df	SS	F	p-value
G1Wa	1	0.0560	11.63	0.0010
caste	1	0.0146	3.04	0.085
caste*G1Wa	1	0.00582	1.21	0.27
error	96	0.462		

Table 18.1. *Wasp ancova on head width*

that after adjusting for size (TW) and caste, there is neglible evidence for an association between G1Wa and HW. Indeed, there is still strong evidence for head width differences between castes after adjusting for size and gonadium width.

Figure 18.5(b) displays head width after removing size. That is, the responses are the residuals from regression of HW on TW. The dotted parallel lines were fitted ignoring TW, which turns out to be incorrect. More properly, it is better to perform ancova of the HW residuals on TW, G1Wa and caste, which yields the solid lines evaluated at the mean size. Table 18.3 shows the improper and proper ancova on residuals. The degrees of freedom and

source	df	SS	F	p-value
G1Wa on TW				
TW	1	0.00267	16.83	0.0001
caste	1	0.0358	225.34	0.0001
error	97	0.0154		
HW on TW and G1Wa				
TW	1	0.230	93.21	0.0001
G1Wa	1	0.00176	0.71	0.40
caste	1	0.349	141.21	0.0001
error	96	0.237		

Table 18.2. *Wasp two-covariate ancova*

source	df	SS	F	p-value
Improper:				
G1Wa	1	0.0103	3.90	0.051
caste	1	0.412	156.52	0.0001
error	97	0.255		
Proper:				
TW	1	0.0179	7.27	0.0083
G1Wa	1	0.00176	0.71	0.40
caste	1	0.349	141.21	0.0001
error	96	0.237		

Table 18.3. *Wasp ancova on residuals*

adjusted sums of squares and test statistics for the proper ancova agree with those in Table 18.1. Note that the adjusted sum of squares for TW is not zero; the slope for TW is modified by the other covariate G1Wa and the factor caste. ◇

An alternative approach uses **stepwise discriminant analysis** (e.g. proc stepdisc in SAS) to decide first on a subset of the p multiple responses which adequately discriminate among the a groups. Formally, this is equivalent to the forward selection procedure for stepwise multiple regression when there are only $a = 2$ groups. More generally, the first step finds that response which has the most significant univariate F test for group differences. Subsequent tests find the response which has the largest

F test of the hypothesis that group means are equal after adjusting for all other responses already included. Stepwise discriminant analysis packages provide convenient summaries of this process. However, the details can be traced by recasting the problem in terms of the **analysis of covariance**.

Suppose that the responses $\mathbf{y}_1, \cdots, \mathbf{y}_k$ have already been included in earlier steps. (That is, relabel the responses so that the first k have been chosen already.) The next response to be considered would be the one ($\mathbf{y}_l$) that has the largest test of significance of the hypothesis $H_0 : \alpha_i = 0$ for the model

$$y_{ijl} = \mu + \alpha_i + \sum_{l=1}^{k-1} \beta_l y_{ijl} + e_{ij} .$$

If no further responses are significant, the stepwise procedure halts. Stepwise discriminant analysis is usually followed by canonical discriminant analysis to examine group differences in detail.

Example 18.6 Size: The scientist might first reduce the dimensionality by stepwise discriminant analysis, ending up with one size measurement and two shape measurements which contribute significantly. Canonical discriminant analysis following this reduction shows that the first canonical variate is essentially the size measurement, while the second is approximately an average of the two shape measurements. Analysis with the one size and the average of the two shape measurements shows that they do an adequate job of discriminating among groups. The data analysis suggests the scientist need only take three measurements. ◊

The empirical selection of a subset can greatly reduce the dimensionality and, ideally, enhance interpretation. However, the choice of a subset of multiple responses that are highly correlated is somewhat arbitrary. The same issues that arise in stepwise multiple regression recur here. For instance, a backward elimination procedure might result in a different subset of responses.

Scientists (and statisticians) can easily be confused by the effect of correlation among multiple responses on the weights b_{lk} in the linear combinations. These weights b_{lk} indicate the degree of necessity, or unique contribution, of a response $\mathbf{y}_k$ to the canonical variate $\mathbf{v}_l$, but they are not directly interpretable as estimates. Further, they can change dramatically if a different subset of multiple responses is used.

Sometimes, easy interpretation is not possible. There is no guarantee that multivariate analysis leads to an understandable explanation. In fact, it could lead to two or more contradictory models. However, this is the nature of scientific investigation. The challenge involves designing a new experiment to sort things out next time.

18.4 Causal models

Most experiments are built on knowledge gained from earlier experiments under somewhat similar conditions. Further, scientists often draw on several types of evidence about a natural process. While formal statistical data analysis cannot attribute cause and effect, it can be guided by evidence from other studies which suggest causal relationships. This section explores causal models using analysis of covariance and step-down analysis. Connections to methods examined in the previous section are briefly made to illustrate how path analysis can be used to display relationships graphically and model assumptions.

Example 18.7 Path: Analysis of covariance can be interpreted in terms of cause and effect if there is outside information about the way factor and covariate influence response. The ancova model can be written in terms of path coefficients as

$$Y = p_a A + p_x X + p_e E .$$

However, interpretation depends on the relationship between A and X as depicted in Figure 18.6. If the design is balanced in the sense presented in the previous chapter, then interpretation is unambiguous, as in Figure 18.6(a). In terms of path coefficients, the proportion of explained variation can be partitioned as

$$1 - p_e^2 = p_a^2 + p_x^2 .$$

However, interpretation in unbalanced analysis of covariance requires some care. One approach, displayed in Figure 18.6(b), considers the association between factor A and covariate X to be undirected. The explained variation includes

$$1 - p_e^2 = p_a^2 + p_x^2 + 2p_a r_{ax} p_x$$

in which r_{ax} is a measure of correlation between A and X. The first two terms connote the variation explained respectively by A or X after adjusting for the other, while the last term $2p_a r_{ax} p_x$ represents the proportion of variation explained by the synergy (if positive) or antagonism (if negative) between A and X. Depending on the experiment, it may be convenient to consider the factor as driving the covariate, as in Figure 18.6(c). In this case, the explained variation is

$$1 - p_e^2 = p_a^2 + p_x^2 + 2p_a p_{ax} p_x$$

in which p_{ax} is the path coefficient from A to X. That is, p_{ax} is the proportion of variation in X explained by the factor A, using a one-factor analysis of variance. If the covariate is measured before the experiment, it may drive the response and much of the factor effect, as in Figure 18.6(d). In terms of explained variation,

$$1 - p_e^2 = p_a^2 + p_x^2 + 2p_x p_{xa} p_a$$

in which $p_x(p_x + 2p_{xa}p_a)$ is the variation in Y explained by the regression of Y on X alone. In other words, p_x^2 is the proportion of variation explained directly by X while $2p_x p_{xa} p_a$ is attributed to the indirect effect of X on Y through factor A. The values of r_{ax}, p_{ax} and p_{xa} are identical; the difference lies in the interpretation in terms of cause and effect. ◇

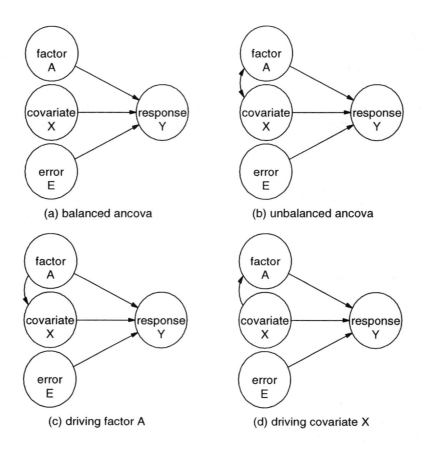

Figure 18.6. *Path diagrams for analysis of covariance. Factor* A *and covariate* X *may affect response* Y *in a variety of ways depending on the balance of the covariate (a,b) and whether the factor or covariate is driving the other (c,d).*

Designed experiments involving two or more responses could be interpreted as simply having correlated responses. Alternatively, consider arbitrarily interpreting the first response as a driving covariate. The effect of the factor on the second response would be examined after adjusting for the first as if it were a covariate. The effect of the first response could be exam-

CAUSAL MODELS

ined using analysis of variance, as shown in Figure 18.7. This corresponds to the two models

$$y_{ij2} = \mu_{i2} + \beta y_{ij1} + e_{ij2},$$
$$y_{ij1} = \mu_{i1} + e_{ij1}.$$

Here, there is a distinct sense of cause and effect. That is, the first response y_1 measures a process which may be affected by the group (μ_{i1}). This first process may in turn affect the second process, as measured by the second response y_2. In addition, the second response may be further influenced by the group (μ_{i2}).

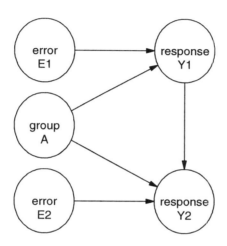

Figure 18.7. *Path diagram for step-down analysis. Factor A can affect both responses; in addition the first response drives the second. Note that errors are independent.*

Example 18.8 Path: Step-down ancova, shown in Figure 18.7, has path coefficients and models

$$Y_1 = p_{a1}A + p_{e1}E_1$$
$$Y_2 = p_{a2}A + p_{12}Y_1 + p_{e2}E_2.$$

The explained variation for the responses is

$$1 - p_{e1}^2 = p_{a1}^2$$
$$1 - p_{e2}^2 = p_{a2}^2 + 2p_{a1}p_{12}p_{a2}$$

with p_{12} the path coefficient from Y_1 to Y_2. The interpretation of the paths concerning Y_2 is identical to analysis of covariance. ◊

This idea can be generalized to consider p multiple responses if there is a natural order of cause and effect for the responses, based on outside knowledge from other studies. The relative contribution of each response can be considered in the specified order after adjusting for earlier measurements. This **step-down analysis** is strongly dependent on the scientist's belief about cause and effect. However, statistical methods cannot test this belief – they can only use it as a model assumption to guide data analysis. That is, consider the following models in order,

$$y_{ijp} = \mu_{ip} + \sum_{k=1}^{p-1} \beta_k y_{ijk} + e_{ijp}$$
$$\vdots$$
$$y_{ij2} = \mu_{i2} + \beta y_{ij1} + e_{ij2}$$
$$y_{ij1} = \mu_{i1} + e_{ij1}$$

Note that the errors between models are assumed to be independent. The natural approach to investigating group effects in a step-down analysis is to use **backward elimination**. That is, consider the response $\mathbf{y}_p$ in an analysis of covariance by group adjusting for the other $p-1$ responses as covariates. If the pth response is not significant in the presence of the other $p-1$, it is dropped and the $(p-1)$th is considered as response. The last step is a univariate anova on the first response by group.

Questions arise about how to assess significance in this form of multistage testing in order to control Type I error. These issues are similar to ones encountered in multiple comparisons and model selection. The step-down approach could be enhanced somewhat by examining whether some intermediate responses are important to overall assessment of group differences; however, it is not obvious how this affects significance levels in formal inference.

This approach appears on the surface to be identical to discriminant analysis considered in the previous section. There are in fact close connections between the mathematical models, although the formal criteria for selection or elimination at each step could be different. However, conceptually they are quite distinct. Discriminant analysis assumes no causal relationships, letting the data suggest which responses are important and in which order. The step-down analysis assumes a specific ordering for responses from the start.

Example 18.9 Path: The effect of experimental factors may only reach the responses indirectly through latent variables, as in Figure 18.8. In this case, the relationships may be represented as

$$\begin{aligned} V &= p_{av}A + p_{ev}E \\ Y_1 &= p_{v1}V + p_{e1}E_1 = p_{av}p_{v1}A + p_{ev}p_{v1}E + p_{e1}E_1 \\ Y_2 &= p_{v2}V + p_{e2}E_2 = p_{av}p_{v2}A + p_{ev}p_{v2}E + p_{e2}E_2 \ . \end{aligned}$$

PROBLEMS

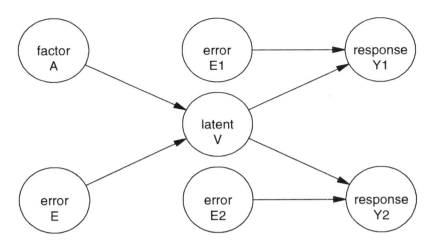

Figure 18.8. *Path diagram for latent variable. Factor A affects (unobservable) latent variable V which in turn affects multiple responses. Thus correlation among responses is modeled through the latent variable.*

The correlation between the responses is thus $p_{v1}p_{v2}$, which can be partitioned as that explained by the factor $(p_{v1}p_{av}^2 p_{v2})$ and unexplained latent error $(p_{v1}p_{ev}^2 p_{v2})$. Since errors are assumed to be independent, the explained variation for the ith response is $(p_{av}p_{vi})^2$. Thus these models have nice interpretability. ◇

Structural models using latent variables have become increasingly popular in recent years (Bollen 1989). The concept involves supposing that the correlation among multiple responses can be attributed to some 'latent' variables that cannot be directly observed. Readers interested in more information may want to consult Bollen (1989) and learn about the LISREL software package (Jöreskog and Sörbom 1988; Byrne 1989).

18.5 Problems

18.1 Wasp: Your job is to provide statistical evidence to support the claim that queens are not simply larger workers – they likely differed in development at a very early stage. The first task is to reduce the 'dimensionality' of the data.
(a) Overall test: Conduct univariate analyses on all responses to assess worker-queen differences. Conduct a multivariate anova for same. Report test results.
(b) Use a stepwise discriminant analysis procedure (e.g. `proc stepdisc` or

proc ancova in SAS) to a subset of significant responses as discriminators. Briefly summarize these results.

(c) Head width (HW) and gonadium width (e.g. G1Wa) appear to be the most significant discriminators, as found in (b). Plot them against each other to note that one is larger while the other is smaller in queens than workers. Use analyses so far, along with correlations and partial correlations (adjusting for group) among the multiple responses, to argue that these two responses appear to capture the main features.

(d) Conduct a canonical discriminant analysis to examine the 'best' linear combination of responses to discriminate between workers and queens. How much better does it do than any one response? What are the canonical variate correlations (see Total Canonical Structure in the SAS listing) with the two responses considered in (c)? Why is one positive and one negative?

(e) Plot the canonical variate against the two responses in (c). Briefly interpret in light of (d) above.

18.2 Wasp: The scientist thought it might be important first to remove 'size' in order to isolate 'shape' differences between workers and queens. He suggested using TL, as it is one of the poorest discriminators – that is, the distributions of TL substantially overlap for queens and workers.

(a) Conduct an analysis of covariance of the two responses HW and G1Wa on size (TL) and group. Are there group differences after adjusting for size?

(b) Remove the effect of size and conduct a stepwise discriminant analysis. Compare results with Problem 18.1(b).

(c) Conduct canonical discriminant analysis for the residuals with size removed. How do the canonical variate correlations compare with Problem 18.1(d)?

(d) Plot the canonical variates and the residuals for the two responses from Problem 18.1(c) against each other. Briefly compare and contrast with earlier plots.

PART G

Deciding on Fixed or Random Effects

Are effects fixed or random? What information in an experiment suggests one or the other? How does this affect data analysis? This part examines the nature of random effects on their own and in conjunction with fixed effects in designed experiments.

The experimental process involves several sources of variation, or random effects (Chapter 19). While certain components of this variation are embedded in the unexplained 'error', some model factors may have randomly selected levels, or elements, drawn from a larger population rather than fixed by the scientist. Questions arise as to the proper measures of variation to use for particular comparisons. In some cases, this depends on how the scientist views the experiment at hand. Are certain factors of inherent interest, or are they best viewed as a random sample?

Random effects models can get rather complicated. However, Chapter 20 presents some ways to organize what is known about the experiment. This helps simplify both presentation and analysis. The main ideas are presented in terms of sums of squares, to draw analogy to earlier developments in this book. A general treatment must ultimately appeal to other approaches such as maximum likelihood.

Most experiments have a mix of fixed and random effects. The formal development of mixed effects models in Chapter 21 provides some guidelines which will be used heavily in the later chapters. Care is needed, as the presence of some random effects can dramatically alter the form of inference for fixed effects.

CHAPTER 19

Models with Random Effects

This chapter introduces different sources of random variation, examining how these arise from design considerations and how they affect the model form and data analysis. A survey of milk production across the state of Wisconsin might select a few counties at random, and then several farms at random from lists for each county. Any measurements on milk production per cow depend on county to county variability as well as farm to farm variability within counties. This approach differs markedly from an experiment designed to compare milk production in preselected counties.

Thus it is important to distinguish between a **random factor**, comprising a random sample of levels from a population of possible levels, and a **fixed factor**, in which levels are selected by a non-random process or encompass the entire population of possible levels.

The models considered up to this point have consisted of fixed effects and one random factor, unexplained error variation. That is, a random sample of experimental units for each factor combination level provides information about those levels. The key questions for **fixed effects** models concern factor means and effects, possibly in the form of main effects and interactions.

This chapter introduces **random effects models**, in which all factors (except the grand mean) are random. The marginal mean of the random sample of levels for a factor varies from sample to sample around some expected value (the grand mean). Interest lies in how variable the response is across the sampled levels, examining the population variance for that factor. If this variance is small, then this factor is probably negligible. In other words, a random factor with a small **variance component** is not very sensitive and may be ignored in design of future experiments. Some studies (e.g. genetics) may mainly concern estimation of these variance components.

The single-factor random model for an observed response leads to a partition of variation amenable to an anova table very similar to the single-factor fixed model considered earlier. The principal difference lies in interpretation, which is embodied formally in hypotheses and the expected mean squares, $E(MS)$. The latter suggests natural tests and estimates of variance components and raise some non-intuitive issues about estimating the grand mean which have implications for sub-sampling.

This chapter concerns a model with one random effect besides error. Section 19.1 develops the single-factor random model. Section 19.2 develops a test for variation among classes, drawing analogies to comparison of group means in the fixed model. The next section examines the distribution of the partitioned sums of squares for error and for classes, leading to tests in Section 19.4. Estimation of variance components and the intraclass correlation is considered briefly in the next section, followed by a discussion of estimating the grand mean and predicting class means or effects.

For clarity, the text uses Greek letters (μ, α, β) for fixed effects and Latin letters (m, a, b) for random effects, with the exception that errors are still signified by e. Levels of random effects are called classes to distinguish them from fixed factor levels.

19.1 Single-factor random model

The one-factor random model includes two random mechanisms which may affect the response. It is important first to understand the experimental design, how randomness is introduced, in order to grasp the types of questions and methods of analysis appropriate to such a design.

Example 19.1 Random: Consider a population of **classes**, say the counties of Wisconsin, with an arbitrary class in this population labeled by u. This class represents a sub-population of **elements**, say the population of farms in county u. If this class is selected for the experiment, an independently selected random sample of its elements (farms) would be selected for measurement, for instance of annual milk production per farm. The key questions concern estimating the average response across the population (state) and understanding the sources of variation in this measurement. Note that a different way to run the experiment would be to take a random sample of all elements (all farms in the state), without regard to which class (county) they come from. This latter design structure is completely randomized, while the former has two stages of randomization. Substantial variation among classes (counties) would be confounded with element-to-element variation (among farms) in the latter design. $\Diamond$

Consider the response y of an element drawn at random from a class u. The expected response could depend on the class,

$$E(y|\text{ class } u) = M(u) ,$$

with variance among elements within the class

$$V(y|\text{ class } u) = \sigma^2(u) = \sigma^2 .$$

SINGLE-FACTOR RANDOM MODEL

For simplicity, assume equal variance across classes. This suggests a class means model for the response,

$$y = M(u) + e, \quad e \sim N(0, \sigma^2).$$

It is helpful for modelling purposes to consider the class random effects, $A(u) = M(u) - \mu$. A randomly chosen element of class u would have response y which could be partitioned as

$$y = \mu + A(u) + e.$$

The one-factor random model of a response looks very similar to the one-factor fixed model developed earlier. However, in that setting, the factor level i and its mean response $\mu_i = \mu + \alpha_i$ were of direct interest to the scientist. Here, the u is chosen at random, with class mean $M(u)$, and the focus is on variation among the population of classes.

Consider drawing a single class at random from a population of classes. The mean response for this random class U varies around some grand mean μ for the population of all elements, with some variance σ_A^2,

$$E[M(U)] = \mu, \quad V[M(U)] = \sigma_A^2.$$

Now suppose there is a mechanism for randomly choosing a class U and its corresponding class mean $M(U)$. With appropriate assumptions, this random class mean is normally distributed,

$$M(U) \sim N(\mu, \sigma_A^2).$$

For instance, suppose the class labels u are real numbers having a uniform distribution and that $M(U)$ is randomly drawn from $N(\mu, \sigma_A^2)$. If $U \sim U(0,1)$ and Φ is the cumulative distribution function of the standard normal distribution, then

$$M(U) = \mu + \sigma_A \Phi^{-1}(U) \sim N(\mu, \sigma_A^2).$$

This argument needs to be modified somewhat if the population of class labels is finite. Essentially, suppose that the distribution of $M(U)$ can be reasonably approximated by normality.

Example 19.2 Random: Consider simulated normal data of ten classes with ten elements per class, with $\mu = 100$, $\sigma = 5$ and $\sigma_A = 20$. A histogram of the 100 elements looks rather non-normal, until one decomposes it into class means and residual elements, as shown in Figure 19.1. Some other representations using box-plots and trees are shown in Figure 19.2. ◇

If the experiment involves taking a simple random sample of classes, then the grand mean is the 'average' of class means,

$$\mu = E[M(U)] = \bar{M}(\cdot) = \sum_{\text{class } u} M(u)/\{\text{number of classes}\}$$

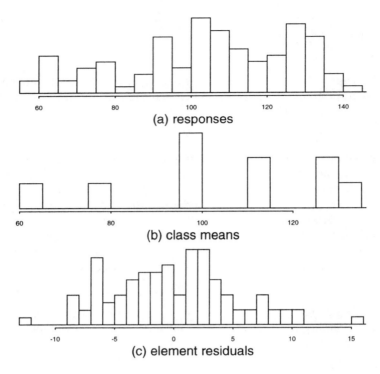

Figure 19.1. *Random model histograms for single factor. Plotting the raw responses (a) can give a misleading idea about distributions. The class means (b) show considerable spread relative to the element residuals (c).*

for a finite population of classes, or

$$\mu = E[M(U)] = \bar{M}(\cdot) = \int M(u)du$$

for an infinite population indexed by $u \in (0,1)$.

An experiment in this framework would draw classes at random and then draw elements at random from the selected classes. That is, pick a random sample of classes $\{U_i, i = 1, \cdots, a\}$. For class U_i, the mean response is

$$M_i = M(U_i) \ .$$

For each i, independently select n_i elements at random from class U_i. Let y_{ij} be the response for the jth element from the ith class. Write the **one-factor random model** as

$$y_{ij} = M_i + e_{ij} \ , \text{ independent } M_i \sim N(\mu, \sigma_A^2), \ e_{ij} \sim N(0, \sigma^2) \ ,$$

SINGLE-FACTOR RANDOM MODEL

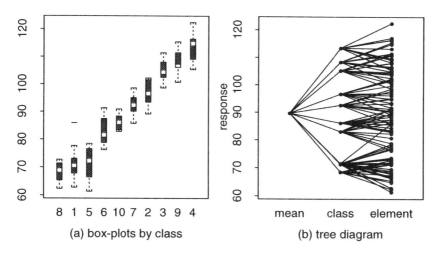

Figure 19.2. *Random model box-plots and tree diagram. Separate box-plots by class ordered by median (a) highlight spread of classes and elements within classes. Tree diagram (b) shows similar information, with lines connecting elements to classes and classes together to the grand mean.*

or in terms of random effects, with $A_i = A(U_i)$,

$$y_{ij} = \mu + A_i + e_{ij} \text{ , independent } A_i \sim N(0, \sigma_A^2), \ e_{ij} \sim N(0, \sigma^2) \ ,$$

for $i = 1, \cdots, a$, and $j = 1, \cdots, n_i$, with independent class random effects and random errors. Note that $E(A_i) = \bar{A}(\cdot) = 0$, but in general $\sum A_i \neq 0$.

Normality is not required in the random mechanism, but it plays a central role for inference. Inferential tools are built on sums of squares (more generally, quadratic forms) which have χ^2 distributions (more generally, weighted sums of non-central χ^2s) if the random effects are normally distributed. Mild violations of the **normal assumption** are usually not a problem, but major violations would require transformations or rethinking of the inferential framework.

Care is needed to determine the **variance** of the response y_{ij} as it depends on two variances,

$$V(y_{ij}) = V(\mu + A_i + e_{ij}) = V(A_i) + V(e_{ij}) = \sigma_A^2 + \sigma^2 \ .$$

While observations from different classes are uncorrelated, those from the same class are dependent, with covariance

$$\text{cov}(y_{i1}, y_{i2}) = \text{cov}(A_i + e_{i1}, A_i + e_{i2}) = V(A_i) = \sigma_A^2 \ .$$

The distribution of **sample class means** in random models is rather different from those in fixed models. They have expectation $E(\bar{y}_{i\cdot}) = \mu$ and

source	df	SS	MS = SS/df	F
A	$a-1$	$\sum_i n_i(\bar{y}_{i\cdot} - \bar{y}_{\cdot\cdot})^2$	MS_A	$MS_A/\hat{\sigma}^2$
error	$n. - a$	$\sum_i \sum_j (y_{ij} - \bar{y}_{i\cdot})^2$	$MS_E = \hat{\sigma}^2$	–
total	$n. - 1$	$\sum_i \sum_j (y_{ij} - \bar{y}_{\cdot\cdot})^2$	–	–

Table 19.1. *Random one-factor anova*

variance
$$V(\bar{y}_{i\cdot}) = V(A_i) + V(\bar{e}_{i\cdot}) = \sigma_A^2 + \sigma^2/n_i \ .$$
The **sample grand mean** has expectation $E(\bar{y}_{\cdot\cdot}) = \mu$ and variance
$$V(\bar{y}_{\cdot\cdot}) = V\left(\sum_i n_i A_i / n.\right) + V(\bar{e}_{\cdot\cdot}) = \sigma_A^2 \left(\sum_i n_i^2 / n_\cdot^2\right) + \sigma^2/n. \ ,$$
which simplifies for equal sample sizes $(n_i = n)$ to
$$V(\bar{y}_{\cdot\cdot}) = \sigma_A^2/a + \sigma^2/na \ .$$

19.2 Test for class variation

Comparison of class means involves questions about the magnitude of class variance σ_A^2 relative to the error variance σ^2, rather than questions of effects. Nevertheless, the partition of total variation is exactly the same for the one-factor random model as for the one-factor fixed model. Variation explained by classes is mathematically the same as variation explained by factor levels in the one-factor fixed model.

SS_A and SS_E are **statistically independent** in the random model in much the same way as in the fixed model, with the assumption of independence of random effect A_i and error e_{ij}. Following an identical geometric argument, the response space is partitioned into a mean space (of dimension 1), a class space (dimension $a-1$) and the error space (dimension $n.-a$). The squared length of the vector between **y** and its projection into the class+mean space is SS_E, while the squared length of the further projection down to the mean space is SS_A.

The **anova table** is exactly the same for fixed and random effects (Table 19.1). The difference between **hypothesis testing** in fixed effects and random effects models lies in the interpretation. The F test for factor in the

one-factor fixed model tests whether the treatment means are the same,
$$H_0 : \mu_i = \mu \ .$$
This is reflected in the **expected mean square** for the factor
$$E(MS_A) = \sigma^2(1 + \delta^2/(a-1))$$
with non-centrality parameter
$$\delta^2 = \sum_i n_i(\mu_i - \tilde{\mu}.)^2/\sigma^2$$
where $\tilde{\mu}. = \sum_i n_i\mu_i/n.$ as before.

The class means in the random model all have the same expectation. Questions revolve around whether there is any significant variation among class means in the population,
$$H_0 : \sigma_A^2 = 0 \ .$$
This hypothesis is consistent with expected mean square,
$$E(MS_A) = \sigma^2 + n_0\sigma_A^2 \ ,$$
with $n_0 = n$ for equal sample size, or in general,
$$n_0 = [n. - \sum_i n_i^2/n.]/(a-1) \ .$$
The total variation sum of squares can be partitioned again as
$$\sum_i \sum_j (y_{ij} - \bar{y}..)^2 = \sum_i n_i(\bar{y}_{i.} - \bar{y}..)^2 + \sum_i \sum_j (y_{ij} - \bar{y}_{i.})^2 \ .$$
The spaces are orthogonal by construction and the variates SS_A and SS_E are quadratic forms involving normal variables.

The usual **pivot statistic** $F = MS_A/MS_E$ for testing the hypothesis $H_0 : \sigma_A^2 = 0$ in the random model is
$$F \sim F_{a-1, n.-a} \text{ under } H_0 \ .$$
Under the null hypothesis of no level effect, for either fixed or random effects model, the distribution of the F statistic is $F_{a-1, n.-a}$. However, the distribution under any alternative is radically different for random effects than what was found for fixed effects. Detailed derivations are given for the random model in the next two sections after summarizing the key findings.

The distribution of F in the **fixed** model under an alternative of unequal group means was proportional to a non-central F distribution,
$$F \sim \sigma^2 F_{a-1, n.-a; \delta^2} \ ,$$
with non-centrality parameter δ^2 depending on the alternative as defined earlier. However, in the **random** model under the alternative of some class

variation, the distribution of F is proportional to a central F statistic only if the sample sizes are equal

$$F \sim \frac{(\sigma^2 + n\sigma_A^2)}{\sigma^2} F_{a-1,a(n-1)} \text{ provided } n_i = n \; .$$

When sample sizes are unequal, F has a bizarre distribution if $\sigma_A^2 > 0$. As shown in Section 19.3, MS_A is proportional to a weighted sum of χ^2 variates, which may be approximated by a non-central χ^2 in practice using the Satterthwaite method. Thus F can be approximated by a central F if the imbalance is not too great.

Discussion of **power** in practice is usually rather tentative and gets rather complicated for unbalanced experiments. For balanced experiments, power considerations for the fixed model revolve around the non-central F distribution, while power for the random model concerns the central F distribution, with a proportional term which is greater than unity. Note the tradeoff between spread in the factor levels ($\mu_i - \bar{\mu}.$ or σ_A^2, respectively), sample size per level (n) and error variance (σ^2). Quite often, discussions of power can be expressed in terms of the difference in mean response for two levels of the factor (whether fixed or random) relative to their standard error.

Power can be investigated for the unbalanced case in a certain way. Consider the weighted mean square for the factor

$$WMS_A = \sum_i w_i (\bar{y}_{i \cdot} - \hat{\mu}_B)^2 / (a-1) \; ,$$

with weights being the inverse of the variance of the sample class means, $w_i = n_i / (\sigma^2 + n_i \sigma_A^2)$, and $\hat{\mu}_B$ the best linear unbiased estimator (see Section 19.5). The statistic $F^* = \sigma^2 WMS_A / MS_E$ has an $F_{a-1, n. -a}$ distribution regardless of σ_A^2, and it reduces to F under the hypothesis $H_0 : \sigma_A^2 = 0$. Note that the statistic F^* changes as the hypothesis about σ_A^2 changes. This is a different approach from pivot statistics, which remain the same while the underlying distribution changes. See Searle, Casella and McCulloch (1992) for further details and references.

19.3 Distribution of sums of squares

This section uses a general result about sums of squares, developed in Example 4.13, to examine properties of their distribution. While the error sum of squares always has a nice form, the distribution of the model sum of squares can be quite complicated in the case of unbalanced data.

Under either the fixed or random model, the **error sum of squares** only concerns the errors e_{ij}. Thus the distribution of SS_E is related to a central χ^2,

$$SS_E \sim \sigma^2 \chi^2_{n. -a} \; .$$

DISTRIBUTION OF SUMS OF SQUARES   305

To show this, note that for each i, $e_{ij} \sim N(0, \sigma^2)$. Hence, using the above result with equal weights,

$$\sum_j (y_{ij} - \bar{y}_{i\cdot})^2 = \sum_j (e_{ij} - \bar{e}_{i\cdot})^2 \sim \sigma^2 \chi^2_{n_i - 1} .$$

Since these a terms are independent and the sum of independent χ^2 variables has a χ^2 distribution,

$$SS_E = \sum_i \sum_j (y_{ij} - \bar{y}_{i\cdot})^2 \sim \sigma^2 \chi^2_{n\cdot - a} ,$$

or, the error variance has distribution

$$MS_E = \hat{\sigma}^2 \sim \sigma^2 \chi^2_{n\cdot - a} / (n\cdot - a) ,$$

which has mean $E(\hat{\sigma}^2) = \sigma^2$ and variance $V(\hat{\sigma}^2) = 2\sigma^4/(n\cdot - a)$. This argument is identical for fixed and random models, provided the underlying data are normally distributed.

The **model sum of squares** SS_A is proportional to a central χ^2_{a-1} variate if the design is balanced and/or the null hypothesis of no class variation holds. Under the alternative of class variation ($\sigma^2_A > 0$) the distribution of SS_A is a weighted sum of χ^2 variates which can hopefully be approximated by a central χ^2.

Under the null hypothesis for the fixed or random model, the model sum of squares is proportional to a χ^2 variate,

$$SS_A \sim \sigma^2 \chi^2_{a-1} \text{ under } H_0 .$$

This is where the similarity ends. Under a fixed model, the distribution of SS_A depends on a non-central χ^2. In general, the distribution of SS_A for the random model is proportional to a weighted sum of χ^2, reducing in the balanced case to a central χ^2.

The distribution of SS_A in the random effects model with **equal sample sizes** is related instead to a central χ^2 distribution,

$$SS_A \sim (\sigma^2 + n\sigma^2_A) \chi^2_{a-1} .$$

The variance σ^2_A affects only the magnitude of proportionality. This follows from observing that the class means are distributed as

$$\bar{y}_{i\cdot} \sim N(\mu, \sigma^2_A + \sigma^2/n) ,$$

yielding a centered sum of squares,

$$SS_A = n \sum_i (\bar{y}_{i\cdot} - \bar{y}_{\cdot\cdot})^2 \sim (\sigma^2 + n\sigma^2_A) \chi^2_{a-1} .$$

Thus for a balanced experiment,

$$E(MS_A) = \sigma^2 + n\sigma^2_A \text{ and } V(MS_A) = 2(\sigma^2 + n\sigma^2_A)^2/(a-1) .$$

The distribution of SS_A in the random model is more complicated with **unequal sample sizes**. In general, SS_A is a quadratic form which is distributed as a weighted sum of central χ^2, with weights depending on sample sizes. It is possible to verify that SS_A is not in general proportional to a χ^2_{a-1} by computing the moments. The mean is

$$\begin{aligned} E(SS_A) &= \sum_i n_i V[\bar{y}_{i.}] - n.V[\bar{y}_{..}] \\ &= \sum_i n_i(\sigma_A^2 + \sigma^2/n_i) - n.(\sigma_A^2 \sum_i n_i^2/n_.^2 + \sigma^2/n.) \\ &= (a-1)(\sigma^2 + n_0\sigma_A^2) \ . \end{aligned}$$

After tedious calculations, the variance can be written as

$$V(SS_A) = 2(a-1)(\sigma^2 + n_0\sigma_A^2)^2 + 2v\sigma_A^4 \ ,$$

with $v = 0$ for the balanced case, but in general v is a complicated function of the sample sizes n_i,

$$v = \sum n_i^2 - 2\sum n_i^3/n. + \left(\sum n_i^2\right)^2 /n_.^2 - n_0^2 \ .$$

An experiment with only slight imbalance should have d near zero. Hence, the first two moments of SS_A would roughly agree with a variate proportional to χ^2_{a-1}. Thus the central χ^2 serves as a rough approximation using the **Satterthwaite** method. While some better approximations have been proposed (see Johnson and Kotz 1972), nothing seems entirely satisfactory. The degrees of freedom are close to $a-1$ for mildly unbalanced experiments, or more precisely

$$df_A \approx \frac{2[E(SS_A)]^2}{V(SS_A)} = a - 1\left[1 - v/\left(v + (a-1)(n_0 + \sigma^2/\sigma_A^2)^2\right)\right] \ ,$$

with an estimate usually made by substituting variance component estimates as discussed in the next section. Again, this approximation is generally poor for heavily unbalanced designs.

19.4 Variance components

Quick and easy estimates of variance components for random effects such as σ_A^2 are possible with the **method of moments**, or anova method, by simply matching up expectations. While anova estimators are unbiased and agree with more complicated estimators for balanced experiments, they tend to have unacceptably large variance for unbalanced experiments. In practice, **maximum likelihood** or **restricted maximum likelihood** are preferred for their superior properties, although they in general require intense computation.

The distribution of sums of squares for model and error suggests estimates of variance components. The unbiased estimate of error variance σ^2 is

naturally
$$\hat{\sigma}^2 = MS_E \sim \sigma^2 \chi^2_{n.-a}/(n.-a) ,$$
which has minimum variance if the data are normal. A confidence interval for σ^2 is derived in the same way as for the fixed effects model,
$$\left(\hat{\sigma}^2(n.-a)/\chi^2_{\alpha/2;n.-a},\ \hat{\sigma}^2(n.-a)/\chi^2_{1-\alpha/2;n.-a}\right) .$$

In practice, interval estimates of the error variance may not be a central concern. However, they can indicate the **precision**, or number of **significant digits**, that should be reported for estimates. For instance, the standard error of the variance, $\sigma^2\sqrt{2/(n.-a)}$, indicates that the error variance is known within a fraction of $\sqrt{2/(n.-a)}$. Thus if $n.-a$ is at least 200, then the SD is known to within 10%, or one digit. In practice, it is a good idea to report one extra digit: one digit for 1–20 degrees of freedom, two digits for 21–200 degrees of freedom and three digits for more than 200 degrees of freedom for error.

An unbiased estimate of class variance σ^2_A is
$$\hat{\sigma}^2_A = (MS_A - MS_E)/n_0 ,$$
which is minimum variance for normal data provided the experiment is balanced ($n_i = n_0 = n$). Note that this **anova**, or method of moments, estimator may be negative, particularly if the class variance is small relative to σ^2. Such a situation often (but not always) corresponds to a non-significant F test, which fails to reject the hypothesis $H_0: \sigma^2_A = 0$. A confidence interval for σ^2_A requires some approximation, even in the case of balanced data. The **Satterthwaite** method matches the first two moments of a χ^2 variate. Here the variance of the estimate of class variance is
$$V(\hat{\sigma}^2_A) = [V(MS_A) + V(MS_E)]/n_0^2 = 2\sigma^4_A/r .$$
The r approximate degrees of freedom for an approximate χ^2 statistic are
$$r = \frac{(a-1)(n.-a)n_0^2\sigma^4_A}{(a-1)\sigma^4 + (n.-a)(\sigma^2 + n_0\sigma^2_A)^2} .$$
The degrees of freedom r could be approximated by using variance component estimates $\hat{\sigma}^2$ and $\hat{\sigma}^2_A$ in the above equation. Alternatively, if n_0 is large (or all the n_i are large), r is approximately $a-1$. An approximate $(1-\alpha)$ confidence interval for σ^2_A can be constructed as
$$\left(\hat{r}\hat{\sigma}^2_A/\chi^2_{\alpha/2;\hat{r}},\ \hat{r}\hat{\sigma}^2_A/\chi^2_{1-\alpha/2;\hat{r}}\right) .$$

Several other estimators of σ^2_A have been proposed (see Searle, Casella and McCulloch (1992) for detailed reviews and derivations). Here we simply note some highlights. A maximum likelihood estimator might seem attractive, but it is biased and is not minimum variance for unbalanced experiments. Two other methods have had success, although they can be computer intensive. Both coincide with the anova estimator for balanced data.

The **minimum norm quadratic unbiased estimator (MINQUE)**, minimizes a squared norm (squared distance measure, or generalized variance). It has the advantage of not requiring the normal assumption. If normality is assumed, then MINQUE is also minimum variance, or **MIVQUE**. Unfortunately, MINQUE depends heavily upon initial estimates of variance components and can yield negative estimators. It is only unbiased when the initial estimates correspond to the true values! In practice, the initial estimates are taken as the method of moments estimates. Most packages do not handle this estimation adequately. SAS appears to work well in most situations.

The MIVQUE estimator can be improved by iteration (using MIVQUE estimates of variance components as initial estimates, etc.). It has been shown that iterating in this way leads ultimately to the **restricted maximum likelihood**. REML estimators of variance components have the nice property of being most probable given the data, subject to the restriction that variance component estimators be nonnegative. The approach requires the iterative solution of a nonlinear system of equations. REML estimators are unbiased but can be very computer intensive. However, in the balanced data case, REML agrees with anova and there is no costly iteration. Recent research (Jiang 1996a, b) has shown that REML and ML estimators have nice large-sample properties even if the data are not normally distributed.

Both REML and MIVQUE are available in some packages such as SAS. The SAS procedure `varcomp` only provides one iteration step for MIVQUE and uses this as the starting estimate for REML iteration. It also provides method of moments estimators for the anova approach (called `Type I`). The `random` statement in `proc glm` with the `test` option performs Satterthwaite approximations. The most satisfactory approach to date appears to be `proc mixed`, which uses REML for components declared in the `random` statement (see Littell et al. 1996). There are some concerns about the reliability of some iteration procedures employed for REML and MIVQUE when the design is severely out of balance (see Milliken and Johnson 1992, ch. 19, or Searle Casella and McCulloch 1992).

Reporting of class variances $\hat{\sigma}_A^2$ should reflect the precision, or number of significant digits. Many packages provide an estimate of the variance of the class variance, which can be used as with the standard error of $\hat{\sigma}^2$ discussed earlier to decide on the number of reportable digits. Roughly speaking, the variance is known to within a fraction of $\sqrt{2/\hat{r}}$.

The **intraclass correlation** between elements of the same class,

$$\rho = \mathrm{cov}(y_{i1}, y_{i2})/V(y_{i1}) = \sigma_A^2/(\sigma_A^2 + \sigma^2) \, ,$$

is the proportion of variation due to classes. This correlation is used quite heavily in genetics and to a lesser degree in other disciplines which employ

GRAND MEAN

random and mixed effects models. An unbiased **estimate of intraclass correlation** ρ based on the method of moments estimators of variance components is

$$\hat{\rho} = \hat{\sigma}_A^2/(\hat{\sigma}^2 + \hat{\sigma}_A^2)$$
$$= (MS_A - MS_E)/[MS_A + (n_0 - 1)MS_E] \,,$$

which would be negative whenever $\hat{\sigma}_A^2$ is negative. Improved estimators would use MIVQUE, or better yet, REML.

19.5 Grand mean

A natural estimate of the grand mean μ in a **balanced** experiment is

$$\hat{\mu} = \bar{y}_{..} \sim N(\mu, (\sigma^2 + n\sigma_A^2)/an) \,,$$

with corresponding pivot statistic for testing or for confidence intervals,

$$T = (\hat{\mu} - \mu)/\sqrt{MS_A/an} \sim t_{a-1} \,.$$

The naïve estimator of the grand mean for the **unbalanced** one-factor random model has variance

$$V(\hat{\mu}) = V(\bar{y}_{..}) = \frac{\sigma^2}{n_.} + \frac{\sigma_A^2(n_. - (a-1)n_0)}{n_.} \,.$$

A method of moments (anova) estimator of this variance is

$$\hat{V}(\hat{\mu}) = MS_A/n_. + (MS_A - MS_E) \times (a/n_. - 1/n_0) \,.$$

Note that if there is only slight imbalance, then $n_. \approx an_0$, and the second term is negligible. However, it is possible to get a negative estimate. The distribution of the pivot statistic

$$T = (\hat{\mu} - \mu)/\sqrt{\hat{V}(\hat{\mu})}$$

is complicated for unbalanced experiments. Again, it is possible to use the method of moments and Satterthwaite approximations as in the previous section on variance components.

While $\hat{\mu}$ is unbiased for μ, it is not the **best linear unbiased estimate (BLUE)**, in the sense that it does not have the smallest possible variance. The BLUE is a weighted mean of sample class means,

$$\hat{\mu}_B = \sum_i w_i \bar{y}_{i\cdot} / \sum_i w_i \,,$$

with weights proportional to the inverse of the variances of the sample class means $w_i = 1/(\sigma_A^2 + \sigma^2/n_i)$. Its variance, $V(\hat{\mu}_B) = 1/\sum w_i$, is the harmonic mean of $\sigma_A^2 + \sigma^2/n_i$, $i = 1, \cdots, a$, divided by the number of classes, a. Unfortunately, the variance components, and hence the weights, are unknown in practice and must be estimated. Naturally, estimating the weights using variance component estimates results in an estimator of μ which is no longer linear in y_{ij}, unbiased, or minimum variance! While it is possible to iterate toward a solution which maximizes the likelihood, this is very computationally intensive and is generally avoided. Instead, recent work has been investigating the use of Monte Carlo methods to improve estimation while reducing computational cost.

The a random effects A_i, or corresponding class means M_i, are specific to the randomly selected classes. However, experiments tend to build on previous knowledge. Breeding studies, for instance, may consider lines as random effects for the present study but may, based on significant variation among classes, choose particular classes for further study based on their high (or low) class mean of response. These class levels would be considered as fixed effects in subsequent experiments.

Example 19.3 Random: Consider the days to budding for selected plant varieties in the mustard family. An initial experiment takes a random sample of offspring from a cross of two varieties and looks at variation in response. A subsequent experiment uses selected offspring from the first experiment, based on observed characteristics. These offspring would serve as parents for the next generation. These new parents would be viewed as fixed factor levels rather than as a random sample. ◇

Thus there may be considerable interest in predicting the mean response for future experiments with particular levels. It can be shown that $\bar{y}_{i\cdot}$ is an unbiased estimator since $E(\bar{y}_{i\cdot}) = \mu$. Further, it is unbiased for the conditional mean of a chosen class since $E(\bar{y}_{i\cdot}|\text{ class } u = U_i) = M_i$. However, $\bar{y}_{i\cdot}$ is not the **best linear unbiased predictor (BLUP)** of class mean M_i, and $\bar{y}_{i\cdot} - \bar{y}_{\cdot\cdot}$ is not the BLUP of A_i. Instead, the best predictors depend on the unknown variances σ^2, σ_A^2 and on the sample sizes n_i.

Essentially, it is most appropriate to consider a weighted least squares approach as in the BLUE, with the same weights. Thus the class effects BLUPs are

$$\hat{A}_i = \sigma_A^2 w_i (\bar{y}_{i\cdot} - \hat{\mu}_B) ,$$

and the class means BLUPs are

$$\hat{M}_i = \hat{\mu}_B + \hat{A}_i .$$

Note that since the variances are unknown, the weights cannot be determined and must be estimated. Once this is done, as is common practice, the predictors are no longer linear, unbiased, nor best. These can be improved using restricted maximum likelihood. The interested reader should consult Searle (1987) and Searle, Casella and McCulloch (1992, ch. 7) and other recent literature to learn further details. Kennedy (1991) provides a historical perspective on Henderson's development of BLUPs.

19.6 Problems

19.1 Random: Consider the simulated data introduced in Example 19.2, or simulate a new set of data with those characteristics.

(a) Make this real for yourself. That is, imagine it as an experiment in your field. Identify model components in your own language. For instance, consider a random sample of hens, and from each a random sample of eggs that are measured for shell thickness. Think up a different example.

(b) Perform an analysis of variance and report test results in the language of your experiment.

(c) Estimate variance components and their standard errors. Justify the choice of method, either on practical or theoretical grounds.

(d) Construct confidence intervals for the variance components. Justify the use of either the normal or χ^2 approximation in your own words.

(e) Estimate the grand mean along with its standard error for the one-factor fixed and random models. That is, interpret this experiment both ways. How and why do the estimates and/or standard errors differ?

19.2 Random: Reconsider the previous problem, but drop one observation so that the design is unbalanced.

(a) Estimate variance components and their standard errors using at least two different methods. Present results in a small table and indicate exactly where you got the values. Briefly compare these methods numerically and in terms of their properties. Justify the choice of one method to use in practice (either on practical or theoretical grounds).

(b) Estimate the grand mean along with its standard error for the one-factor fixed and random models. Discuss why the estimate of the population grand mean (in the fixed effects model, computed in `proc glm`) differs from the best linear unbiased estimator (in the random effects model, computed in `proc mixed`), and why both differ from the sample grand mean (reported as Y Mean on `proc glm` printout).

CHAPTER 20

General Random Models

This chapter considers more complicated models which have random effects. The first section considers the balanced two-factor random model. General random models are presented in Section 20.2, where the unbalanced two-factor random model is briefly considered as a special case. For a more detailed approach see Searle (1987) and Searle, Casella and McCulloch (1992).

20.1 Two-factor random models

Two-factor random models can be viewed as beginning with a two-factor fixed effects model with all levels of the two factor populations. In the random model, random samples of levels from each factor are drawn. Again, the mean response can be partitioned into four parts, the grand mean plus two terms involving each main effect plus an interaction. However, now the main effects and interactions are independent random variates rather than fixed but unknown values.

Example 20.1 Random: Consider a population of seed packets and a population of fields in Dane County. The experiment consists of taking a random sample of seed packets and a random sample of fields, then placing several seeds from each seed packet in each field. Individual plants grown from seed are measured for yield at the end of the growing season. ◇

Let $M(u, w)$ be the mean response (yield) for class (u, w), that is seed packet u in field w. The grand mean for all seeds over all fields is $\mu = \bar{M}(\cdot, \cdot)$. The main effect of packet u averaged over the population of fields is $A(u) = \bar{M}(u, \cdot) - \mu$, while the main effect of field w averaged over the population of seed packets is $B(w) = \bar{M}(\cdot, w) - \mu$. The part of the mean response (yield) for packet u in field w which is not explained by main effects is the interaction, $C(u, w) = M(u, w) - \bar{M}(u, \cdot) - \bar{M}(\cdot, w) + \mu$. Thus the mean response can be written as

$$M(u, w) = \mu + A(u) + B(w) + C(u, w)$$

with $\bar{A}(\cdot) = \bar{B}(\cdot) = \bar{C}(u, \cdot) = \bar{C}(\cdot, w) = 0$.

Now consider a **random class** from these factor populations. Let U be a random class from the seed packet population and W a random class from the population of fields. Arguing as for the single-factor model, the main effects are on average zero,

$$E[A(U)] = \bar{A}(\cdot) = 0 \text{ and } E[B(W)] = \bar{B}(\cdot) = 0.$$

In addition, for any seed packet class u, $E[C(u,W)] = \bar{C}(u,\cdot) = 0$. Hence the random interaction is also centered,

$$E[C(U,W)] = 0.$$

The random main effects $A(U)$ and $B(W)$ are independent by construction. It turns out that the random interaction is uncorrelated with the main effects. To see this, note that for any class u,

$$A(u)E[C(u,W)|u] = A(u)\bar{C}(u,\cdot) = A(u) \times 0 = 0.$$

Thus

$$\operatorname{cov}(A(U), C(U,W)) = E[A(U)C(U,W)]$$
$$= E\{A(U)E[C(U,W)|U]\} = 0.$$

and similarly for $\operatorname{cov}(B(W), C(U,W))$. This agrees with the earlier fixed effects model interpretation of interaction as that part of the mean response which cannot be explained by, and hence uncorrelated with, main effects.

Typically, for purposes of inference, the population of mean responses $M(u,w)$ is believed to follow a **normal** distribution. Hence the marginal mean responses $\bar{M}(u,\cdot)$ and $\bar{M}(\cdot,w)$ follow a normal as well. It follows that if U and W are independent uniform variates, then $A(U)$, $B(W)$ and $C(U,W)$ are mutually independent and normally distributed, with zero mean and respective variances denoted by σ_A^2, σ_B^2 and σ_{AB}^2.

A designed experiment might draw random samples of classes from these two populations, $\{(U_i, W_j), i = 1, \cdots, a; j = 1, \cdots, b\}$, with class means of

$$M(U_i, W_j) = M_{ij} = \mu + A_i + B_j + C_{ij}$$

where $A_i = A(U_i)$, $B_j = B(W_j)$ and $C_{ij} = C(U_i, W_j)$ are uncorrelated (independent if normal) random variates. Random samples of elements are drawn from each class (U_i, W_j). The **two-factor random model** can be written as

$$y_{ijk} = M_{ij} + e_{ijk}$$
$$= \mu + A_i + B_j + C_{ij} + e_{ijk} .$$

Note that in a two-factor **additive random model**, the mean response would be

$$M_{ij} = \mu + A_i + B_j$$

TWO-FACTOR RANDOM MODELS

which could be viewed as taking the sum of μ plus independent random draws from $N(0, \sigma_A^2)$ and $N(0, \sigma_B^2)$. The two-factor random model with interaction would have a further dimension for interaction.

Questions of interest would focus on determining the magnitude of class variances for main effects and interactions and perhaps on inference for the grand mean. Since the levels are a random sample, comparison of factor levels would only by done on a post-hoc basis to select levels for subsequent experiments. Genetics and breeding experiments perform such comparisons on a regular basis.

General treatment of unbalanced design structures is relegated to the next section, as a number of issues emerge that need to be addressed. If the design is **balanced**, the expected mean squares and distributions of MS can be determined directly with a Type I approach. For instance, the sum of squares for A can be expressed as

$$SS_A = bn \sum_i (\bar{y}_{i..} - \bar{y}_{...})^2$$
$$= bn \sum_i [(A_i - \bar{A}_.) + (\bar{C}_{i.} - \bar{C}_{..}) + (\bar{e}_{i..} - \bar{e}_{...})]^2 .$$

Its expected value is readily apparent, using an approach similar to the previous section and noting that cross-products between different random effects have zero expectation. This yields

$$E(MS_A) = bn[\sigma_A^2 + \sigma_{AB}^2/b + \sigma^2/bn]$$
$$= \sigma^2 + n\sigma_{AB}^2 + bn\sigma_A^2 .$$

All MS for random effects in balanced factorial arrangements are independent and are proportional to χ^2 variates.

The anova in Table 20.1 summarizes the key features of sums of squares and expected mean squares. Note that expected mean squares for both main effects depend on the random interaction variance σ_{AB}^2. The interaction terms do not cancel out in expected mean squares for random effects factorial arrangements. Instead the variation in higher-order interactions persists when examining main effects.

Hypothesis testing requires a statistic whose distribution under the null hypothesis does not depend on any model parameters. Examining the expected mean squares and noting that MS are independent, leads to the following pivot statistics.

The test for **interaction** is operationally the same as for the fixed model. The hypothesis
$$H_0 : \sigma_{AB}^2 = 0$$
has corresponding pivot

$$F = MS_{AB}/MSE \sim \frac{\sigma^2 + n\sigma_{AB}^2 + bn\sigma_A^2}{\sigma^2 + n\sigma_{AB}^2} F_{a-1,(a-1)(b-1)} ,$$

Source	SS	$E(MS)$
A	$bn \sum_i (\bar{y}_{i..} - \bar{y}_{...})^2$	$\sigma^2 + n\sigma^2_{AB} + bn\sigma^2_A$
B	$an \sum_j (\bar{y}_{.j.} - \bar{y}_{...})^2$	$\sigma^2 + n\sigma^2_{AB} + an\sigma^2_B$
$A*B$	$n \sum_i \sum_j (\bar{y}_{ij.} - \bar{y}_{i..} - \bar{y}_{.j.} + \bar{y}_{...})^2$	$\sigma^2 + n\sigma^2_{AB}$
Error	$\sum_i \sum_j \sum_k (y_{ijk} - \bar{y}_{ij.})^2$	σ^2
Total	$\sum_i \sum_j \sum_k (y_{ijk} - \bar{y}_{...})^2$	

Table 20.1. *Random two-factor anova*

which reduces exactly to a central F under the hypothesis. Note that under the alternative, the distribution is proportional to a central F, as was found for the balanced single-factor random model.

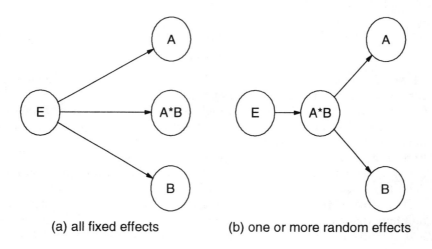

(a) all fixed effects   (b) one or more random effects

Figure 20.1. *Random and fixed model errors for tests. Directed lines connect appropriate errors for factors in (a) fixed effect and (b) random or mixed effect models.*

Main effects tests run counter to intuition gained from fixed effects models (Figure 20.1). Here the interaction is the appropriate denominator for F tests, since σ^2_{AB} appears in the expected mean square for both main

effects. The hypothesis for A is
$$H_0 : \sigma_A^2 = 0 ,$$
and is associated with the pivot statistic
$$F = MS_A/MS_{AB} \sim \frac{\sigma^2 + n\sigma_{AB}^2 + bn\sigma_B^2}{\sigma^2 + n\sigma_{AB}^2} F_{b-1,(a-1)(b-1)}$$
with a similar construction for testing σ_B^2. Note that under the hypothesis of no main effect, the pivot statistic has a central F distribution and does not depend on model parameters.

Method of moments estimators for **variance components** correspond in the balanced case to MINQUE. However, in general, only the error variance estimator, $\hat{\sigma}^2 = MSE$, has a central χ^2 distribution. The other variance component estimators,

$$\begin{aligned}
\hat{\sigma}_A^2 &= (MS_A - MS_{AB})/bn , \\
\hat{\sigma}_B^2 &= (MS_B - MS_{AB})/an , \\
\hat{\sigma}_{AB}^2 &= (MS_{AB} - MSE)/n ,
\end{aligned}$$

have rather messy distributions which could be approximated using the Satterthwaite approach outlined in the previous chapter or through simulations.

Example 20.2 Random: Two seed packets were drawn at random. Four farms were selected at random from a list of cooperators. Seeds from each packet were used on all four farms, with two replicates per farm. The anova in Table 20.2, derived from proc glm, shows that the interaction is significant but main effects are not. The restricted maximum likelihood estimates of variance components as reported by proc mixed

```
              Covariance Parameter Estimates (REML)
  Cov Parm        Ratio     Estimate  Std Error      Z Pr>|Z|
  seed         0.00000000 0.00000000         .       .   .
  farm         2.79663323 5.04481852 7.25613273   0.70 0.4869
  seed*farm    3.21863303 5.80605971 4.76466416   1.22 0.2230
  residual     1.00000000 1.80388993 0.90194497   2.00 0.0455
```

do not inspire great confidence. This is not surprising given the small sample sizes. ◊

Inference for the **grand mean** μ can follow in an approximate way by matching moments. Its estimator has distribution

$$\hat{\mu} = \bar{y}_{...} \sim N(\mu, (\sigma^2 + bn\sigma_A^2 + an\sigma_B^2 + n\sigma_{AB}^2)/abn)$$

which does not correspond to any one MS as was the case in the single-

source	df	SS	MS	F	p-value
seed	1	4.81	4.81	0.30	0.62
farm	3	100.79	33.60	2.06	0.28
seed*farm	3	48.85	16.28	9.03	0.0060
error	8	14.43	1.80		
total	15	168.88			

Table 20.2. *Random two-factor anova*

factor random model. Instead use the pivot statistic
$$T = (\hat{\mu} - \mu)/\sqrt{(MS_A + MS_B - MS_{AB})/abn} ,$$
which has roughly a t distribution. The estimated degrees of freedom follow from a Satterthwaite approximation,
$$\begin{aligned} r &= 2[E(MS_A + MS_B - MS_{AB})]^2/V(MS_A + MS_B - MS_{AB}) \\ &= \frac{[E(MS_A+MS_B-MS_{AB})]^2}{E(MS_A)^2/(a-1)+E(MS_B)^2/(b-1)+E(MS_{AB})^2/(a-1)(b-1)} , \end{aligned}$$
and substituting the actual MS for $E(MS)$. It's not pretty, but it can be done. Note that this approach is *not* the same as viewing the class structure as a one-factor random effect, nor would the latter be appropriate in the presence of two random effects.

In general, it is more appropriate to use ML or REML methods for careful calculations, keeping in mind that some statistical packages may have numerical instabilities. It is a good idea when possible to double-check calculations between two or more statistical packages for complicated models.

Example 20.3 Random: The best linear unbiased predictors of the combinations of seed*farm are shown in Figure 20.2 connected by lines. Notice how the BLUPs are shrunk toward the mean slightly in comparison to the ordinary least squares estimates (closed circles). ◇

20.2 Unbalanced two-factor random model

Consider the two-way random model with interactions and unequal numbers per class combination but no missing cells,
$$y_{ijk} = \mu + A_i + B_j + C_{ij} + e_{ijk} ,$$
with $n_{ij} > 0$ samples per class combination (i,j) and independent $A_i \sim N(0, \sigma_A^2)$, $B_j \sim N(0, \sigma_B^2)$ and $C_{ij} \sim N(0, \sigma_{AB}^2)$. The Type I sum of squares

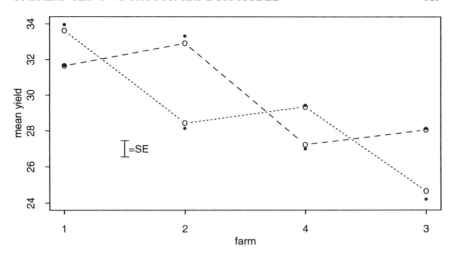

Figure 20.2. *Random two-factor interaction plot. Best linear unbiased predictors (BLUPs) by* **farm** *and* **seed** *packet connected by lines with open circles. Ordinary least squares estimates are shown with closed circles. Standard errors are for individual BLUPs but not for comparisons.*

for σ_A^2 alone is

$$SS_A = \sum_{i=1}^{a} n_{i\cdot}(\bar{y}_{i\cdot\cdot} - \bar{y}_{\cdot\cdot\cdot})^2 \ .$$

Consider the columns of the design matrix for each of the random effects separately. For instance, substitute the appropriate column of **Z** for **y** in SS_A and sum the resultant quadratic forms for each random effect. For the random effect A, this yields

$$\sum_{l=1}^{a} \sum_{i} n_{i\cdot}(\delta_{il} - n_{l\cdot}/n_{\cdot\cdot})^2 \ ,$$

with the Kronecker delta, $\delta_{il} = 1$ if $i = l$ and 0 otherwise. After some algebra, this reduces to $n_{\cdot\cdot} - \sum_{i} n_{i\cdot}^2/n_{\cdot\cdot}$. Arguing similarly for the contribution to $E(SS_A)$ from the other random main effect B yields

$$\sum_{j=1}^{b} \sum_{i} n_{i\cdot}(n_{ij}/n_{i\cdot} - n_{\cdot j}/n_{\cdot\cdot})^2 = \sum_{ij} n_{ij}^2/n_{i\cdot} - \sum_{j} n_{\cdot j}^2/n_{\cdot\cdot} \ ,$$

and from the random interaction $C = A * B$,

$$\sum_{l=1}^{a} \sum_{ij} n_{i\cdot}[n_{lj}(\delta_{il}/n_{l\cdot} - 1/n_{\cdot\cdot})]^2 = \sum_{ij} n_{ij}^2(1/n_{i\cdot} - 1/n_{\cdot\cdot}) \ .$$

Thus the Type I expected mean square for σ_A^2 is

$$\begin{aligned} E(SS_A) &= (a-1)\sigma^2 + [n_{..} - \sum_i n_{i.}^2/n_{..}]\sigma_A^2 \\ &+ [\sum_i \sum_j n_{ij}^2/n_{i.} - \sum_j n_{.j}^2/n_{..}]\sigma_B^2 \\ &+ [\sum_i \sum_j n_{ij}^2(1/n_{i.} - 1/n_{..})]\sigma_{AB}^2 \;, \end{aligned}$$

which reduces in the balanced case to

$$E(SS_A) = (a-1)[\sigma^2 + nb\sigma_A^2 + n\sigma_{AB}^2] \;.$$

The Type III sum of squares for A given the random effects B and C is

$$SS_{A|B,A*B} = \sum_i n_{i.}(\hat{M}_{i.} - \hat{M}_{..})^2 \;,$$

with $\hat{M}_{i.} = \sum_j \sum_k y_{ijk}/bn_{ij}$ and $\hat{M}_{..} = \sum_i \sum_j \sum_k y_{ijk}/abn_{ij}$ the estimated marginal means for the random sample of class levels. Arguing as above, the contribution of σ_A^2 to the Type III expected sum of squares is

$$\sum_{l=1}^{a} \sum_i n_{i.}(\delta_{il} - 1/a)^2 = (a-1)n_{..}/a \;,$$

which is considerably cleaner than the Type I contribution. The random effect B contributes nothing,

$$\sum_{ij} n_{i.}(1/b - 1/b)^2 = 0 \;,$$

while the random interaction C adds

$$\sum_{l=1}^{a} \sum_{ij} n_{i.}(\delta_{il}/b - 1/ab)^2 = (a-1)n_{..}/ab \;.$$

Thus the Type III expected mean square for σ_A^2 is

$$E(MS_{A|B,A*B}) = \sigma^2 + (n_{..}/a)\sigma_A^2 + (n_{..}/ab)\sigma_{AB}^2 \;,$$

which reduces to the right value for the balanced case.

Now consider the expected mean square for interaction, which is the same for all four approaches. The sum of squares is

$$SS_{B*A|A,B} = \sum_{ij} n_{ij}(\bar{y}_{ij.} - \bar{y}_{i..} - \bar{y}_{.j.} + \bar{y}_{...})^2 \;.$$

An argument similar to those above demonstrates the lack of contribution from the random main effects, and that the contribution from the interaction random effects C is

$$\sum_{l=1}^{a} \sum_{m=1}^{b} \sum_{ij} n_{ij}(\delta_{li}\delta_{mj} - \delta_{li}/b - \delta_{mj}/a + 1/ab)^2 \;,$$

which reduces to $(b-1)(a-1)n_{..}/ab$. Thus the expected mean square for interaction is

$$E(MS_{A*B|A,B}) = \sigma^2 + (n_{..}/ab)\sigma_{AB}^2 ,$$

which matches coefficients with the σ^2 and σ_{AB}^2 terms from the expected mean square for Type III main effects.

Problems arise for unequal cell sizes. Terms involving variance of other factors do not drop out, and $E(MS)$ may not balance out. Usually, weighted sums and/or differences of MS yield approximate inference for most variance components besides interaction.

Unfortunately, the MS for random effects in the unbalanced case are not proportional to χ^2, requiring a Satterthwaite or some other approximation to develop a pivot statistic. The Type I MS for σ_A^2 is independent of MS_{AB}, but its expected mean square contains a term for the other main effect and the coefficient for σ_{AB}^2 does not match. On the other hand, the Type III MS for main effects are not independent of MS_{AB}. This suggests using

$$F = MS_{A|B,A*B}/MS_{A*B|A,B}$$

as the pivot statistic. It is a ratio of dependent quadratic forms that are each weighted sums of χ^2, even under $H_0 : \sigma_A^2 = 0$. Each MS is roughly a central χ^2 with degrees of freedom from the Satterthwaite approximation,

$$df \approx 2[E(MS)]^2/V(MS) .$$

The F is approximately $F_{dfA,dfAB}$ under H_0 provided the design is not too unbalanced.

This same Satterthwaite approximation idea can be used to get quick **variance component estimates**. Better estimates may be found using the ML or REML method with the SAS procedures varcomp or mixed. Arguments follow similar lines to the previous section but with the added complications already noted.

The variance of the grand mean is

$$V(\hat{\mu}) = \left(\sigma^2 + \sigma_A^2 \sum_i n_{i.}^2/n_{..} + \sigma_B^2 \sum_j n_{.j}^2/n_{..} + \sigma_{AB}^2 \sum_{ij} n_{ij}^2/n_{..}\right)/n_{..} ,$$

which is rather messy and would have an even messier estimator for the unbalanced case.

Again, Satterthwaite approximations provide a quick method for inference concerning pivot statistics based on weighted sums of quadratic forms. There are two problems with this, as already encountered above. The numerator and denominator may not be statistically independent, and either or both may not be close to central χ^2. Thus such approaches should be used with caution.

If the imbalance is great and the sample size is large, a likelihood ratio test works well for comparing the full versus the reduced model. Unlike the fixed model case, the likelihood ratio test for $H_0 : \sigma_A^2 = 0$ does not correspond to the F test. For one thing, the reduced model has one fewer parameters, and the likelihood ratio statistic is compared to a χ_1^2 distribution. The likelihood ratio test is presented in Milliken and Johnson (1992, ch. 20). Note that if sample size is moderate and the design is rather unbalanced, then neither the likelihood ratio test nor the Satterthwaite approximate F test performs very well. This situation would require Monte Carlo simulation to investigate properties.

20.3 General random model

Unbalanced two-factor random models were referred to repeatedly in the previous section with guarded tones. When the design is unbalanced, the SS for main effects and for interaction are no longer independent. The test for random interaction works fine and is as described above for the balanced design. The problems arise in trying to test MS which have more than one variance component in their $E(MS)$. The pragmatic approach is to try to match up variance components by taking linear combinations of $E(MS)$. Different types of SS (I, II, III and IV) might suggest different linear combinations of those types. In general, there appears to be no clear-cut choice of type of SS to use, as there is for fixed effects models.

This section examines the general random model, in which all effects are random but the design may be rather unbalanced. The unbalanced two-factor random model is discussed at various stages. The development is largely in terms of matrix algebra, building on linear fixed effects model examined in Chapter 12. For further information on the general random model, see Searle (1987, ch. 13).

Consider a linear random model in matrix notation. Let $\mathbf{y}$ be the n-vector of reponses, μ the overall mean and $\mathbf{1}$ an n-vector of all 1s. The random components are separated into pure error $\boldsymbol{\epsilon}$ and a random effect $\mathbf{u}$ of length q whose relation to response is summarized in the $n \times q$ matrix $\mathbf{Z}$. Together, the model can be written as

$$\mathbf{y} = \mathbf{1}\mu + \mathbf{Zu} + \boldsymbol{\epsilon} \ .$$

Random effects and random errors are assumed to be mutually independent and normally distributed. The variance of the response $\mathbf{y}$ is

$$V(\mathbf{y}) = \mathbf{V} = \mathbf{Z}V(\mathbf{u})\mathbf{Z}^\mathrm{T} + \sigma^2 \mathbf{I} \ .$$

If the random effects $\mathbf{u}$ are independent with a common variance σ_A^2 as in the one-factor random model, the variance reduces to $\mathbf{V} = \sigma_A^2 \mathbf{Z}\mathbf{Z}^\mathrm{T} + \sigma^2 \mathbf{I}$. A multivariate approach would consider arbitrary covariances for $\mathbf{u}$. However, it is often possible to partition the random effects in a natural way and examine **variance components**.

The random effects design matrix $\mathbf{Z}$ can be partitioned into several smaller matrices corresponding to independent random effects, such as those associated with main effects and interactions. Quite often, these matrices consist of 0s and 1s, with exactly one 1 in each row. In other words,

$$\mathbf{Zu} = \mathbf{Z}_1\mathbf{u}_1 + \mathbf{Z}_2\mathbf{u}_2 + \cdots + \mathbf{Z}_r\mathbf{u}_r$$

and, if the random effects $\{\mathbf{u}_k\}_{k=1}^r$ are uncorrelated with each other,

$$V(\mathbf{y}) = \mathbf{V} = \sum_{k=1}^r \mathbf{Z}_k V(\mathbf{u}_k)\mathbf{Z}_k^\mathrm{T} + \sigma^2\mathbf{I} .$$

If in addition each random effect has independent elements, say $\mathbf{u}_k \sim N(\mathbf{0}, \sigma_k^2\mathbf{I})$,

$$V(\mathbf{y}) = \mathbf{V} = \sum_{k=1}^r \sigma_k^2 \mathbf{Z}_k\mathbf{Z}_k^\mathrm{T} + \sigma^2\mathbf{I} .$$

That is, description and inference may reduce to consideration of the grand mean μ and the variance components $\{\sigma_k^2\}_{k=1}^r$ and σ^2.

20.4 Quadratic forms in random effects

Inference about random effects can be made in terms of quadratic forms under suitable conditions. That is, if the data are normal then quadratic forms in $\mathbf{y}$ are related in some way to chi-square variates. An understanding of general quadratic forms for the partitioned random model developed above can offer insight into how to address inferential questions about variance components.

Consider a quadratic form involving a projection matrix $\mathbf{H}$,

$$SSH_0 = \mathbf{y}^\mathrm{T}\mathbf{H}\mathbf{y} = \sum_i \sum_j y_i h_{ij} y_j ,$$

The matrix $\mathbf{H}$ must be idempotent, and the columns of $\mathbf{H}$ should be orthogonal to the constant space spanned by $\mathbf{1}$, implying that

$$\mathbf{1}^\mathrm{T}\mathbf{H}\mathbf{1} = \sum_i \sum_j h_{ij} = 0 .$$

All the sum of squares for model effects commonly considered are of this form. An associated null hypothesis $H_0 : V(\mathbf{HZu}) = \mathbf{0}$ implies that $\mathbf{Hy} \sim N(\mathbf{0}, \sigma^2 \mathbf{H})$. That is, certain variance components must be zero. Thus under the null hypothesis, the quadratic form divided by the variance, $\mathbf{y}^\mathrm{T}\mathbf{H}\mathbf{y}/\sigma^2$ has a chi-square distribution with $\mathrm{tr}(\mathbf{H})$ degrees of freedom.

The expected sum of squares for the quadratic form in general is

$$E[\mathbf{y}^\mathrm{T}\mathbf{H}\mathbf{y}] = \mathrm{tr}[\mathbf{H}E(\mathbf{y}\mathbf{y}^\mathrm{T})] = \mathrm{tr}(\mathbf{H}\mathbf{V}) = \sigma^2\mathrm{tr}(\mathbf{H}) + \mathrm{tr}(\mathbf{H}\mathbf{Z}V(\mathbf{u})\mathbf{Z}^\mathrm{T}) .$$

The first equality can be verified by noting that the diagonal entries are

$\{\mathbf{Hyy}^T\}_{ii} = \sum_j h_{ij} y_i y_j$. For the partitioned random model,

$$E[\mathbf{y}^T\mathbf{Hy}] = \sigma^2 \text{tr}(\mathbf{H}) + \sum_{k=1}^r \sigma_k^2 \text{tr}(\mathbf{HZ}_k\mathbf{Z}_k^T)$$
$$= \sigma^2 \text{tr}(\mathbf{H}) + \sum_k \sum_j \sigma_k^2 \mathbf{z}_{kj}^T \mathbf{Hz}_{kj} .$$

with $\mathbf{z}_{kj}$ the jth column of $\mathbf{Z}_k$. The quadratic forms $\mathbf{z}_{kj}^T \mathbf{Hz}_{kj}$ are similar to $\mathbf{y}^T\mathbf{Hy}$. Thus the expected mean square can be derived without explicitly knowing $\mathbf{H}$ by successively replacing $\mathbf{y}$ by each of the columns of $\mathbf{Z}$. Further, since $\mathbf{Z}$ typically has 0s and 1s, the computations are quick. In balanced designs, all but one of the traces of $\mathbf{HZ}_k\mathbf{Z}_k^T$ may be zero, focusing attention on one variance component. However, there is no guarantee of this simplification for unbalanced experiments.

It is not necessary to begin with a projection matrix. Consider a hypothesis $H_0 : V(\mathbf{AZu}) = \mathbf{0}$. Let $\mathbf{P} = (\mathbf{I} - \mathbf{J}/n)$ be the projection matrix that annihilates the fixed mean μ. Then the corresponding projection matrix $\mathbf{H}$ for the hypothesis is

$$\mathbf{H} = \mathbf{PA}^T(\mathbf{APA}^T)^{-1}\mathbf{AP} .$$

Notice in particular that the choice $\mathbf{A} = \mathbf{I}$ yields the matrix $\mathbf{E} = \mathbf{P}$ which annihilates all the random effects except the residual error: $\mathbf{y}^T\mathbf{Py} = \sum_i (y_i - \bar{y}_.)^2$.

In other words, proper care in writing the design matrix can help unravel the expected mean squares and corresponding questions of inference. Packages such as SAS can perform these calculations with appropriate options. Recall from the fixed effects model that different types of SS result in different quadratic forms. Regardless of the type of SS, expected mean square can be computed. The key consideration when forming ratios for **hypothesis testing** using F pivot statistics is to select quadratic forms which are independent of one another and which differ in $E(MS)$ only for the variance component being examined. Ideally the quadratic forms are proportional to χ^2 variates or nearly so.

20.5 Application to two-factor random model

This general approach is illustrated below with the two-factor random model. Suppose that $\mathbf{H}$ measures within-group variation and the columns of $\mathbf{Z}_k$ identify group membership (i.e., are indicators for that group random effect) then, for balanced designs,

$$\text{tr}(\mathbf{Z}_k^T \mathbf{HZ}_k) = 0 ,$$

and the kth term drops out of $E(MS)$ for $\mathbf{y}^T\mathbf{Hy}$.

Example 20.4 Random: Consider the additive random model

$$y_{ij} = \mu + A_i + B_j + e_{ij} ,$$

APPLICATION TO TWO-FACTOR RANDOM MODEL

which is made explicit as

$$\begin{bmatrix} y_{11} \\ y_{12} \\ y_{13} \\ y_{21} \\ y_{22} \\ y_{23} \end{bmatrix} = \begin{bmatrix} 1 & 1 & 0 & 1 & 0 & 0 \\ 1 & 1 & 0 & 0 & 1 & 0 \\ 1 & 1 & 0 & 0 & 0 & 1 \\ 1 & 0 & 1 & 1 & 0 & 0 \\ 1 & 0 & 1 & 0 & 1 & 0 \\ 1 & 0 & 1 & 0 & 0 & 1 \end{bmatrix} \begin{bmatrix} \mu \\ A_1 \\ A_2 \\ B_1 \\ B_2 \\ B_3 \end{bmatrix} + \begin{bmatrix} \epsilon_{11} \\ \epsilon_{12} \\ \epsilon_{13} \\ \epsilon_{21} \\ \epsilon_{22} \\ \epsilon_{23} \end{bmatrix}.$$

Now assign matrix notation,

$$\mathbf{y} = \begin{bmatrix} \mathbf{1} & : & \mathbf{Z}_1 & : & \mathbf{Z}_2 \\ 6 \times 1 & & 6 \times 2 & & 6 \times 3 \end{bmatrix} \begin{bmatrix} \mu \\ A_1 \\ A_2 \\ B_1 \\ B_2 \\ B_3 \end{bmatrix} + \boldsymbol{\epsilon} .$$

The symmetric matrix to pick up variation among levels of A is

$$\mathbf{H} = \frac{1}{6} \begin{bmatrix} 1 & 1 & 1 & -1 & -1 & -1 \\ 1 & 1 & 1 & -1 & -1 & -1 \\ 1 & 1 & 1 & -1 & -1 & -1 \\ -1 & -1 & -1 & 1 & 1 & 1 \\ -1 & -1 & -1 & 1 & 1 & 1 \\ -1 & -1 & -1 & 1 & 1 & 1 \end{bmatrix},$$

with corresponding quadratic form

$$\mathbf{y}^\mathrm{T}\mathbf{H}\mathbf{y} = 3 \sum_{i=1}^{2} (\bar{y}_{i\cdot} - \bar{y}_{\cdot\cdot})^2$$

having expected sum of squares

$$E[\mathbf{y}^\mathrm{T}\mathbf{H}\mathbf{y}] = \sigma_A^2 \sum_{j=1}^{2} \mathbf{z}_{1j}^\mathrm{T} \mathbf{H} \mathbf{z}_{1j} + \sigma^2 tr(\mathbf{H}) = 3\sigma_A^2 + \sigma^2 .$$

To show this, notice that $\mathbf{H}$ picks up the two A effects

$$\mathbf{z}_{1j}^\mathrm{T} \mathbf{H} \mathbf{z}_{1j} = 3 \sum_{i=1}^{2} (1 - 1/2)^2 = 3/2 ,$$

and annihilates the three B effects

$$\mathbf{z}_{2j}^\mathrm{T} \mathbf{H} \mathbf{z}_{2j} = 3 \sum_{i=1}^{2} (1/3 - 1/3)^2 = 0 .$$

These could be computed directly using $\mathbf{H}$ or from the quadratic form $3 \sum_{i=1}^{2} (\bar{y}_{i\cdot} - \bar{y}_{\cdot\cdot})^2$, substituting the columns of $\mathbf{Z}$ for $\mathbf{y}$. ◇

20.6 Problems

20.1 Random: Consider the two-way random experiment labeled simply as row and column treatments.
(a) Describe in words an experiment resulting in these data. That is, identify realistic treatments instead of 'row' and 'column'. Justify your choice as a random rather than fixed model.
(b) Write down the model with all assumptions. Indicate in a small table the sample sizes. Test whether there is any variation among row and column treatments, and any evidence of row by column interaction.
(c) Verify the Type I estimates of the four variance components from `proc varcomp` using the `random` phrase information in `proc glm`. Use this to guide you in calculating Type II or Type III (you only need to do one type) estimates of variance components. Why are the estimates of main effects variance components negative? Is this consistent with the results of tests from part (c)?
(d) Estimate the interaction and error variance components using ML or REML (say, using `proc varcomp` or `proc mixed`). Comment briefly on the ML or REML estimates of main effects variance components. Briefly justify your choice of method – are you selecting it because it is easy to explain to the scientist or because it has some (specified) 'optimal' properties?
(e) Find the BLUE for the grand mean and its standard error. Why are the estimates for `proc glm` incorrect?
(f) Find the BLUPs and their standard for the six row by column combinations. Present the BLUPs in an interaction plot (draw by hand!). Add the cell means estimates of these six combinations (see `proc glm` output) and connect corresponding points. Comment briefly on the difference between BLUPs and cell mean estimates.

CHAPTER 21

Mixed Effects Models

Most experiments combine elements of fixed effects and random effects models. Section 21.1 introduces two-way **mixed effects models**, with a general treatment in Section 21.2. For a more detailed approach see Searle (1987) and Searle, Casella and McCulloch (1992).

21.1 Two-factor mixed models

Many models contain both fixed and random effects in addition to random error. Usually mixed models have a factorial arrangement. with one or more main effects being random and one or more main effects being fixed. Interactions involving random effects are assumed to be random effects. Thus the only fixed effects interactions are those involving fixed main effects.

The **two-factor mixed model** with one fixed and one random effect illustrates the key features. Consider

$$y_{ijk} = M_{ij} + e_{ijk},$$

with the random mean part M_{ij} independent of random error e_{ijk}. Suppose factor A has fixed treatment levels $i = 1, \cdots, a$, while factor B has class levels w drawn at random from a population. The mean response for treatment level i of A and class w from B may be denoted by $M(i, w)$. Let μ_i be the mean response for level i of A averaged across all possible class levels of B. The grand mean is $\mu = \bar{\mu}_. = \bar{M}(\cdot, \cdot)$.

The **fixed main effect** for level i of factor A, averaged across the population of levels B, is

$$\alpha_i = \mu_i - \mu = \bar{M}(i, \cdot) - \mu.$$

The **random main effect** for class w from B, averaged across the fixed levels of A, is

$$B(w) = \bar{M}(\cdot, w) - \mu.$$

Any part of the random mean $M(i, w)$ which is not explained by the grand mean or main effects comprises the **random interaction**,

$$C_i(w) = M(i, w) - \bar{M}(i, \cdot) - \bar{M}(\cdot, w) + \mu.$$

Thus the mean response can be partitioned as $M(i,w) = \mu + \alpha_i + B(w) + C_i(w)$, with $\bar{B}(\cdot) = \bar{C}_i(\cdot) = 0$ and some suitable side condition on $\boldsymbol{\alpha}$ such as $\bar{\alpha}. = 0$. Some authors, such as Scheffé (1959), assume that $\bar{C}.(w) = 0$. However, this introduces unnatural restrictions on the interactions and is not the approach used in most packages today.

Consider a **random class** W from population B. Arguing as for the random effects models, with suitable assumptions, such as normality for inference, $E(B(W)) = 0$ and $V(B(W)) = \sigma_B^2$. This implies that $E(C_i(W)) = 0$, and it is natural to suppose that $C_i(W)$ is normally distributed with some variance $V(C_i(W)) = \sigma_{AB}^2$. Note that assuming $\bar{C}.(W) = 0$ leads to correlation among the interaction random effects which could be problematic when trying to develop inferential tools. No such assumption is pursued below. A conditional argument similar to that given for random models shows that $B(W)$ and $C_i(W)$ are independent. Thus, the random effects are usually

$$B(W) \sim N(0, \sigma_B^2) \text{ and } C_i(W) \sim N(0, \sigma_{AB}^2) .$$

A **designed experiment** for such a mixed model has selected fixed levels $i = 1, \cdots, a$, of factor A and a random sample $\{W_j, j = 1, \cdots, b\}$ from the possible levels of B. The mean response is

$$M(i, W_j) = M_{ij} = \mu + \alpha_i + B_j + C_{ij} ,$$

with $B_j = B(W_j)$ and $C_{ij} = C_i(W_j)$. The random effects **b** and **c** are mutually uncorrelated (and independent if normally distributed). As before, consider a random sample of size n_{ij} for each treatment/class combination. The model for observed responses is

$$y_{ijk} = M_{ij} + e_{ijk} = \mu + \alpha_i + B_j + C_{ij} + e_{ijk} .$$

The formal overall tests for main effects and interactions are the same as for the two-factor random effects model presented in the previous chapter (see the directed graph of Figure 20.1(b)). The real difference lies in the interpretation for the fixed effect A. Recalling arguments for SS from fixed and random effects models, tedious calculations can show that for a balanced design the sums of squares

$$\begin{aligned}
SS_A &= nb \sum_i [\alpha_i + (\bar{C}_{i\cdot} - \bar{C}_{\cdot\cdot}) + (\bar{e}_{i\cdot\cdot} - \bar{e}_{\cdots})]^2 \\
SS_B &= na \sum_j [(B_j - \bar{B}.) + (\bar{C}_{\cdot j} - \bar{C}_{\cdot\cdot}) + (\bar{e}_{\cdot j\cdot} - \bar{e}_{\cdots})]^2 \\
SS_{AB} &= n \sum_i \sum_j [(C_{ij} - \bar{C}_{\cdot j} - \bar{C}_{i\cdot} + \bar{C}_{\cdot\cdot}) + (\bar{e}_{ij\cdot} - \bar{e}_{i\cdot\cdot} - \bar{e}_{\cdot j\cdot} + \bar{e}_{\cdots})]^2
\end{aligned}$$

TWO-FACTOR MIXED MODELS

lead to expected mean squares

$$E(MS_A) = \sigma^2 + n\sigma_{AB}^2 + nb\sum_i \alpha_i^2/(a-1)$$
$$E(MS_B) = \sigma^2 + n\sigma_{AB}^2 + na\sigma_B^2$$
$$E(MS_{AB}) = \sigma^2 + n\sigma_{AB}^2 .$$

Tests for $H_0 : \sigma_{AB}^2 = 0$ and $H_0 : \sigma_B^2 = 0$ and estimates of these variance components are the same as those derived for the random effects model. For the fixed effect, $E(MS_A)$ has elements of both fixed and random effects models. This affects the hypothesis test for $H_0 : \alpha_i = 0$, leading to the pivot statistic

$$F = MS_A/MS_{AB} \sim F_{b-1,(b-1)(a-1);\delta_A}$$

with non-centrality parameter $\delta_A^2 = nb\sum \alpha_i^2/(\sigma^2 + n\sigma_{AB}^2)$ now depending on the fixed effects and on the random interaction variance. If there were no interaction random effect, $\sigma_{AB}^2 = 0$, this would profoundly change inference for α_i.

The distribution of the **grand mean** in the mixed model differs from both the fixed and random models,

$$\hat{\mu} = \bar{y}_{...} \sim N(\mu, (\sigma^2 + na\sigma_B^2 + n\sigma_{AB}^2)/nab) .$$

This suggests the pivot statistic

$$T = (\hat{\mu} - \mu)/\sqrt{MS_B/nab} \sim t_{a-1}$$

for hypothesis tests or confidence intervals.

The **marginal means** for factor A are somewhat complicated, with distribution

$$\hat{\mu}_i = \bar{y}_{i..} \sim N(\mu_i, (\sigma^2 + n\sigma_{AB}^2 + n\sigma_B^2)/nb) ,$$

suggesting a pivot statistic

$$T = (\hat{\mu}_i - \mu_i)/\sqrt{[MS_{AB} + (MS_B - MS_{AB})/a]/nb} .$$

The distribution of this statistic is at best approximate whether using Satterthwaite or another approach.

Contrasts of main effects for factor A are more tractable. The variance of the difference

$$\hat{\alpha}_1 - \hat{\alpha}_2 = \bar{y}_{1..} - \bar{y}_{2..} \sim N(\alpha_1 - \alpha_2, 2(\sigma^2 + n\sigma_{AB}^2)/nb)$$

has a natural estimator

$$\hat{V}(\hat{\alpha}_1 - \hat{\alpha}_2) = 2MS_{AB}/nb .$$

Similar issues arise for mixed models to those found for fixed and random models when there are unequal sample sizes. Some guiding principles are outlined in the next section. For unbalanced designs, Type III SS can be

source	degrees of freedom	sums of squares
A	$a-1$	$(\sigma^2 + n(\sigma^2_{ABC} + b\sigma^2_{AC}))(1 + \delta^2_A/(a-1))$
B	$b-1$	$(\sigma^2 + n(\sigma^2_{ABC} + a\sigma^2_{BC}))(1 + \delta^2_B/(b-1))$
AB	$(a-1)(b-1)$	$(\sigma^2 + n\sigma^2_{ABC})(1 + \delta^2_{AB}/(a-1)(b-1))$
C	$c-1$	$\sigma^2 + n(\sigma^2_{ABC} + b\sigma^2_{AC} + a\sigma^2_{BC} + ab\sigma^2_C)$
AC	$(a-1)(c-1)$	$\sigma^2 + n(\sigma^2_{ABC} + b\sigma^2_{AC})$
BC	$(b-1)(c-1)$	$\sigma^2 + n(\sigma^2_{ABC} + a\sigma^2_{BC})$
ABC	$(a-1)(b-1)(c-1)$	$\sigma^2 + n\sigma^2_{ABC}$
error	$abc(n-1)$	σ^2

Table 21.1. *Random factor and two fixed factors anova*

used to match $E(MS)$. However, recall that MS are no longer distributed as χ^2 and may not be independent. Further, contrasts of levels of the fixed effect no longer have a simple variance,

$$V(\hat{\alpha_1} - \hat{\alpha_2}) = \sigma^2(1/n_1 + 1/n_2) + \sigma^2_{AB}(\sum n^2_{1j}/n^2_{1\cdot} + \sum n^2_{2j}/n^2_{2\cdot})$$
$$+ \sigma^2_B \sum (n_{1j}/n_{1\cdot} - n_{2j}/n_{2\cdot})^2 \ .$$

Most methods dealing with unbalanced mixed models are at best approximate.

Three-factor random and mixed effects models present new dilemmas as it may not be possible to match expected mean squares for all terms. Consider a model with one random factor and two fixed factors, leading to the anova in Table 21.1. The non-centrality parameters are the usual ones for fixed effects while the variance components correspond to the random effects. The directed graph in Figure 21.1 depicts the corresponding errors for testing each effect, with dotted lines for the random factor C. The latter has no natural error. Instead, it is necessary to appeal to a Satterthwaite approximation or, more appropriately, to use REML as outlined in the next section.

21.2 General mixed models

The guiding principle for mixed models is to deal separately with the fixed and random parts of the model. Examine the random parts by first removing all the fixed model parts. Quadratic forms for inference concerning the variance components in the random part should all be orthogonal to the fixed part, and hence will be distributed as weighted sums of central χ^2

GENERAL MIXED MODELS 331

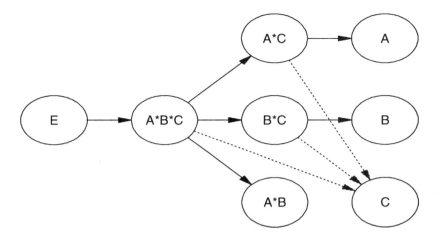

Figure 21.1. *Random factor and two fixed factors. Factor* C *is random while factors* A *and* B *are fixed. Most tests are straightforward in balanced experiments. However, three random factors must be combined for appropriate inference about factor* C *itself.*

variates. The fixed part of the model is composed of main effects and any of their interactions deemed to be present. Quadratic forms appropriate to these have expected mean squares whose distribution may depend on one or more variance components from the random part.

In other words, treat the model as

response = fixed effects + random effects + random error

and use methods already discussed for fixed models and random models. The general layout in matrix notation would be

$$\mathbf{y} = \mathbf{X}\boldsymbol{\beta} + \mathbf{Z}\mathbf{u} + \boldsymbol{\epsilon}$$

or, expanding the random effects in terms of variance components,

$$\mathbf{y} = \mathbf{X}\boldsymbol{\beta} + \sum_{k=1}^{r} \mathbf{Z}_k \mathbf{u}_k + \boldsymbol{\epsilon}$$

with $\mathbf{X}$ and $\boldsymbol{\beta}$ the design matrix and unknown parameters for the fixed effects, $\mathbf{Z} = [\mathbf{Z}_1 : \cdots : \mathbf{Z}_r]$ the design matrix for the random effects and $\mathbf{u}_k$ the random effects with variance components σ_k^2, $k = 1, \cdots, r$. The response has expected value $E(\mathbf{y}) = \mathbf{X}\boldsymbol{\beta}$ and variance

$$V(\mathbf{y}) = \mathbf{V} = \sigma^2 \mathbf{I} + \sum_{k=1}^{r} \sigma_k^2 \mathbf{Z}_k \mathbf{Z}_k^{\mathrm{T}} \ .$$

The least squares estimates for the expected responses are
$$\mathbf{X}\hat{\boldsymbol{\beta}} = \mathbf{X}(\mathbf{X}^T\mathbf{X})^-\mathbf{X}^T\mathbf{y} = (\mathbf{I} - \mathbf{P})\mathbf{y} \ .$$
Estimable functions, which are linear combinations of $\mathbf{X}\boldsymbol{\beta}$, could be estimated in the usual way. However, these estimates are not efficient in the sense that they are not minimum variance. Instead, it would be better to consider
$$\mathbf{X}\hat{\boldsymbol{\beta}} = \mathbf{X}(\mathbf{X}^T\mathbf{V}^{-1}\mathbf{X})^-\mathbf{X}^T\mathbf{V}^{-1}\mathbf{y} = (\mathbf{I} - \mathbf{P}_V)\mathbf{y} \ .$$
Unfortunately, the covariance matrix $\mathbf{V}$ has unknown variance components. The appropriate procedure is to find restricted maximum likelihood estimates of the variance components, and then estimate the fixed effects parameters given these. In practice, this involves projecting down to a random effects model,
$$\mathbf{P}\mathbf{y} = \sum_{k=1}^{r} \mathbf{P}\mathbf{Z}_k\mathbf{u}_k + \mathbf{P}\boldsymbol{\epsilon} \ ,$$
which can be analyzed using methods discussed in the previous chapter. That is, use Satterthwaite approximations, REML, or MINQUE methods, depending on the importance of precision in estimation and testing.

While the estimates $\mathbf{X}\hat{\boldsymbol{\beta}}$ were the best linear unbiased estimates for $\mathbf{X}\boldsymbol{\beta}$ in the fixed model, they no longer have that property in the mixed model. This is because $\mathbf{V}$ is random, and the estimators are no longer linear combinations of responses $\mathbf{y}$. For balanced designs, the BLUEs have simple forms. However, in general the BLUEs are neither best nor linear nor unbiased! In practice, inference on fixed effects proceeds as if the covariance $\mathbf{V}$ were known. Tests and confidence intervals for fixed effects can be developed by matching $E(MS)$, which in general unbalanced designs are very messy. Instead, it is better to consider a full REML approach, as is implemented in the SAS procedure `proc mixed`.

21.3 Problems

21.1 Infer: The two-factor mixed model has been the subject of some controversy. Some authors and some packages add restrictions to the random interactions as specified below. This problem explores some of the issues, but the interested reader should see Samuels, Casella and McCabe (1991) or one of Searle's texts (Searle 1971, ch. 9; Searle, Casella and McCulloch 1992, sec. 4.3).

Consider the two-factor mixed model
$$y_{ijk} = \mu + \alpha_i + B_j^* + C_{ij}^* + e_{ijk}$$
with the restriction that $\sum_{i=1}^{a} C_{ij}^*) = 0$.
(a) Show how B_j^* and C_{ij}^* are related to the random effects B_j and C_{ij} in the unrestricted mixed model.

(b) Suppose all C_{ij}^* have the same distribution. Find their variances and covariances using the relation in (a).

(c) Find the expected mean squares for the fixed effect and two random effects in this model.

(d) Show the relationship between the values in (c) and the expected mean squares for the unrestricted model.

(e) What hypotheses are appropriate for this model? How do they relate to the unrestricted model?

PART H

Nesting Experimental Units

What is the experimental unit? Simply put, it is that item to which a treatment is applied. However, this question can be subtle and is often disregarded in the rush to analyze the data. This part concerns the identification and analysis of designs in which there may be several different experimental units for different aspects of the experiment.

Chapter 22 recasts models from earlier chapters by examining how factors may be nested or crossed with each other. Replication is a simple form of nesting. Blocking and sub-sampling are developed as prototypes of more general nesting. Nesting issues can be subtle, requiring great care uncovering the experimental design and determining appropriate data analysis.

The split plot is the prototype nested design involving two or more factors. Application of different fertilizers may require large fields, while varieties can be planted in much smaller plots on each field. The experimental unit for comparing fertilizers is a field, while the EU for comparing varieties is a plot within a field. Chapter 23 examines this split plot design in some detail.

Chapter 24 briefly considers some general concepts about nested designs. Extensions to split plot and related designs are discussed, along with cautions about unbalanced designs. Covariates can affect response in addition to factors of interest. It may be important to adjust separately for covariates for each size of experimental unit.

CHAPTER 22

Nested Designs

Nested models may arise in experiments restricting randomization or in certain factorial arrangements of combinations of levels. Sometimes restrictions are dictated by pragmatic considerations. It may be possible to hand-plant different varieties in small rows, but fertilizer applications may need a mechanical spreader appropriate to a large field. Manufacturers may want to examine the quality of mechanical parts coming from several plants, but different types of parts may be produced in each plant.

When nesting occurs in the design structure, different factors may be assigned to **different sized experimental units**. When this occurs, each size of experimental unit has an associated source of variation, and inference on factor differences (or variation across classes) must be made relative to that error variation. Thus a random effect must be identified as error variation for each size of experimental unit. However, experiments with nesting only in the treatment structure may still have only one type of experimental unit and hence one identifiable source of error variance.

Sometimes it is very tricky to sort out these different sized experimental units. Pay very close attention to how the experiment was actually conducted, asking questions rather than making assumptions. It is important to understand where randomization is introduced, since it may occur (or perhaps should have occurred!) at several different levels.

Blocking and **sub-sampling** are key features of experiments which have nesting. Many complicated nested designs can be examined in a modular fashion by identifying blocking and sub-sampling for each factor viewed on its own. In a certain sense, nested designs have incomplete blocks, with only some combinations of (design or treatment) factors occurring in each block.

Section 22.1 examines sub-sampling while Section 22.2 develops ideas for blocking. Section 22.3 studies the difference between nesting in the design and treatment structures. Some designs are completely nested and have interesting features in themselves, as developed in Section 22.4. For convenience, this chapter assumes designs are balanced.

22.1 Sub-sampling

Sub-sampling arises naturally when several measurements are taken on the same experimental unit. These measurements do not constitute 'new' experimental units, although they can increase slightly the precision of those already established. Sub-sampling is sometimes known as **pseudo-replication** (Hurlbert 1984). It is quite tempting to consider the added measurements as extra EUs, but this is inappropriate.

Suppose there are a levels of factor A and b experimental units per level. Suppose in addition that there are n measurements per experimental unit. The sub-sampling model for a single factor may be written as

$$y_{ijk} = \mu + \alpha_i + \beta_{ij} + e_{ijk}$$

with $\mu_i = \mu + \alpha_i$ the mean for factor level i, β_{ij} the random effect for the jth experimental unit at level i and e_{ijk} the measurement error. The variance of a response is

$$V(y_{ijk}) = \sigma_B^2 + \sigma^2$$

with variance $\sigma_B^2 = V(\beta_{ij})$ and $\sigma^2 = V(e_{ijk})$ for the (assumed independent) random effects for EU and measurement, respectively. For later consideration of nested designs, it is convenient to denote by $B(A)$ the experimental units. That is, consider the experimental units B as nested within levels of the factor A.

If there was only $n = 1$ measurement per experimental unit, this would simply partition to variance into components for environmental (EU) effects and measurement effects. However, it would not be possible to estimate σ_B^2 and σ^2 separately. With $n = 2$ or more, the variance of the mean response per experimental unit is

$$V(\bar{y}_{ij\cdot}) = \sigma_B^2 + \sigma^2/n \ .$$

That is, there is a common variance component σ_B^2 and a component σ^2/n for the average over the n sub-samples. No matter how many sub-samples are taken, the variance will always be at least σ_B^2. Put another way, sub-sampling can reduce measurement error but cannot affect the inherent error of an experimental unit.

The group means are

$$\bar{y}_{i\cdot\cdot} = \mu + \alpha_i + \bar{\beta}_{i\cdot} + \bar{e}_{i\cdot\cdot}$$

with $\bar{\beta}_{i\cdot} = \sum_{j=1}^{b} \beta_{ij}/b$. The random effect $\bar{\beta}_{i\cdot}$ contributes to the variance of the marginal mean. Focusing on one level of A, the sub-sampling model can be viewed as a one-factor random effects model. The variance of the group means in such a model depends directly on the the experimental unit rather than on the measurement. Thus for level i

$$V(\bar{y}_{i\cdot\cdot}) = (\sigma_B^2 + \sigma^2/n)/b = \sigma_B^2/b + \sigma^2/bn \ .$$

SUB-SAMPLING

source	df	E(MS)
A = factor	$a-1$	$(\sigma^2 + n\sigma_B^2)(1 + \delta^2/(a-1))$
$B(A)$ = EU error	$a(b-1)$	$\sigma^2 + n\sigma_B^2$
sub-sampling error	$ab(n-1)$	σ^2
total	$abn-1$	

Table 22.1. *Sub-sampling expected mean squares*

Table 22.1 summarizes expected mean squares for a factor A in the presence of sub-sampling. The sub-sampling error line drops out if $n = 1$ since it would have no degrees of freedom.

The **pivot statistic** for testing $H_0 : \alpha_i = \bar{\alpha}.$ is

$$F = \frac{MS_A}{MS_{B(A)}} \sim F_{a-1, a(b-1); \delta^2} ,$$

with non-centrality parameter $\delta^2 = bn \sum_i (\alpha_i - \bar{\alpha}.)^2 / (\sigma^2 + n\sigma_B^2)$.

A test of negligible error among experimental units might seem reasonable in practice. The test for the hypothesis $H_0 : \sigma_B^2 = 0$ relies on the pivot statistic

$$F = \frac{MS_{B(A)}}{\hat{\sigma}^2} \sim \frac{\sigma^2 + n\sigma_B^2}{\sigma^2} F_{a(b-1), ab(n-1)} .$$

Negligible variance among experimental units implies that sub-sampling can substantially help improve power and precision. However, as the following example shows, the best way to improve power is to increase the number of experimental units rather than the number of sub-samples.

Example 22.1 Power: The tradeoff between variance components in sub-sampling is sometimes easier to understand with hard numbers. Suppose that measurement variance is $\sigma_B^2 = 1$ while EU variance is $\sigma^2 = 2$, and that two groups are to be compared have a difference of means of $\mu_1 - \mu_2 = 1$. Consider a test to be conducted at significance level 5%. Suppose that experimental units are relatively inexpensive relative to sub-sampling measurements and that a fixed number $nb = 100$ of measurements can be performed per factor level. Table 22.2 shows that as the number of sub-samples per experimental unit n increases, the degrees of freedom decrease and the expected mean square for experimental units $E(MS_{B(A)})$ increases. Both the non-centrality parameter $\delta^2 = 2bn/E(MS_{B(A)})$ and the power decrease as the number of sub-samples n increases.

n	b	df	$E(MS)$	δ^2	power
1	100	99	3	16.7	0.98
2	50	49	5	5.00	0.59
5	20	19	11	0.909	0.14
10	10	9	21	0.238	0.066
20	5	4	41	0.0610	0.052
50	2	1	101	0.00990	0.050
100	1	0	201	0.00248	--

Table 22.2. *Power decrease with sub-sampling*

Figure 22.1 shows the power for selected ratios of measurement to EU variances $\sigma_B^2 : \sigma^2$ from 1:10 to 10:1. Notice that when the measurement error is large, power is low regardless of the number of sub-samples. For a modest ratio, the power decreases quickly after a few sub-samples. This suggests that there is little lost in taking two or three sub-samples to examine measurement error. However, it is harmful to take more than five regardless of the EU to measurement variance ratio. ◊

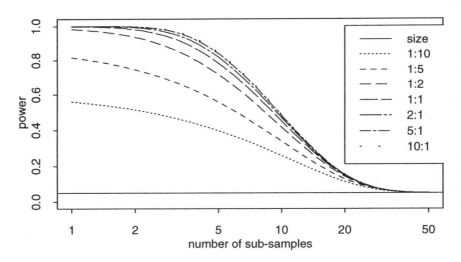

Figure 22.1. *Power decrease with sub-sampling: power to detect mean difference of 1 with 100 total measurements as the number of sub-sample measurements per experimental unit (horizontal axis) and variance ratio $\sigma_B^2 : \sigma^2$ (separate lines) both vary. Based on t test at significance level 0.05.*

22.2 Blocking

The two-factor additive model allows a way formally to present blocking with a single factor in a randomized complete block design. Consider

$$y_{ij} = \mu + \alpha_i + \beta_j + e_{ij}$$

with A_i the factor at level i and B_j the level of the jth block. This provides a mathematical model for the word model

response = mean + factor effect + block effect + error.

The main interest is usually in the hypothesis $H_0 : \alpha_i = \bar{\alpha}.$ for all i, with the block effect a nuisance parameter which much be planned into design and analysis. The blocks may be considered as a fixed or random effect depending on the experiment. Analysis proceeds in the same fashion in either event by removing block effects before examining the factor.

Example 22.2 Nested: Consider again three varieties randomized within each of four rows (= blocks). The model means can be displayed as:

Variety	Block				mean
	1	2	3	4	
A	y_{A1}				$\bar{y}_{A\cdot}$
B					$\bar{y}_{B\cdot}$
C					$\bar{y}_{C\cdot}$
mean	$\bar{y}_{\cdot 1}$	$\bar{y}_{\cdot 2}$	$\bar{y}_{\cdot 3}$	$\bar{y}_{\cdot 4}$	$\bar{y}_{\cdot\cdot}$

The estimator of variety A mean averaged over blocks, $\bar{\mu}_{A\cdot}$, is

$$\hat{\mu}_{A\cdot} = \bar{y}_{A\cdot} \ ,$$

while the estimator of variety A in block 1, μ_{A1}, is

$$\hat{\mu}_{A1} = \bar{y}_{\cdot\cdot} + (\bar{y}_{A\cdot} - \bar{y}_{\cdot\cdot}) + (\bar{y}_{\cdot 1} - \bar{y}_{\cdot\cdot}) = \bar{y}_{A\cdot} + \bar{y}_{\cdot 1} - \bar{y}_{\cdot\cdot} \ .$$

Block means $\hat{\mu}_{\cdot j}$ could be examined for evidence of trend or gradient. ◊

The analysis of variance table for a single factor with balanced blocking is very similar to the anova table for the one-factor model and the two-factor additive model found earlier (Table 22.3). SS_A is calculated in the same way for both completely randomized and randomized complete block designs when they are balanced. The difference lies in removing SS_{BLOCK} from the error sum of squares. The clear advantage is the removal of possible variability due to a known or suspected gradient. Unfortunately, reducing df_{ERROR} lowers the power to detect factor level differences if there is no block effect. Computations are more complicated in unbalanced designs with blocking.

source	df	SS
block	$b-1$	$a\sum_j(\bar{y}_{\cdot j}-\bar{y}_{\cdot\cdot})^2$
A = variety	$a-1$	$b\sum_i(\bar{y}_{i\cdot}-\bar{y}_{\cdot\cdot})^2$
error	$(a-1)(b-1)$	$\sum_{ij}(y_{ij}-\bar{y}_{\cdot j}-\bar{y}_{i\cdot}+\bar{y}_{\cdot\cdot})^2$
total	$ab-1$	$\sum_{ij}(y_{ij}-\bar{y}_{\cdot\cdot})^2$

Table 22.3. *Blocking with one-factor anova*

source	df	SS
block	$b-1$	$an\sum_j(\bar{y}_{\cdot j\cdot}-\bar{y}_{\cdots})^2$
A = variety	$a-1$	$bn\sum_i(\bar{y}_{i\cdot\cdot}-\bar{y}_{\cdots})^2$
error	$abn-a-b-1$	$\sum_{ijk}(y_{ijk}-\bar{y}_{\cdot j\cdot}-\bar{y}_{i\cdot\cdot}+\bar{y}_{\cdots})^2$
total	$abn-1$	$\sum_{ijk}(y_{ijk}-\bar{y}_{\cdots})^2$

Table 22.4. *Blocking with replication in one-factor anova*

If factor levels appear more than once in a block, then it is necessary to expand the model slightly to give

$$y_{ijk}=\mu+\alpha_i+\beta_j+e_{ijk}\ ,$$

with $k=1,\cdots,n_{ij}$. Ideally, the blocking is balanced ($n_{ij}=n$) leading to an anova table of similar form, as shown in Table 22.4. Expected mean squares are also similar, such as

$$E(MS_A)=\sigma^2+nb\sum_i\alpha_i^2/(a-1)\ .$$

Blocking in more complicated designs usually can be presented in the above fashion. That is, remove block effects before examining factor effects. For instance, complicated treatment structures in designs with blocking can be viewed as single factors. After the blocking is understood, the treatment structure could be broken down. Similarly, complicated blocking could be viewed as a single-factor blocking and later dissected.

BLOCKING

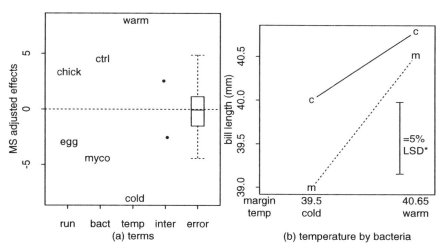

Figure 22.2. *Bacteria interaction plot adjusted for runs. (a) Mean square adjusted effect plot of* run *plus interaction of* temp *and* bact *on* bill *length. (b) Interaction plot of* temp *and* bact *on* bill *length adjusting for effect of* run.

source	df	MS	F	p-value
run	1	8.33	2.10157	0.1488140
bact	1	16.35	4.12617	0.0436328
temp	1	65.06	16.41493	0.0000744
error	188	3.96		

Table 22.5. *Bacteria anova with blocking*

Example 22.3 Bacteria: This experiment on bird development was conducted in two runs separated by several weeks. Several things could have changed in that time, including the mycoplasma culture, seasonal changes of chick growth and food or water conditions. The scientist inoculated eggs in the first run, but decided to switch to inoculating young chicks in the later run. Earlier analysis in the text has assumed that inoculation could be just considered as another factor. Here it is viewed as a blocking factor with no replication. That is, strictly speaking it is not possible to assess the main effect of inoculation method since there is no replication of runs. However, it would be possible to assume that interactions with run were interactions with inoculation.

The tables associated with Example 10.6 can be reinterpreted by considering the inoc term as run, a blocking factor. In particular, the additive model would result in Table 22.5, again with adjusted (Type III) sums of

source	df	SS
B = block	$b-1$	$an\sum_j(\bar{y}_{\cdot j\cdot}-\bar{y}_{\cdots})^2$
A = variety	$a-1$	$bn\sum_i(\bar{y}_{i\cdot\cdot}-\bar{y}_{\cdots})^2$
$A*B$ = interaction	$(a-1)(b-1)$	$n\sum_{ij}(\bar{y}_{ij\cdot}-\bar{y}_{i\cdot\cdot}-\bar{y}_{\cdot j\cdot}+\bar{y}_{\cdots})^2$
error	$ab(n-1)$	$\sum_{ijk}(y_{ijk}-\bar{y}_{ij\cdot})^2$
total	$abn-1$	$\sum_{ijk}(y_{ijk}-\bar{y}_{\cdots})^2$

Table 22.6. *Blocking interaction with one-factor anova*

squares. Note that run should remain in the model and anova table whether or not it is significant. Lack of significance indicates that blocking was not efficient in terms of removing a significant source of variation. However, blocking as two runs was necessary in this experiment given limitations of laboratory space and health considerations. ◊

If the blocks themselves belong to different groups (say, two blocks in each of three locations), it is possible to ignore this distinction initially (say, viewing these as six blocks) and later look for location differences. In addition, separate analyses at each location may help check for consistency across locations. Blocks are usually considered as having additive effects. However, it is sometimes appropriate to consider **interactions** with blocks as in Table 22.6. However, appropriate tests for A now depend on whether blocks are considered fixed or random. That is, if B and hence $A*B$ are random, the appropriate error for a test of $H_0 : \alpha_i = 0$ is MS_{AB}. For an interesting discussion of blocking and mixed models, see the article and discussion of Samuels, Casella and McCabe (1991) and Problem 21.1 above. If there is no replication, a simpler form of interaction such as that suggested by Tukey can be examined as in Chapter 9, Part C.

The **balanced incomplete block design (BIBD)** is used when blocks are too small to contain all factor levels. A balance is achieved by ensuring that every pair of factor levels appears together equally often within blocks, although not necessarily in every block. Suppose there are a levels of the factor of interest, but only $m > 1$ levels can be accommodated in each block. Then there must be a multiple of

$$b = \binom{a}{m} = \frac{a(a-1)\cdots(a-m+1)}{m(m-1)\cdots 1}$$

source	df	$m = a-1$	$m = a-2$
block	$b-1$	$a-1$	$a(a-1)/2 - 1$
factor	$a-1$	$a-1$	$a-1$
error	$b(m-1) - a + 1$	$a(a-3) + 1$	$a(a^2 - 4a + 1)/2 + 1$
total	$bm - 1$	$a(a-1) - 1$	$a(a-1)(a-2)/2 - 1$

Table 22.7. *BIBD degrees of freedom*

source	df	SS	MS	F	p-value
B	2	2.00	1.00	555	0.030
A	2	3.25	1.62	904	0.024
error	1	0.0018	0.0018		
total	5	4.24			

Table 22.8. *Biotron one-factor BIBD anova*

incomplete blocks to have balance. That is, there are b ways to choose m out of a factor levels. The degrees of freedom in the anova table for such a BIBD for a few values of $a - m$ are shown in Table 22.7. Of course, the levels should be randomly assigned to units within each block.

Analysis would proceed as for randomized complete block designs except that it is important to use adjusted (Type III) sums of squares. The argument for using adjusted sums, and for obtaining least squares estimates by factor levels, is exactly the same as that given for additive models with missing cells in Chapter 11, Part D.

Example 22.4 Biotron: A pilot experiment to investigate the effect of calcium on sturdiness of potatoes used a small chamber which could only handle two pots at a time. However, the investigator wanted to compare three levels of calcium. Therefore, three runs were performed, each lasting two weeks, with two calcium levels at a time. The levels in the three blocks were 1,2; 1,3; 2,3. Following the experiment, an anova table was constructed as shown in Table 22.8. The unadjusted (marginal) and adjusted (least squares) means are shown in Table 22.9. The most notable adjustment is for `calcium` level 1, reflecting the higher responses on average for blocks 1 and 2. ◇

calcium	unadjusted	adjusted
1	0.58	0.11
2	0.98	1.28
3	2.03	2.19

block	unadjusted	adjusted
1	1.02	1.52
2	1.76	1.80
3	0.81	0.28

Table 22.9. *Biotron one-factor BIBD means*

22.3 Nested and crossed factors

All the designs considered in this book could be viewed as partially nested designs. Often this is overlooked, but sometimes it can play an important role in understanding possible avenues for analysis. This section defines nesting and considers the implication of nesting for both the design and treatment structures.

Factor B is **nested** within factor A if every level of B appears with only one level of A. The levels of B thus fall into a disjoint sets. Conversely, if every level of B appears with every level of A then A and B are **completely crossed**. Factors which are neither nested nor completely crossed are termed **partially crossed**. These definitions can be applied to factors which are part of the design structure or the treatment structure and whether effects are considered to be fixed or random.

Nesting can be viewed as a special form of **incomplete blocking**. Suppose B is nested within A and A is a blocking factor. Each block has distinct levels of factor B, whereas in complete blocking all levels of B would appear in all blocks. Nested designs are not fully **connected**, and certain comparisons of marginal effects are not possible. The difference in response at levels of B nested within different levels of A depends on both factors. Marginal comparisons of levels of A may or may not be meaningful, depending on the interpretation of the nested levels of B. Thus nested designs may share some of the characteristics of unbalanced designs with missing cells.

Example 22.5 Nested: Suppose one applied some projects to classrooms in several schools which spanned two districts. Measured responses indicate something about the interpersonal intelligence training in each classroom. Classrooms are nested within schools, which are nested within districts. One might expect the response to depend on district philosophies (and

NESTED AND CROSSED FACTORS 347

funding) of education, as well as on fine differences among schools.

Perhaps schools chosen in one district were large while those in the other district were all small. It makes sense to compare schools within a district, but it may be misleading to compare districts, or to compare schools across districts, even adjusting for a 'district' effect. ◇

Example 22.6 Nested: Consider a breeding experiment in which a scientist wants to investigate a trait (say, days to ripening of tomatoes) and has two different genetic varieties of interest. The scientist selects seed from each variety and develops separate 'lines' from each seed. The offspring of these lines are eventually grown to maturity and measured for ripening time.

The offspring (replicates) are nested within lines, which are nested within variety. If the scientist is interested in what line is best, then a broad comparison across lines would be appropriate. This could be useful for selecting lines for future breeding. Alternatively, the scientist may want to know which variety performs best. Another area of investigation might be whether there is any variability among lines, accounting for varietal differences. ◇

Replication in a completely randomized design is nested within factor combination (say, $A * B$). This can be acknowledged by labelling each experimental unit uniquely or by referring to replicates within combinations of factor levels (say, $R(A * B)$). Thus replicates can be divided into ab disjoint sets.

Common **notation** for nesting of B within A is $B(A)$, although some disciplines use B/A. Subscript notation for nesting follows similar conventions, such as

$$\mu_{ij} = \mu + \alpha_i + \beta_{j(i)} \ .$$

While the main effect for levels of factor A can be compared, there is no possible separation of a 'main effect' term for B from interaction. Thus $B(A) = B + B * A$ and the main effects and interactions involving B are usually meaningless and inestimable.

The **simple nested model** may be written as

$$y_{ijk} = \mu + \alpha_i + \beta_{j(i)} + e_{k(ij)} \ .$$

The explicit nesting notation for the random errors is used here just to emphasize that it can be interpreted in this manner. Some disciplines use this form, while others stick with the more compact e_{ijk}. The effects α_i and $\beta_{j(i)}$ may be fixed or random, with appropriate assumptions and/or side conditions.

The SAS language treats explicit nesting differently from interactions. Always make sure that the degrees of freedom match hand calculations.

Unique identifiers on chambers, while not necessary, do help remind the scientist that chambers are nested within temperature levels.

Completely nested designs are somewhat interesting in their own right. The next two sections briefly consider complete nesting with all fixed effects or some random effects to illustrate further the differences between fixed and random effects in nesting.

22.4 Nesting of fixed effects

Nesting in the **treatment structure** is typically in the form of fixed effects. For instance, a breeder may have selected specific lines from certain varieties and be primarily interested in comparing the lines. Experiments with nesting in the treatment structure may have only one size of experimental unit (the offspring).

Consider two fixed factors that are nested, as in the model

$$y_{ijk} = \mu + \alpha_i + \beta_{j(i)} + e_{k(ij)} .$$

Note that a further side conditions are needed on $\beta_{j(i)}$ in order to solve the normal equations. While the natural choice in a balanced design is $\bar{\beta}_{.(i)} = 0$, it is worthwhile to follow this term, especially in unbalanced experiments.

The population marginal means for levels of factor A are

$$\bar{\mu}_{i.} = \mu + \alpha_i + \bar{\beta}_{.(i)}$$

with $\bar{\beta}_{.(i)} = \sum_{j=1}^{b} \beta_{j(i)}/b$. The marginal means $\bar{\mu}_{i.}$ are averaged over the levels of B which are nested within A_i. If the fixed levels of $B(A)$ are not comparable between different levels of A, this may introduce bias, leading to a confounding of the effect of factor A with the levels of B within A.

It is important to use an adjusted (Type III) approach for unbalanced designs. Otherwise, hypotheses can depend on sample sizes. For instance, the sample marginal means have expectations

$$E(\bar{y}_{i..}) = \mu + \alpha_i + \sum_{j=1} n_{ij}\beta_{j(i)}/n_{i.}$$

and hence inference based on these (as in Type I and Type II) may concern hypotheses that depend on the sample sizes. This is analogous to the situation with two crossed factors in unbalanced designs considered in Chapter 10, Part D.

Inference for a nested factor $B(A)$ involves comparing levels after adjusting for the main effect of A in the usual fashion. It is important to keep in mind that different levels of $B(A)$ are associated with the different levels of A. Thus comparisons of B levels should usually be performed within a given level of A. Sometimes it may be appropriate to compare B levels across different levels of A, being aware of possible bias due to the incom-

NESTING OF FIXED EFFECTS

source	df	Type III $E(MS)$
A	$a-1$	$\sigma^2 + \sum n_{i\cdot}(\bar{\mu}_{i\cdot} - \bar{\mu}_{\cdot\cdot})^2/(a-1)$
$B(A)$	$a(b-1)$	$\sigma^2 + \sum n_{ij}(\mu_{ij} - \bar{\mu}_{i\cdot})^2/a(b-1)$
$E(B*A)$	$ab(n-1)$	σ^2
total	$abn-1$	—

Table 22.10. *Two-factor nesting with fixed effects*

plete block structure. Expected mean squares for A and $B(A)$ are shown in Table 22.10.

The **pivot statistic** for testing $H_0 : \alpha_i = 0$ is

$$F = \frac{MS_A}{\hat{\sigma}^2} \sim F_{a-1, ab(n-1); \delta^2}$$

with non-centrality parameter $\delta_A^2 = bn \sum_i (\alpha_i - \bar{\alpha}_\cdot)^2 / \sigma^2$. Tests of $H_0 : \beta_{j(i)} = 0$ rely on the pivot statistic

$$F = \frac{MS_{B(A)}}{\hat{\sigma}^2} \sim F_{a(b-1), ab(n-1); \delta^2}$$

with non-centrality parameter $\delta_{B(A)}^2 = n \sum_{ij} (\beta_{j(i)} - \bar{\beta}_{\cdot(i)})^2 / \sigma^2$.

Example 22.7 Nested: Consider cities (A) nested within counties (B) nested within states (C). Suppose one is interested in all cities with between 100,000 and 150,000 residents and wants to compare driving skills. A random sample of people from each of these cities is chosen. If all cities are chosen, then cities, counties and states would be fixed effects. ◇

The **completely nested three-factor fixed model** would be

$$y_{ijkm} = \mu_{ijk} + e_{ijkm}$$

with model effects

$$\mu_{ijk} = \mu + \alpha_i + \beta_{j(i)} + \gamma_{k(ij)}$$

and corresponding expected mean squares summarized in Table 22.11. Estimable parameters include the city means μ_{ijk}, the county means

$$\bar{\mu}_{ij\cdot} = \mu + \alpha_i + \beta_{j(i)} + \bar{\gamma}_{\cdot(ij)} ,$$

the state means

$$\bar{\mu}_{i\cdot\cdot} = \mu + \alpha_i + \bar{\beta}_{\cdot(i)} + \bar{\gamma}_{\cdot(i\cdot)}$$

and the population grand mean

$$\bar{\mu}_{\cdot\cdot\cdot} = \mu + \bar{\alpha}_\cdot + \bar{\beta}_{\cdot(\cdot)} + \bar{\gamma}_{\cdot(\cdot\cdot)} .$$

source	df	E(MS)
A	$a-1$	$\sigma^2(1+\delta_A^2/(a-1))$
$B(A)$	$a(b-1)$	$\sigma^2(1+\delta_{B(A)}^2/a(b-1))$
$C(A*B)$	$ab(c-1)$	$\sigma^2(1+\delta_{C(A*B)}^2/ab(c-1))$
$E(A*B*C)$	$abc(n-1)$	σ^2

Table 22.11. *Three-factor nesting with fixed effects*

Completely nested models involving fixed effects are relatively rare in practice. More often nesting arises by the way in which randomization is employed in an experiment. Thus the **completely nested random design** merits some examination.

22.5 Nesting of random effects

The **design structure** typically encompasses the random effects in the model for a designed experiment. The designed randomization in fact leads to the justification of the normal approximation for random effects. A completely randomized design has only one random effect, the random error. The split plot design, considered in the next chapter, has two random effects, one for randomized assignment of whole plots and the second for randomization on subplots. Thus a careful understanding of the design of an experiment, in particular where randomization is introduced, can reveal the nature of nesting in the design structure. Design structures with nesting usually have smaller experimental units nested within larger ones.

Consider two random factors that are nested

$$y_{ijk} = \mu + A_i + B_{j(i)} + e_{k(ij)}$$

with $A_i \sim N(0, \sigma_A^2)$ and $B_{j(i)} \sim N(0, \sigma_B^2)$ independent, or a random effect nested within a fixed effect,

$$y_{ijk} = \mu + \alpha_i + B_{j(i)} + e_{k(ij)}$$

with side constraints on α_i to solve the normal equations. The nested random effect $B(A)$ contributes to the variance of main effect A (whether it is fixed or random). Within each level of A, the measurement model is a one-factor random model. The variance of that level mean depends primarily on the random effect $B(A)$ rather than on the random error $E(B(A))$. If A is fixed and the design is balanced, then level i of A has mean $E(\bar{y}_{i..}) = \mu + \alpha_i$

NESTING OF RANDOM EFFECTS

source	df	random A	fixed A
A	$a-1$	$\sigma^2 + n\sigma_B^2 + nb\sigma_A^2$	$(\sigma^2 + n\sigma_B^2)(1 + \delta_A^2/(a-1))$
$B(A)$	$a(b-1)$	$\sigma^2 + n\sigma_B^2$	$\sigma^2 + n\sigma_B^2$
$E(B*A)$	$ab(n-1)$	σ^2	σ^2
total	$abn-1$		

Table 22.12. *Two-factor nesting with random or mixed effects*

and variance
$$V(\bar{y}_{i..}) = \sigma_B^2/b + \sigma^2/nb = E(MS_{B(A)})/nb \ .$$
For a random factor A, the mean is $E(\bar{y}_{i..}) = \mu$ with variance
$$V(\bar{y}_{i..}) = \sigma_A^2 + \sigma_B^2/b + \sigma^2/nb = E(MS_A)/nb \ .$$
Inference about levels of a fixed effect A is done relative to the random effect $B(A)$ rather than the random error. In fact, this was encountered already with sub-sampling. Table 22.12 summarizes expected mean squares for a fixed or random A with random nested $B(A)$.

The **pivot statistic** for testing the fixed A hypothesis $H_0 : \alpha_i = \bar{\alpha}.$ is
$$F = \frac{MS_A}{MS_{B(A)}} \sim F_{a-1,a(b-1);\delta^2} \ ,$$
with non-centrality parameter $\delta_A^2 = bn\sum_i(\bar{\mu}_{i.} - \bar{\mu}_{..})^2/(\sigma^2 + n\sigma_B^2)$. For random A, the hypothesis $H_0 : \sigma_A^2 = 0$ has pivot
$$F = \frac{MS_A}{MS_{B(A)}} \sim \frac{\sigma^2 + n\sigma_B^2 + nb\sigma_A^2}{\sigma^2 + n\sigma_B^2} F_{a-1,a(b-1)} \ .$$
Power for fixed vs. random A differs in subtle ways, analogous to issues already discussed for mixed models in Part G. Tests of $H_0 : \sigma_B^2 = 0$ rely on the pivot statistic
$$F = \frac{MS_{B(A)}}{\hat{\sigma}^2} \sim \frac{\sigma^2 + n\sigma_B^2}{\sigma^2} F_{a(b-1),ab(n-1)} \ .$$

Example 22.8 Nested: Consider a study to examine a certain brand of pain relieving pills for effectiveness, following complaints from the public. Suppose these pills have already reached the stores in their boxes. Perhaps distribution centers in general service several states. Imagine taking a random sample of citys within a state and a random sample of stores

source	df	E(MS)
A	$a-1$	$\sigma^2 + n\sigma_C^2 + cn\sigma_B^2 + bcn\sigma_A^2$
$B(A)$	$a(b-1)$	$\sigma^2 + n\sigma_C^2 + cn\sigma_B^2$
$C(A*B)$	$ab(c-1)$	$\sigma^2 + n\sigma_C^2$
$E(A*B*C)$	$abc(n-1)$	σ^2

Table 22.13. *Three-factor nesting with random effects*

within those citys. Further, take random samples of pillboxes within these stores.

Let u be a city from the population of citys in a selected state (A), $v|u$ be a store from the population of stores in city u $(B(A))$ and $w|u,v$ be a pillbox from the population of pillboxes in store v of city u $(C(A*B))$. Thus a pillbox must be identified by all three classes, as (u,v,w) from the joint population of all three factors. It would have effectiveness (mean response) $m(u,v,w)$. Denoting the grand mean by $\mu = \bar{m}(\cdot,\cdot,\cdot)$, the city effect is $A(u) = \bar{m}(u,\cdot,\cdot) - \bar{m}(\cdot,\cdot,\cdot)$, the store effect is $B(u,v) = \bar{m}(u,v,\cdot) - \bar{m}(u,\cdot,\cdot)$ and the pillbox effect given city and store is $C(u,v,w) = m(u,v,w) - \bar{m}(u,v,\cdot)$. This partitions the mean response as

$$m(u,v,w) = \mu + A(u) + B(u,v) + C(u,v,w) ,$$

with side conditions on the conditional populations implying $\bar{A}(\cdot) = \bar{B}(u,\cdot) = \bar{C}(u,v,\cdot) = 0$.

Draw a random element from the joint population by first drawing a city U, then a store (U,V) and finally a pillbox (U,V,W). Let

$$A(U) \sim N(0,\sigma_A^2)$$
$$B(U,V) \sim N(0,\sigma_B^2)$$
$$C(U,V,W) \sim N(0,\sigma_C^2)$$

It is possible to prove $A(U), B(U,V), C(U,V,W)$ are uncorrelated using a conditional argument as was used earlier for two-factor random models.

An experiment would take a random sample from populations, leading to the expected response

$$\begin{aligned} m_{ijk} &= m(U_i, V_j, W_k) \\ &= \mu + A(U_i) + B(U_i, V_j) + C(U_i, V_j, W_k) \\ &= \mu + A_i + B_{j(i)} + C_{k(ij)} \end{aligned}$$

NESTING OF RANDOM EFFECTS   353

The experimenter would then take a random sample of pills from the pillbox for evaluation of their effectiveness. Sources of variation could be examined at each level of the process by appealing to Table 22.13.   ◇

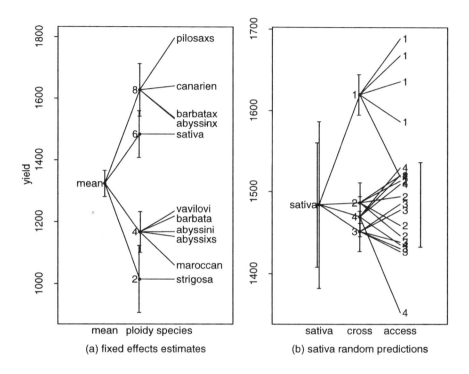

Figure 22.3. *Wheat BLUEs and BLUPs with complete nesting. (a) Best linear unbiased estimators of fixed effects for* ploidy *and* species *nested within* ploidy *show an increasing* yield *with* ploidy. *(b) Best linear unbiased predictors for four randomly chosen* crosses *for species* sativa *each had five* accessions. *Notice that* cross *1 appears to have higher spread among its* accessions *than do the other three. The bars are 95% confidence intervals. The two bars for* sativa *are from the entire data set (left) and from the* sativa *data alone (right).*

Example 22.9 Wheat: An agronomist (Katsiotis 1993) studied the effect of ploidy (number of copies of DNA chromosomes – humans have 2) on the yield in several species of alfalfa. The ploidy ranged from 2 to 8, with one to five species per ploidy. Within each species several matings (known as crosses) were arranged. Each mating produced about five accessions. Roughly 15 offspring from each accession were planted and later harvested to measure yield. This completely nested design has a mix of fixed effects (`ploidy, species`) and random effects (`crosses, accessions, offspring`). Part of this relationship is depicted in the tree diagram of Figure 22.3. ◇

There may be situations where a sequential (Type I) approach would be appropriate for a completely nested random design that is unbalanced. Hypotheses should be checked carefully by writing out models and performing appropriate averaging. The sequential approach leads to SS which are independent and can be combined in ways previously described for random models to construct pivot statistics which are approximately distributed as central F. Variance components could be estimated as well. However, if sample sizes are sufficiently large, the restricted maximum likelihood approach should probably be employed.

22.6 Problems

22.1 Power: Determine the power as a function of the number of subsamples for a 1:4 ratio of EU to measurement variances (as in Figure 22.1) for comparing two groups when the number of measurements per group is only 12. This can be done either assuming the variances are known or that the ratio of estimates is 1:4.

22.2 Bacteria: Refer to Example 22.3 and earlier presentations for details of this experiment. Measurements were taken every 1–2 days during duckling development. The data available here are `bill` length and `leg` length at 21 or 22 days for each duckling. Complete the following for both measurements.
(a) Determine the experimental design (what is the experimental unit for each factor?).
(b) Write out an appropriate model with assumptions.
(c) Test for main effects and interactions, reducing the model accordingly.
(d) Use tables and/or interaction plots of estimates (with appropriate SEs) to illustrate your findings.

22.3 Tomato: The original experiment covered two years (`yr`), with 93 unique plant entries (`entry`). There were several replications each year. Unfortunately, the data now available to the scientist consist of the mean fruit weight (`mfw`), averaged across the replicates for each year. (The raw data are in notebooks halfway around the world!) The scientist believes a $\log_{10}$ transformation (`mfwlog`) is reasonable, and has presented the data in that way.

Nevertheless there are still 2 years of data for most entries. The marker `tg430` can be used to classify entries into one of three categories, 1 = parent A, 3 = parent B, 2 = hybrid of A and B (and . = missing marker value). The

PROBLEMS   355

scientist is particularly interested in the 'additive effect' (parent A minus parent B) and the 'dominance effect' (Hybrid minus mean of parents). Note that if the dominance is zero, then the hybrid would be halfway between the two parents.

Your task is to compare the three marker 'genotypes' over the two years. Are they different? Is the pattern consistent over years? As usual, include enough information so that someone else could reproduce your results. Use graphics to complement formal tests of hypotheses and/or interval estimates as appropriate.

(a) Perform a separate analysis for the second year. [Note that the first year's data are presented in various parts of the book. You may use these as guides.] What do you conclude about the additive and dominance effects? Are the results similar for the first year presented in this text?

(b) Comment briefly on the appropriateness of log transformation for these data. Show evidence (e.g. graphics) to support your statements. (And use plot symbols!)

(c) Discussions with the scientist indicate there were three replicates for each entry arranged in a randomized complete block design. Lay out the model and anova table for one year's complete data, including df and $E(MS)$ and assuming there were no missing data.

22.4 Oocyte: The data for the problem introduced in Example 10.1 were actually collected over several **dates**. Redo the analysis, adjusting for the **dates**. Be sure to include relevant plots (interaction, margin, residual, effect) to support the results and interpretation.

22.5 Wheat: This design involves complete nesting – every factor is nested within the factor before it. Your task is to help the scientist understand the sources of variation and draw inference about the relative performance at different levels of nesting. Since this is a complicated experiment (which took a long time to conduct of course!), it is broken down into manageable modules.

Pretend for the moment that there is simple replication within species. That is, suppose there are just 4-5 independent replicates for each **species** within each **ploidy**, ignoring **cross** and **accession** for now. These might be approximated by averaging across plants within each accession.

(a) Explain in words (briefly) why **species** are nested within **ploidy**. In addition, argue whether **species** and **ploidy** should be considered fixed or random effects.

(b) Write down a model for **species** and **ploidy**, stating assumptions clearly.

(c) Analyze the fixed model as if it were correct. Perform and report on appropriate tests.

(d) Estimate means and standard errors for **ploidy** and **species** levels.

Indicate exactly how these would be calculated in a balanced design.
(e) It may seem wise to perform weighted least squares using weights being the number of plants averaged to get one measurement per cross within species. Why is this NOT a good idea here? You may want to verify for yourself that estimates are worse as compared to Problem 22.7(d).

22.6 Wheat: Now consider only one species. You have an example of the diploid (ploidy = 6) named sativa shown in Figure 22.3. Your task is to do something similar for the hexaploid (ploidy = 2) named strigosa.
(a) Argue briefly the relative merits of using an approximate Satterthwaite approach versus a REML or ML approach to inference for the variance components for cross and accession.
(b) Report on inference for the variance components.
(c) To my knowledge the estimates and standard errors presented in proc mixed are correct. Select one cross within species sativa and verify the relationship between BLUPs of cross means, the BLUE for species mean and the BLUPs of random effects for cross.
(d) The denominator degrees of freedom for estimators and predictors (DDF) are determined in proc mixed using the 'containment method'. That is, they are the smallest degrees of freedom among the random effects that contribute to an effect. This agrees with the interpretation under balanced designs. Justify this in your own words. Verify that the DDFs from proc mixed actually do this.

22.7 Wheat: Now consider the full model with all species for the four ploidy levels and the random effects for cross and accession.
(a) Write down the full model, indicating nesting explicitly. You may do this with math symbols or word models. Define terms carefully.
(b) Test whether effects of ploidy and species are significant. Interpret your results.
(c) Present REML estimates of variance components along with standard errors. Are all variance components significantly different from zero?
(d) Compare estimates and standard errors of ploidy and species with those found in Problem 22.5. Where is there approximate agreement? Did we lose much by simply averaging? Why do you think this might (or might not) be a big problem in this experiment? (As usual, be brief and explicit.)

22.8 Running: Return to Problem 8.2, now considering subjects as blocks. That is, first argue that this is a randomized complete block design with a three-factor treatment structure, and then conduct appropriate analysis. Present relevant plots and interpretation of results.

CHAPTER 23

Split Plot Design

Experiments with two or more factors and different sized experimental units can be very confusing. The split plot is the prototype nested design, encompassing the primary features of nesting with two or more factors. It helps to examine subsets of the experiment to learn about the design and about how to analyze the data to address key questions. Ideas are drawn from blocking and sub-sampling to develop the analysis of the split plot in stages.

Section 23.1 introduces several views of the split plot. A nested model and inference on main effects and interactions are presented in Section 23.2. Contrasts are developed in Section 23.3 with special attention to different sized experimental units.

23.1 Several views of split plot

Split plot designs arise commonly in agriculture, industry and across the biological sciences. They have a two-way or higher-way treatment structure arranged in incomplete blocks, with one or more treatment factors assigned across blocks and the rest assigned to smaller units within each block. The blocks themselves are considered to be random effects and are the experimental units for factors assigned by 'whole plots'. The smaller units, sometimes called 'split plots' or 'subplots', are the experimental units for factors assigned within block by subplot. Assignments are randomized for each size of experimental unit. The design is hopefully balanced. Unbalanced nested designs are relegated to the next chapter.

Example 23.1 Nested: The name **split plot** comes naturally from agricultural field trials in which one factor (A = fertilizer) is assigned by whole plot and a second factor (B = variety) is assigned by split plot. Suppose three fertilizers were assigned at random to six field plots, and four varieties were planted in random order to rows within each plot. Considering only the factor assigned to field plots, the design for fertilizer is a one-factor completely randomized design with sub-sampling, leading to Table 23.1, with δ_A^2 being the usual non-centrality parameter.

source	df	E(MS)
A = fert	2	$(\sigma^2 + 4\sigma_P^2)(1 + \delta_A^2/2)$
plot error	3	σ^2
subplot error	18	σ^2
total	23	

Table 23.1. *Split plot design as sub-sampling*

source	df	E(MS)
block=plot error	5	$\sigma^2 + 4\sigma_P^2$
B = variety	3	$\sigma^2(1 + \delta_B^2/3)$
subplot error	15	σ^2
total	23	

Table 23.2. *Randomized complete block design*

Viewing the field plots as blocks and ignoring the effect of fertilizer yields a one-factor randomized complete block design for varieties within subplots of field plots (Table 23.2). Note that blocks are viewed here as random effects, since the plots are a random sample from a population. Inference about B is not affected by the interpretation of plots as fixed or random. The expected block mean square has a contribution from the subplot, due to averaging over subplot measurements to arrive at the plot means. This was ignored in the previous table, which had only one measurement by plot.

The primary goal may be to assess the response of varieties to different fertilizers, that is the interaction $A*B$. The assignment of variety by fertilizer factor combination is done per row (subplot). Thus the experimental unit for this interaction is the row. This again is an RCBD, but with a more complicated treatment structure, as shown in Table 23.3. The main effect for A is absorbed in the plot error.

source	df	E(MS)
plot error	5	$\sigma^2 + 4\sigma_P^2$
B = variety	3	$\sigma^2(1 + \delta_B^2/3)$
A * B	6	$\sigma^2(1 + \delta_{AB}^2/6)$
subplot error	15	σ^2
total	23	

Table 23.3. *Split plot design as RCBD*

source	df	E(MS)
A = fert	2=$a-1$	$(\sigma^2 + b\sigma_P^2)(1 + \delta_A^2/(a-1))$
plot error	3=$a(n-1)$	$\sigma^2 + b\sigma_P^2$
B = variety	3=$b-1$	$\sigma^2(1 + \delta_B^2/(b-1))$
A * B	6=$(a-1)(b-1)$	$\sigma^2(1 + \delta_{AB}^2/(a-1)(b-1))$
subplot error	9=$a(b-1)(n-1)$	σ^2
total	23=$abn-1$	

Table 23.4. *Split plot design*

Now put these pieces all together, clearly differentiating the part of the anova table concerned with whole plot as experimental unit from that with the split plot as the experimental unit. An easy way to do this is to draw a line between different sized experimental units. ◊

Example 23.2 Forage: Dhiman et al. 1995) examined the effect of the percentage of alfalfa in forage on dry matter intake (dmi). The initial design called for nine heifers and nine cows for each of the five treatments, but three animals were removed from study. The skewed nature of the data and the intuitive reasoning that spread increased with mean dmi suggested a log transformation (results with raw data and with square root transformations

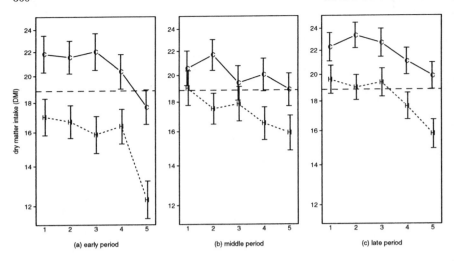

Figure 23.1. *Forage interaction plots by period. Separate interaction plots of dry matter intake (dmi) on heifer/cow and forage treatment (trt) with 95% confidence intervals for (a) early, (b) middle and (c) late periods of lactation cycle. Dashed line is at grand mean.*

turned out to be fairly similar). The scientist was interested in the change in dmi over the course of the lactation cycle for this experiment and divided the study into three periods. That is, the whole plot experimental units are the $(9 + 9) * 5 - 3 = 87$ animals, while the split plot units are the $87 * 3 = 261$ periods per animal. Figure 23.1 shows the (log) averages by period for the five treatments and two ages of animal (heifer/cow). ◊

The split plot experiment can be viewed as a particular form of **mixed model** with two random effects (whole plot and subplot errors) and a two-factor structure of fixed effects. In words, the response is partitioned as

$$\text{response} = (\text{whole plot}) + (\text{split plot}),$$

with each part subdivided into mean model and error,

whole plot = mean + fert + whole plot error
split plot = var + var*fert + subplot error.

Viewing this as a mixed model, consider partitioning the

$$\text{response} = (\text{fixed}) + (\text{random})$$

into components

fixed = mean + fert + var + var*fert
random = whole plot error + subplot error .

SPLIT PLOT MODEL

Put another way, the split plot design has fixed factors A and B crossed. However, the random subplots are nested within random plots. Plots are nested within factor A but are crossed with factor B. subplots are nested within A, B and their interaction, and hence are nested within the fixed factor combinations. Actually, factors A and B need not be fixed, although this is not developed here.

23.2 Split plot model

The **split plot model** in notation is

$$y_{ijk} = \mu_{ij} + r_{ik} + e_{ijk} ,$$

as a mixed model, or, keeping track of different sized experimental units and corresponding random errors,

$$y_{ijk} = \mu + (\alpha_i + r_{ik}) + (\beta_j + \gamma_{ij} + e_{ijk}) .$$

Suitable side conditions identify the fixed effects model parameters,

$$\mu_{ij} = \mu + \alpha_i + \beta_j + \gamma_{ij} .$$

When considering inference, added assumptions are usually made about independent normal errors $r_{ik} \sim N(0, \sigma_P^2)$ and $e_{ijk} \sim N(0, \sigma^2)$.

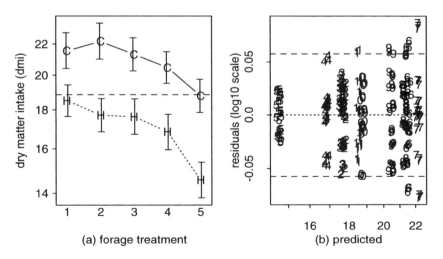

Figure 23.2. *Forage whole plot interaction and residuals. (a) Interaction plot of dry matter intake (*dmi*) on* heifer/cow *and forage treatment (*trt*) with 95% confidence intervals; dashed line at grand mean. (b) Residual plot with symbols (1-5 =* heifer trt*; 6-9, 0 =* cow trt*). Dotted line at zero residual; dashed lines at one SD.*

Example 23.3 Forage: Figure 23.2 shows the whole plot fit (i.e., average over length of study) to the log-transformed dmi. These are appropriate interaction plot and whole plot residuals for examining the main effects of heifer vs. cow based on the average treatment effect and treatment by heifer/cow interactions. However, they are not sufficient if there is evidence of significant interactions found in the split plot (period-within-cow). ◇

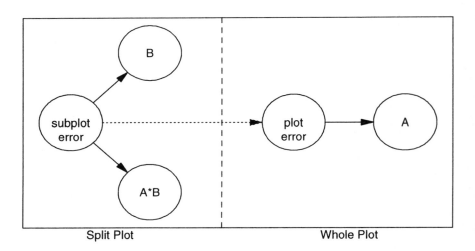

Figure 23.3. *Errors for tests in split plot design. Directional solid lines point from appropriate errors to model parts; directional dotted line suggests informal assessment of whole plot error.*

Inference follows naturally by carefully considering the subparts or by examining the $E(MS)$ now that the design is properly laid out (see Figure 23.3). For fertilizer, assigned by whole plot, the hypothesis $H_0 : \alpha_i = \bar{\alpha}.$ has pivot statistic

$$F = MS_A/MS_{PE} \sim F_{a-1, a(n-1); \delta_A^2} ,$$

with non-centrality parameter $\delta_A^2 = bn \sum_i (\alpha_i - \bar{\alpha}.)^2/(\sigma^2 + b\sigma_P^2)$. Inference for varieties, assigned by split plot, revolves around the hypothesis $H_0 : \beta_j = \bar{\beta}.$, with pivot statistic

$$F = MS_B/MS_{SPE} \sim F_{b-1, a(b-1)(n-1); \delta_B^2} ,$$

and non-centrality parameter $\delta_B^2 = an \sum_j (\beta_j - \bar{\beta}.)^2/\sigma^2$. Interactions similarly have the split plot as the experimental unit. Thus the statistic

$$F = MS_{AB}/MS_{SPE} \sim F_{(a-1)(v-1), a(b-1)(n-1); \delta_{AB}^2} ,$$

addresses the hypothesis $H_0 : \gamma_{ij} = \bar{\gamma}_{i.} - \bar{\gamma}_{.j} - \bar{\gamma}_{..}$ with the non-centrality

SPLIT PLOT MODEL

parameter being

$$\delta^2_{AB} = n \sum_{ij} (\gamma_{ij} - \bar{\gamma}_{i\cdot} - \bar{\gamma}_{\cdot j} + \bar{\gamma}_{\cdot\cdot})^2 / \sigma^2 \ .$$

Example 23.4 Nested: A typical SAS run for a balanced split plot design could use the test statement in the anova or glm procedure, such as:

```
proc anova;
    class A B R;
    model y = A R(A) B A*B;
    test h = A e = R(A);
    means A / lsd e = R(A);
    means B A*B / lsd;
```

Here A and B are the factors of interest, and R(A) identifies each whole plot experimental unit. Analysis of unbalanced split plot can be approximated with proc glm:

```
proc glm;
    class A B R;
    model y = A R(A) B A*B;
    random R(A) / test;
    lsmeans A / stderr pdiff e=R(A);
    means B A*B / stderr pdiff;
```

The random statement occasionally does not work properly when the design is rather complicated and unbalanced. In fact, the newer proc mixed procedure is preferred in practice:

```
proc mixed;
    class A B R;
    model y = A B A*B;
    random R(A);
    lsmeans A B A*B;
```

However, it is wise to check this against proc glm when possible to build confidence in the REML method. ◇

Example 23.5 Forage: Figure 23.4 shows the mean-square adjusted effects as suggested in Part F of this book (see Figure 10.2). Separate plots are needed for the different sized experimental units. Note how the spread in main effects and interactions corresponds to the level of significance of those terms. Formal analyses showing significant high-order effects support presentation of separate results by period such as Figure 23.1. ◇

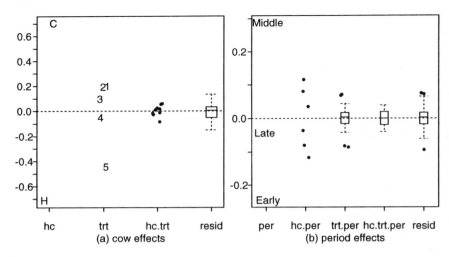

Figure 23.4. *Forage mean square adjusted effects. Effect plots for (a) whole plot and (b) split plot. Size of estimated effects is adjusted so that the sum of squared deviations from 0 (dashed lines) equals the respective mean squares. Thus the spread for a main effect or interaction can be directly compared to the spread of residuals. Box-plots replace individual points when there are many levels to an effect. Note difference in vertical scales.*

Estimation of variance components and fixed effects follows along already developed lines for mixed effects models. There is little new to add here, except to note that this is a very specific type of experiment which is widely used and which illustrates key features of many nested designs. However, **residual plots** should be developed with some care, as shown in the following example.

Example 23.6 Forage: There is some evidence of unequal variance even with the log transform, as seen in the residual plots for the cow as experimental unit (Figure 23.1) and for each period (Figure 23.5). Another way to examine residuals across the levels of nesting is to consider orthogonal linear functionals which span the whole-cow and period-within-cow spaces. That is, there are 86 whole-cow degrees of freedom and 174 for the periods. Figure 23.6 shows the predicted values against residuals for the orthogonal linear functionals corresponding to these degrees of freedom. Again, there is some evidence of unequal variance, but it seems minor and is ignored here. While these predicted values and residuals have nice statistical properties for examining patterns of fit, they are difficult to relate to the observed values. ◇

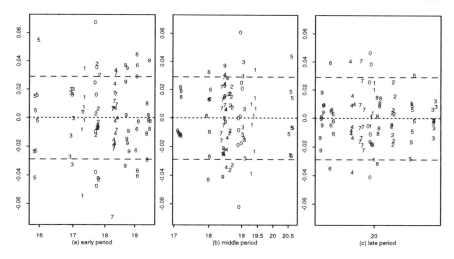

Figure 23.5. *Forage split plot residuals by period. Residual plots for (a)* early, *(b)* middle *and (c)* late *periods. Plot symbols represent forage treatment (1–5 =* heifer trt; *6–9, 0 =* cow trt*). Dotted line at zero residual; dashed lines at one SD.*

23.3 Contrasts in a split plot

Contrasts of fixed main effects naturally depend on the size of experimental unit. For the whole plot,

$$\begin{aligned}
\hat{\mu}_{1\cdot} - \hat{\mu}_{2\cdot} &= \mu_{1\cdot} - \mu_{2\cdot} + (\bar{r}_{1\cdot} - \bar{r}_{2\cdot}) + (\bar{e}_{1\cdot\cdot} - \bar{e}_{2\cdot\cdot}) \\
&= \bar{y}_{1\cdot\cdot} - \bar{y}_{2\cdot\cdot} \\
&\sim N(\mu_{1\cdot} - \mu_{2\cdot}, 2E(MS_{PE})/bn),
\end{aligned}$$

which leads to confidence intervals, for instance,

$$\bar{y}_{1\cdot\cdot} - \bar{y}_{2\cdot\cdot} \pm t_{\alpha/2;a(n-1)}\sqrt{2MS_{PE}/bn}.$$

Contrasts at the subplot level,

$$\begin{aligned}
\hat{\mu}_{\cdot 3} - \hat{\mu}_{\cdot 4} &= \mu_{\cdot 3} - \mu_{\cdot 4} + (\bar{e}_{\cdot 3\cdot} - \bar{e}_{\cdot 4\cdot}) \\
&= \bar{y}_{\cdot 3\cdot} - \bar{y}_{\cdot 4\cdot} \\
&\sim N(\mu_{\cdot 3} - \mu_{\cdot 4}, 2E(MS_{SPE})/an),
\end{aligned}$$

lead to confidence intervals involving the split plot variance,

$$\bar{y}_{\cdot 3\cdot} - \bar{y}_{\cdot 4\cdot} \pm t_{\alpha/2;a(b-1)(n-1)}\sqrt{2MS_{SPE}/an}.$$

There is more **power** (in terms of degrees of freedom) for contrasts in the split plot than in the whole plot. This has profound implications for

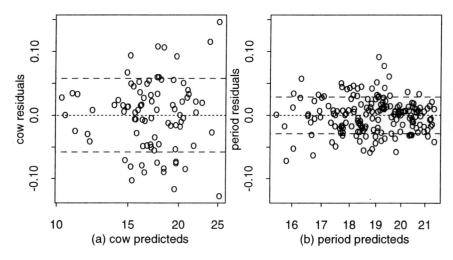

Figure 23.6. *Forage orthogonal residuals. Plots contain one point for each degree of freedom in the (a) whole plot (*cow*) or (b) split plot (*period*). Horizontal lines at zero (dotted) and one SD (dashed).*

when and how to use split plot designs when there is a choice in the matter. It is best to put the factor of most interest in the split plot. If both factors have equal interest, then it is better to avoid split plot designs and use a completely randomized design (or RCBD), if possible.

Comparison of cell means must be done with care. Some comparisons represent pure interactions ($\mu_{13} - \mu_{14} - \mu_{23} + \mu_{24}$),

$$\bar{y}_{13\cdot} - \bar{y}_{14\cdot} - \bar{y}_{23\cdot} + \bar{y}_{24\cdot} \pm t_{\alpha/2;a(b-1)(n-1)}\sqrt{4MS_{SPE}/n} ,$$

but some concern main effects in the whole plot ($\mu_{1j} - \mu_{2j}$) while others involve main effects comparisons in the subplot ($\mu_{i3} - \mu_{i4}$). Others mix up all model aspects, for example comparing arbitrary means $\mu_{13} - \mu_{24}$ as in a one-factor model interpretation. Determine what random effects remain in differences and ascertain appropriate variance estimates. Some comparisons, like the last one, might be approximated using a Satterthwaite approach, or more appropriately, employing REML.

Linear and other polynomial contrasts can be used to examine trends in fixed factors. Explicitly extract linear, quadratic and cubic components as contrasts or use regressors and Type I SS. However, take some care how this is done in the presence of random effects. Milliken and Johnson (1992, ch. 24) have a nice example.

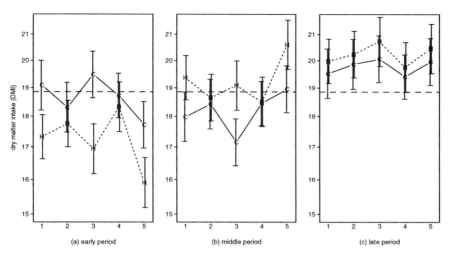

Figure 23.7. *Forage split plot interactions by period. Separate interaction plots with 95% confidence intervals for (a)* early, *(b)* middle *and (c)* late period *of lactation cycle. Whole plot effects of* heifer/cow, *forage treatment and animal have been removed. Grand mean (dashed line) has been added back for easy comparison.*

Example 23.7 Forage: The change in dmi over the course of the lactation cycle for this experiment can be examined with the residual within period. That is, subtract out the cow effect and examine the variation across the three periods. For convenient interpretation, the plots in Figure 23.7 add back the overall mean. Notice the reversal of heifers and cows from the early to later periods and the higher response than average for the middle period. Another interpretation involves noting the common treatment and heifer/cow effects (Figure 23.2) with evidence of complications as shown in either Figure 23.1 or 23.7. ◇

23.4 Problems

23.1 Brassica: This experiment is described in Example 24.2. Here, consider only the first year of data, ignoring crop 2. Carefully write down the statistical model for the effect of crop and pyrethrum rate on yield (with words or symbols), defining all terms and clearly stating assumptions. State hypotheses, carry out analysis and report results including any relevant interaction plots. A column of plotting codes (trt) is provided for convenience.

23.2 Tomato: Conduct a split plot analysis over both years. (See Problem 22.3 and examples in the book for details of the experiment.) You may reduce to a balanced subset (only entries with data for both years).

(a) What are the whole plots? What are the subplots? Be exact in your description (in math or words) of the model and assumptions. Conduct formal tests and summarize results graphically.

(b) What more would you have done with 'all' the data? You may assume there were four replicates per entry in each year and no missing data. Describe the design, making educated guesses if needed, and sketch the anova table (indicate sources and degrees of freedom).

CHAPTER 24

General Nested Designs

General designs are scary because there is so much going on. Some factors may be nested, others crossed, still others only partially crossed. Further, there may be covariates at several levels of nesting. It can be very helpful to examine subsets of the design as if they were the whole experiment. Once these are understood, they can often be put together in a logical fashion.

The split plot design embodies most of the key features that arise when there is more than one random error, associated with more than one size of experimental unit. Mixed models provide the machinery for separately handling fixed and random components of an experimental design. That is, use an approach which fits the fixed parts of the model first, then sorts out the random parts conditional on the fixed part. This insures that SS for fixed effects are orthogonal to their corresponding SS for error.

Some natural extensions of split plot designs are noted in Section 24.1. Unbalanced nested designs are briefly examined in Section 24.2. Covariates in nested designs are addressed in Section 24.3. Overall summaries such as explained variation are discussed briefly in Section 24.4.

24.1 Extensions of split plot

The split plot design detailed in the previous chapter had a completely randomized design for whole plots and a randomized complete block design for subplots. More complicated arrangements are possible at each level. The key lies in recognizing the split plot nature of the assignment of factor levels.

Several variations on the split plot model readily present themselves. There are split split plot models. Consider a different twist in which the whole plots are assigned with an RCBD. Here are just a few.

Example 24.1 Nested: Suppose there are n locations and a blocks per location. The a fertilizers may be assigned at random to blocks (field plots) and assignment of varieties as before. The anova table is shown in Table 24.1. There are three sizes of experimental units, although no factors are assigned to the largest one (location). This RCBD on the whole plots does not alter how factor levels assigned on subplots would be analyzed. ◇

source	df	E(MS)
$R = $ loc	$n - 1$	$\sigma^2 + b\sigma_P^2 + ab\sigma_R^2$
$A = $ fert	$a - 1$	$(\sigma^2 + b\sigma_P^2)(1 + \delta_A^2/(a-1))$
plot error	$(a-1)(n-1)$	$\sigma^2 + b\sigma_P^2$
$B = $ variety	$b - 1$	$\sigma^2(1 + \delta_B^2/(b-1))$
$A * B$	$(a-1)(b-1)$	$\sigma^2(1 + \delta_{AB}^2/(a-1)(b-1))$
subplot error	$a(b-1)(n-1)$	σ^2
total	$abn - 1$	

Table 24.1. *Split plot with blocking*

Example 24.2 Brassica: Miller (1993) examined the effects of rate of application of a certain herbicide (pyridate) on a range of closely related crops. The experiment was conducted over two years, with three blocks each year. The two years were conducted in roughly the same location, but there was separate randomization in each year. Three herbicide rates (pyrrat = 0, 1, 2) were randomly assigned to separate rows within each crop plot. Recorded values consist of yield per row for each crop*pyrrat combination within each block for each year. Since crop 2 was not used in the second year, it is dropped from further consideration.

This is essentially a split split plot design with randomized blocking in the whole plot. Figure 24.1 shows the essential features of the data analysis. Figure 24.1(a) shows the split plot interaction of crop and year. Notice that year 2 appeared to be better for one high-yield crop (4) while year 1 seemed better for lower-yielding crops (3, 8). Figure 24.1(b) focuses on deviations from mean crop yield associated with pyridate, showing that only two crops suffered in the presence of the herbicide (4, 6) while the others maintained or improved slightly. The LSD bars are meant for vertical comparisons at each level of the horizontal axis; other comparisons might use different length bars.

It is easier to understand this experiment by considering one year at a time. However, Table 24.2 shows a comprehensive analysis. Formal F tests are given for the random effects, although these should be interpreted as advisory, showing the efficiency of blocking. The year*crop and crop*pyrrat are both highly significant, supporting evidence for Figure 24.1. ◊

EXTENSIONS OF SPLIT PLOT 371

source	df	SS	F	p-value
year	1	36.132	9.93	0.035
block(year)	4	14.555	7.60	0.0001
crop	6	518.284	32.49	0.0001
year*crop	6	222.782	13.97	0.0001
block*crop(year)	24	63.804	5.55	0.0001
pyrrat	2	3.904	4.08	0.022
year*pyrrat	2	0.490	0.51	0.60
crop*pyrrat	12	25.758	4.48	0.0001
year*crop*pyrrat	12	7.814	1.36	0.21
error	56	26.819		
total	125	920.341		

Table 24.2. *Brassica split split plot anova*

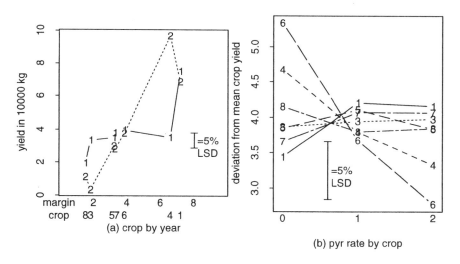

Figure 24.1. *Brassica split split plot anova. (a) Margin plot for* year*crop *interaction with symbols for* year *(1 = 1991, 2 = 1992). (b) Interaction plot on deviations from mean* crop *yield by* pyridate *rate with symbols for* crop*).*

24.2 Strip plot

A design which is superficially similar to the split plot is the **strip plot** in which plots are divided into subplots in different ways for different factors. This leads to rather strange designs, with different sized (and shaped) experimental units for different factors.

Example 24.3 Nested: Consider a two-way treatment structure in which one randomly assigns the levels of factor A (say, the amount of a herbicide) by row and the levels of factor B (say, plant variety) by column. This may be the only economical way to set up an experiment using large machines. The EU for the main effect of A is a row, while a column is the EU for the main effect of B. Treatment combinations are actually assigned by specifying both row and column. Thus interactions of $A * B$ have row*column (cells) as their experimental units. There are three different sized experimental units in this experiment. Note that one can draw inference about main effects and interaction only if the strip plot is replicated. ◇

A strip plot model for a design with replication by having different sets (with different randomization of rows and columns in each set) could be written as

$$y_{ijk} = \mu + (s_k) + (\alpha_i + r_{ik}) + (\beta_j + c_{jk}) + (\gamma_{ij} + e_{ijk}) ,$$

with $i = 1, \cdots, a$, $j = 1, \cdots, b$ and $k = 1, \cdots, n$. The fixed effects are, as usual,

$$\mu_{ij} = \mu + \alpha_i + \beta_j + \gamma_{ij} .$$

There are four random effects, corresponding to sets (or blocks) $s_k \sim N(0, \sigma_S^2)$, rows $r_{ik} \sim N(0, \sigma_R^2)$, columns $c_{jk} \sim N(0, \sigma_C^2)$ and cells $e_{ijk} \sim N(0, \sigma^2)$. The design, with its different sized experimental units and random effects, suggests a natural partition of sources of variation (Table 24.3).

Inference follows along similar lines to the split plot design, with obvious adjustments arising from the further complications in this design. The same cautions about estimates and contrasts of cell means arise here. For instance, the **grand mean** is estimated by

$$\bar{y}_{...} \sim N(\mu, E(MS_{\text{SET ERROR}})/abn) ,$$

while **main effects contrasts** for A are estimated by

$$\bar{y}_{1..} - \bar{y}_{2..} \sim N(\bar{\mu}_{1.} - \bar{\mu}_{2.}, 2E(MS_{\text{ROW ERROR}})/bn) .$$

Main effects contrasts for B follow similarly, depending on $E(MS_{\text{COL ERROR}})$. Interaction contrasts get messy as for the split plot design and must be done with care.

source	df	$E(MS)$
set error	$3=n-1$	$\sigma^2 + b\sigma_R^2 + a\sigma_C^2 + ab\sigma_S^2$
A	$2=a-1$	$(\sigma^2 + b\sigma_R^2)(1 + \delta_A^2/(a-1))$
row error	$6=(a-1)(n-1)$	$\sigma^2 + b\sigma_R^2$
B	$1=b-1$	$(\sigma^2 + a\sigma_C^2)(1 + \delta_B^2/(b-1))$
col error	$3=(b-1)(n-1)$	$\sigma^2 + a\sigma_C^2$
$A*B$	$2=(a-1)(b-1)$	$\sigma^2(1 + \delta_{AB}^2/(a-1)(b-1))$
cell error	$6=(a-1)(b-1)(n-1)$	σ^2
total	$23=abn-1$	

Table 24.3. *Strip plot design*

24.3 Imbalance in nested designs

Nested factors can usually be addressed with an adjusted (Type III) approach. Partially crossed factors may have missing cells and require a balanced subsets (Type IV) approach. Sometimes it may be preferable to combine nested factors as a single factor to simplify analysis and/or multiple comparisons. Thus it may not be possible simply to use one approach for all factors in an experiment which is unbalanced in some manner.

Imbalance in nested designs is more critical if it occurs with the smaller experimental units. For instance, in split plot designs, if there are unequal numbers of whole plots, this only affects comparisons of factors assigned to subplots. However, if there are unequal numbers of subplots per whole plot, this affects comparisons for factors assigned to whole plots and subplots.

A Type III approach for the fixed factors addresses these concerns as far as overall F tests (provided there are no missing cells for fixed effects). However, contrasts of cell means in unbalanced nested designs may be problematic, requiring separate handling of each contrast. Milliken and Johnson (1992, ch. 28) consider two examples, one with imbalance in the whole plot and one with imbalance in the split plot.

Note also that the SAS `proc glm` procedure does not produce correct `lsmeans` for mixed models which are unbalanced. In fact, the standard errors for balanced nested designs are incorrect even though the least squares means may be correct. This is because `proc glm` treats all effects as fixed except for results using the `random` statement. These problems have been largely overcome with the newer `proc mixed`. That is, LS means and their standard errors appear to be correct. However, there is some ambiguity when it comes to inference, in particular in deciding on degrees of freedom! The interested reader should carefully read the latest documentation and supporting literature. In practice, it can be handy to use both procedures when possible, to compare results.

24.4 Covariates in nested designs

Covariates can be incorporated into random effects and nested models. However, it is best to proceed cautiously, as interpretation can be tricky. This chapter illustrates some issues which arise with covariates in a **split plot** design and with random effects in another context.

Several questions surface when covariates are part of analysis of a split plot design. What is the experimental unit for the covariate? Was it measured on the whole plot or the split plot? If measured on the whole plot, how does it affect comparisons of whole plot factors and/or split plot factors? If measured on the split plots, does the covariate affect analysis of the whole plot?

Consider first a covariate measured on the **whole plot**. The model would be (parallel line case)

$$y_{ijk} = \mu + \alpha_i + \delta(x_{ik} - \bar{x}_{..}) + r_{ik} + \beta_j + \gamma_{ij} + e_{ijk} ,$$

with the usual assumptions, including sum-to-zero side conditions. The cell mean response,

$$\mu_{ij} = \mu + \alpha_i + \delta(\bar{x}_{i.} - \bar{x}_{..}) + \beta_j + \gamma_{ij} ,$$

depends on the covariate, suggesting the need for some adjustment. Whole plot contrasts for factor A are of the form

$$\mu_{1.} - \mu_{2.} = (\alpha_1 - \alpha_2) + \delta(\bar{x}_{1.} - \bar{x}_{2.}) ,$$

which depend on the covariate. However, split plot contrasts for factor B,

$$\mu_{.3} - \mu_{.4} = \beta_3 - \beta_4 ,$$

do not. Similarly, pure interaction contrasts depend only γ_{ij},

$$(\mu_{13} - \mu_{14}) - (\mu_{23} - \mu_{24}) = \gamma_{13} - \gamma_{14} - \gamma_{23} + \gamma_{24} .$$

Thus a covariate which is measured on the whole plot can only affect comparisons and tests for the whole plot. Effectively, removing the whole plot effect as a 'block', in order to analyze the split plot, removes the covariate as well. The whole plot analysis is identical to that for parallel lines.

The ancova for such a situation would be as in Table 24.4. There are na unique values of the covariate, each repeated b times. In other words, operationally one would define $x_{ijk} = x_{ik}$, or alternatively analyze the whole plot ancova on its own by first averaging over the split plot experimental units. The split plot analysis is unaffected by the whole plot covariate.

source	df	$E(MS)$
$X\|A$	1	$(\sigma^2 + b\sigma_P^2)(1 + \delta_X^2)$
$A\|X$	$a-1$	$(\sigma^2 + b\sigma_P^2)(1 + \delta_{(}A^2/(a-1))$
WP error	$an - a - 1$	$\sigma^2 + b\sigma_P^2$
B	$b-1$	$\sigma^2(1 + \delta_B^2/(b-1))$
$A*B$	$(a-1)(b-1)$	$\sigma^2(1 + \delta_{AB}^2/(a-1)(b-1))$
SP error	$a(b-1)(n-1)$	σ^2
Total	$abn - 1$	—

Table 24.4. *Ancova in whole plot of nested design*

Here is some SAS code (assuming X is properly set up already).

```
proc glm;
   class A B R;
   model Y = X A R(A) B A*B / solution;
   random R(A) / test;
   lsmeans A / e=R(A) stderr pdiff;
   lsmeans B A*B / stderr pdiff;
proc sort; by R A;
proc means noprint; by R A;
   var Y X;
   output out=means m=MY MX;
proc plot;
   plot MY*MX=A;
```

A covariate which is measured on the **split plot** experimental unit, however, should be handled somewhat differently. It will usually have a component which can be attributed to the whole plot and another which can be attributed to the split plot. Here is a model,

$$y_{ijk} = \mu + \alpha_i + \gamma \bar{x}_{i \cdot k} + r_{ik} + \beta_j + (\alpha\beta)_{ij} + \eta(x_{ijk} - \bar{x}_{i \cdot k}) + e_{ijk},$$

with the usual assumptions, including sum-to-zero side conditions. The cell mean response,

$$\mu_{ij} = \mu + \alpha_i + \gamma \bar{x}_{i \cdot \cdot} + \beta_j + (\alpha\beta)_{ij} + \eta(\bar{x}_{ij \cdot} - \bar{x}_{i \cdot \cdot}),$$

depends profoundly on the covariate. Whole plot contrasts for factor A are of the form

$$\mu_{1 \cdot} - \mu_{2 \cdot} = (\alpha_1 - \alpha_2) + \gamma(\bar{x}_{1 \cdot \cdot} - \bar{x}_{2 \cdot \cdot}),$$

which depend on the covariate in the same way as for covariates only in

source	df	$E(MS)$
$W\|A$	1	$(\sigma^2 + b\sigma_P^2)(1 + \delta_{W\|A}^2)$
$A\|W$	$a - 1$	$(\sigma^2 + b\sigma_P^2)(1 + \delta_{A\|W}^2/(a-1))$
WP error	$an - a - 1$	$\sigma^2 + b\sigma_P^2$
$Z\|A, AB$	1	$\sigma^2(1 + \delta_{Z\|B,AB}^2)$
$B\|Z, AB$	$b - 1$	$\sigma^2(1 + \delta_{B\|Z,AB}^2/(b-1))$
$AB\|Z, B$	$(a-1)(b-1)$	$\sigma^2(1 + \delta_{AB\|B,Z}^2/(a-1)(b-1))$
SP error	$a(b-1)(n-1) - 1$	σ^2
total	$abn - 1$	—

Table 24.5. *Ancova in whole plot and subplot*

the whole plot. Now, split plot contrasts for factor B,

$$\mu_{\cdot 3} - \mu_{\cdot 4} = \beta_3 - \beta_4 + \eta(\bar{x}_{\cdot 3 \cdot} - \bar{x}_{\cdot 4 \cdot}) ,$$

depend on the covariate as well. In addition, pure interaction contrasts do not escape the covariate,

$$(\mu_{13} - \mu_{14}) - (\mu_{23} - \mu_{24}) = (\alpha\beta)_{13} - (\alpha\beta)_{14} - (\alpha\beta)_{23} + (\alpha\beta)_{24}$$
$$+ \eta \times (\bar{x}_{13\cdot} - \bar{x}_{14\cdot} - \bar{x}_{23\cdot} + \bar{x}_{24\cdot}) .$$

Thus a covariate which is measured on the split plot can affect all comparisons and tests for both split plot and whole plot. The whole plot analysis is again analogous to that for parallel lines. The split plot analysis has to be adjusted for the covariate.

The ancova for such a situation is shown in Table 24.5. Operationally, there are two covariates. One (call it W) has na unique values of the covariate X averaged over the split plot, each repeated b times, $w_{ijk} = \bar{x}_{i \cdot k}$. The other (call it Z) which has the deviations $z_{ijk} = x_{ijk} - \bar{x}_{i \cdot \cdot}$.

Here is some SAS code that should do the trick. Remember that Type I and Type III interpretations of the covariate may both be important. It is also important to look at residual plots, separating the whole plot residuals from the split plot as described in earlier chapters of this part of the book.

```
proc sort data=a; by R A;
proc means noprint; by R A;
   var X;
   output out=b m=W;
data c; merge a b; by R A;
   Z = X - W;
```

```
proc glm;
   class A B R;
   model Y = W A R(A) Z B A*B / solution;
   random R(A) / test;
   lsmeans A / e=R(A) stderr pdiff;
   lsmeans B A*B / stderr pdiff;
/* look at whole plot and split plot together */
proc plot;
   plot Y*X=A;
   plot Y*X=B;
/* look at whole plot only */
proc sort; by R A;
proc means noprint; by R A;
   var Y X;
   output out=d m=MY MX;
proc plot;
   plot MY*MX=A;
```

24.5 Explained variation in nested designs

Since this section concerns a prototype design for nested models with more than one source of variation, it is appropriate to examine the interpretation of the **explained variation** or (multiple correlation) coefficient of determination, R^2. When there are two or more sources of error variation and several sizes of experimental units, the meaning of R^2 is not entirely clear. Packages automatically display R^2, and it is frequently quoted to indicate the 'goodness of fit'.

In RCBDs, interest typically focuses on factor effects, with the blocking being necessary to control some known or suspected variation. Therefore, it usually does not make sense to include block variation in the explained variation R^2, particularly if this is to be compared with other experiments using the same or similar factor levels. Large block effects raise two counter-intuitive problems if included in R^2. First, the block variation may mask variation explained by factors. Second, large block effects ought to raise suspicions about the results and about a simple model of additive block effects. Therefore it seems prudent to examine **explained variation after removing blocking**,

$$R^2 = (SS_{\text{MODEL}} - SS_{\text{BLOCK}})/(SS_{\text{TOTAL}} - SS_{\text{BLOCK}}) \ .$$

Similar arguments can be made regarding sub-sampling.

In split plot designs, part of the variation 'explained' is the whole plot error. Thus

$$R^2 = SS_{\text{MODEL}}/SS_{\text{TOTAL}}$$

summarizes explained variation only in models with exactly one term for unexplained variation. Blocking and whole plot error are mechanically considered part of the 'model' for most packages, even if they allow subsequent tests using different errors. What should one do with more than one error term? The default package approach for the split plot design yields

$$R^2 = (SS_A + SS_{PE} + SS_B + SS_{AB})/SS_{\text{TOTAL}}$$
$$= (SS_{\text{TOTAL}} - SS_{SPE})/SS_{\text{TOTAL}}.$$

Alternatively, lump all random errors together and consider only that part of the variation explained by model mean terms,

$$R^2 = (SS_A + SS_B + SS_{AB})/SS_{\text{TOTAL}}$$
$$= (SS_{\text{TOTAL}} - (SS_{PE} + SS_{SPE}))/SS_{\text{TOTAL}}.$$

This is not entirely satisfactory, since it fails to distinguish how well the model does on whole versus split plot parts. Perhaps it would be better to consider separate R^2s for each size of EU. For the whole plot,

$$R_P^2 = SS_A/(SS_A + SS_{PE}),$$

while for the split plot

$$R_{SP}^2 = (SS_B + SS_{AB})/(SS_B + SS_{AB} + SS_{SPE}).$$

24.6 Problems

24.1 Diet: The experiment described in Problem 4.2 actually was a randomized block design. That is, `cows` were `blocked` by time, the first six `cows` were randomly assigned among the six diets, and so on. 'Proper' analysis should take account of this, along with the initial capacity of each animal (its `covariate` of dry matter intake at three weeks) and possibly weighting by the number of `weeks` each `cow` was on trial.

24.2 Bacteria: Determine whether duckling `bill` length at 21 or 22 days adequately explains differences in `leg` length. That is, are ducklings affected by `mycoplasma` in the same way in different parts of their body during development?

24.3 Brassica: The scientist would like to chronicle the year-to-year variation in crop yield when no herbicide is applied. That is, do the crops have different yields (in average kilograms per hectare `yldkga`), and is this consistent across years? There are some missing data from one year to the next. Be sure to account for this in your analysis.
(a) State exactly how you handle missing data and how this affects your model and hypotheses. That is, what questions can you ask, and what questions are nonsensical?

PROBLEMS 379

(b) Discuss briefly how the missing data affect analysis. Be specific to this problem, rather than providing a 'cook-book' approach to a generic problem. Illustrate by documenting your analysis of this problem.
(c) Notice that an analysis over years for all the data has identical Type I, II, III and IV sums of squares for rep(year) and year*crop, but not for year and crop. Why? [Hint: what is special about the pattern of missing data here?] Does this mean that you do not have to worry about these four approaches? [Hint: Think about the hypotheses you wish to test, and the pattern of missing data. Where is the balance? Try using algebra to examine hypotheses in terms of cell means, or use the e1, e2, e3, e4 options to the model statement. But examine output on the screen and save paper as much as you can!]

24.4 Brassica: Comment on the assumption of unequal variance for these data.
(a) Is there any evidence to suggest a need for transformation or some other method to address problems with assumptions? Be careful in examining residual plots: are there any cautions for this type of design?
(b) What happens to interactions when these data are transformed?
(c) The scientist wants to report results in the original (untransformed) units. There are good reasons for this, in terms of communicating with peers. Does analysis of transformed data lead to different results which might question the scientist's wish? It is important for this problem to be specific and brief in your comments. Support any claim with results and/or plots.

24.5 Brassica: (a) Now examine both years together. Lay out a comprehensive model. What can you say about year-to-year differences? In other words, can the scientist simply combine information across the two years? Include a careful examination of interactions with appropriate plots.
(b) Does the herbicide act in a linear fashion? How can you examine this with the limited information provided? (There is some relevant information: where?) What further analysis would you propose? Again, be sure to pay attention to the experimental design.
(c) How would design and analysis change if plots for crops were not randomized between years? That is, suppose crops and rates were assigned once but data were recorded for two years. Lay out the model across the two years and set up the anova table. It is not necessary to conduct formal analysis, but include degrees of freedom and sums of squares using values from the above experiment.

24.6 Drying: A dairy scientist (Hoffman 1996) wished to examine the effect of forage sample drying technique on ruminal disappearance of dry matter and crude protein. He gathered 30 uniquely different forage (silage) samples, roughly ten (9–11) per period. Each sample was divided each into sub-samples for drying by three different methods (microwave, freeze dryer and convection oven). There were three Holstein cows available with canulas (openings into the stomach for experimental observation), but each cow's rumen could only hold about 150 grams of forage. The scientist wanted to make sure that each sample was exposed to all nine combinations of the three cows and three drying techniques. Therefore, each drying sub-sample was further divided into 5 gram portions which were randomly assigned to the cows. The percentages of dry matter (dm) and crude protein (cp) remaining after canulation were measured, although some portions were inadvertently destroyed.

(a) Consider dry matter remaining in the first period. Write down a model with assumptions. State key questions and analyze the data. Report findings concisely.

(b) Consider all three periods. What is the design relationship (e.g. crossed or nested) between the factors period and cow? sample and period? sample and cow? What is the experimental unit for drying technique? period? cow*drying interaction?

(c) Write down a model for the full experiment. Lay out the anova table with degrees of freedom and expected mean squares.

(d) Analyze the full data set for dry matter remaining. Report results using interaction plot(s) with appropriate LSD bar(s) or some other graphical measure of precision.

(e) Check residuals for evidence of violation of assumptions. Is it appropriate to look just at the raw residuals with such a design?

24.7 Drying: Use dry matter remaining after canulation (dm) as a covariate to investigate the adjusted effect of drying technique on crude protein remaining (cp).

(a) First analyze crude protein in a similar fashion to Problem 24.6.

(b) Plot cp against dm using plot symbols to highlight drying technique and period. It may be worthwhile to have separate plots by either of these factors. Comment on any unusual patterns.

(c) Part (b) may be flawed because it does not account for the design. Make appropriate adjustments to deal (approximately) with this and produce new plot(s). Does this alter interpretation?

(d) Conduct a formal analysis of covariance. Note that it may be important to distinguish the effects of the covariate at different levels of nesting. If you use an approximate approach, make this clear to the reader.

(e) Present results in a graphical manner accompanied by a concise interpretation.

PART I

Repeating Measures on Subjects

Many experiments incur large expenses per subject, encouraging the scientist to learn as much as possible about each subject. Further, repeated measurements can provide valuable information about the changing response to treatment. Some experiments are so expensive that scientists must reuse subjects for a sequence of treatments.

Repeated measures present new challenges working with possible correlation among measurements on the same subject. Chapter 25 examines when repeated measures designs can be usefully analyzed as if they were split plots.

In general, the pattern of correlation among repeated measures can raise doubts about the utility of a split plot analysis. If the pattern is not too bad, adjustments to the split plot analysis are reasonably easy, as developed in Chapter 26. Other alternatives range from reducing a subject's measurements to a single contrast over time, or employing general multivariate statistics.

Sometimes, due to limited resources, a scientist may run some or all treatments on each subject, one after another. This cross-over within subjects can lead to efficient designs, provided there is no carry-over from one treatment to the next. Chapter 27 examines some experiments of this nature, including replicated Latin squares, showing how inferential questions relate to those considered in earlier experiments.

CHAPTER 25

Repeated Measures as Split Plot

Experiments in a wide range of disciplines involve selecting a sample of 'subjects', assigning them to treatment groups and over some period of time taking repeated measurements to observe the response to treatment. In many situations, the treatment may result in protracted changes, which might result in rather different short-term and long-term effects. Sometimes the expected treatment effect is dramatic, but the timing of this is unknown, leading to a design which hopes to bracket that critical time. Other experiments are designed to examine gradual changes which might be mediated by treatment.

Section 25.1 provides a broad overview of the subject matter. Section 25.2 lays out the repeated measures model and discusses some initial approaches to analysis. Section 25.3 considers the sphericity conditions which justify using a split plot approach in the presence of correlation over time. Section 25.4 develops expected mean squares while section 25.5 examines contrasts under sphericity.

25.1 Repeated measures designs

Repeated measures designs are essentially split plot designs allowing for correlation within each random effect. Measurements over time (split plot) on the same subject (whole plot) are likely to be highly correlated if the spacing is short. However, stretching out the time between measurements may make it difficult to observe the processes of interest. Thus it is important to examine how possible **correlation** modifies the key questions and the methods of analysis for such a design, since in practice it is often difficult or impossible to ensure complete independence. One reason to take repeated measurements on subjects is to learn about changes over time due to treatment. Another is to save scarce resources by reusing subjects. Experiments which involve sequentially assigning two or more treatments to the same subject and then measuring some response are known as **crossover designs**.

The usual situation in a repeated measures design has treatments applied to subject and several measurements (3–20) taken on each subject. It is not possible to randomize the experimental units (e.g. time intervals) within subject as is done in split plot designs. Interest focuses on the average treatment effect (with subject as EU) and on the effect of treatment on the time course of response (with time interval as EU).

Other types of repeated measurements on a subject besides those over time can lead to complications of correlation. Consider examining components of the subject (body parts, leaves on a plant) or a spatial arrangement within a subject (position on a field, location on printed circuit board). Measurements between subjects may be correlated as well. Subjects such as potted plants on a greenhouse table may be spatially correlated, either directly (e.g. shading) or indirectly (e.g. local variation in light intensity). Correlation through kinship of subjects within 'treatment' groups may also be considered with repeated measures approaches. For the purpose of discussion, attention focuses on a design with repeated measurements over time on subjects.

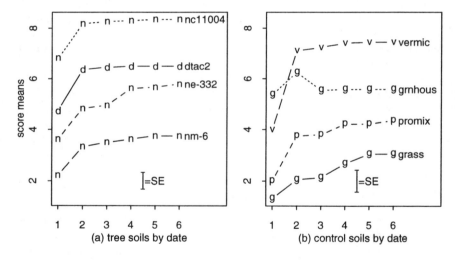

Figure 25.1. *Tree scores over time by soil type. Points are averages over the six replicates for each* **date** *and* **soil**. *The standard errors (SEs) are for* **date**∗**soil** *means, computed separately for (a) tree and (b) control* **soils** *using a split plot (mixed model) design.*

Example 25.1 Tree: A student of plant pathology (Maxwell and Stanosz 1996) studied the 'weed suppressiveness' of tree **soils**. David Maxwell examined the sensitivity of 'Rutgers' variety tomato, known to be sensitive to biological factors, to several different **soils**. Three of the **soils** (**vermix**, **promix** and **grnhous**) are in a certain sense controls, the first two being artificial and the third coming from a controlled environment. A fourth soil contained only grass. The other four are of particular interest, as the scientist wants to know whether soil from tree nm-6 is better (lower score) than those of the other three trees. The score was the number of germinating tomato seeds out of ten planted. In addition, the mean height of seedlings was measured at different times.

For each type of soil, there were six soil samples (reps(soil)). These soil samples were inoculated with pathogen. Then these soil samples were split (acc), with one half autoclaved (sterilized) and the other half not. Thus there were 12 half-samples (acc*reps(soil)). Each of these 12 half-samples was placed in storage and scored on six different dates. In addition, the mean height of seedlings was measured at three of these times. The primary focus is on the half-samples that were not autoclaved. Figure 25.1 shows the average score by (a) tree soil and (b) control soil. For most soils, the score increases between the first two dates and then levels off. However, there are some subtle variations. Are they significant? The SE bars for the plotted mean values on the plots were calculated separately by soil type assuming the experient was a split plot design. (These are appropriate under sphericity for repeated measures designs.) Notice that the standard errors are markedly smaller for the tree soils. ◇

Since this problem arises in many arenas, it has had several names and several approaches. Biomedical disciplines often use the name **longitudinal study** to signify an experiment following subjects under treatment for an extended length of time (months or years). The psychology literature tends to use the term **profile analysis**, which similarly connotes tracking individual profiles over time by periodic laboratory tests. While these and other disciplines focus on different aspects of repeated measures, they have many features in common.

Correlations and measurements over time bring to mind **time series**. Typically, time series analysis requires more than a few repeated measurements. Where there are 50–100 or more measurements per subject, time series methods are very appealing and appropriate. However, experiments with 3–4 or perhaps even 10–20 measurements are less amenable to this approach. Nevertheless, ideas from time series can be employed, such as analyzing the change in response over time intervals, to reduce correlation over time before conducting repeated measures analysis. See Milliken and Johnson (1984) for an example of an autoregressive process.

Sequential testing can involve repeated measurements on subjects. Here, the goal is to stop the experiment early based on results at hand. These arise most notably in large, expensive **clinical trials** which should be halted as soon as results are known, for health and ethical reasons. This approach is not formally considered in this chapter, as it raises a number of statistical issues about multiple comparisons with the same data. Further, in many experiments there may not be enough measurements to have any power for early testing, or it may not possible to process measurements in time to make informed decisions about stopping the experiment.

25.2 Repeated measures model

The **repeated measures model** setup is basically a split plot,

response = grand mean + (subject part) + (time interval part),

with treatments assigned to subject,

subject part = treatment + subject error,

and measurements taken over time intervals on each subject,

time interval part = time + time * treatment + time interval error.

Expressed in mathematical notation, it is

$$y_{ikm} = \mu + (\alpha_i + r_{im}) + (\tau_k + (\alpha\tau)_{ik} + e_{ikm})$$

with $i = 1, \cdots, a$ treatments, $m = 1, \cdots, n_i$ subjects per treatment and $k = 1, \cdots, t$ repeated measurements (times) per subject.

While the number of repeated measurements may vary by subject (r_{im}), this adds considerable complexity and raises questions of estimability and model identification. Up until recently, with the advent of proc mixed in SAS and similar software (lme() and nlme()) in S-Plus, proper analysis of unbalanced repeated measures involved tedious programming. Further, these software additions still require some fine-tuning to handle certain situations. In practice, it is often convenient to restrict attention to balanced subsets with a constant number t of repeated measures.

Treatments are assumed to be fixed effects, although random affects could be examined as well. There may be a systematic time effect (τ_k) which is usually not of interest. Key questions revolve around the overall treatment effect (α_i) and the change in shape of the time course of response due to treatment (($\alpha\tau)_{ik}$).

The mean response, or fixed part of the model,

$$E(y_{ikm}) = \mu_{ik} = \mu + \alpha_i + \tau_k + (\alpha\tau)_{ik} \ .$$

agrees with the split plot model. However, the covariance of responses is more complicated for the repeated measures model.

Distribution assumptions on the random effects are more general for repeated measures than for split plot. Let $\mathbf{r}_i^T = (r_{i1}, \cdots, r_{in_i})$ be the vector of subject errors for treatment i and let $\boldsymbol{\epsilon}_{im}^T = (e_{i1m}, \cdots, e_{itm})$ be the vector of time interval errors for subject m on treatment i. These vectors are assumed to be mutually independent and normally distributed. The split plot approach would have

$$\mathbf{r}_i \sim N(0, \sigma_P^2 \mathbf{I}) \text{ and } \boldsymbol{\epsilon}_{im} \sim N(0, \sigma^2 \mathbf{I})$$

while a multivariate repeated measures approach considers

$$\mathbf{r}_i \sim N(0, \mathbf{U}_i) \text{ and } \epsilon_{im} \sim N(0, \mathbf{V})$$

with subject covariance matrix $\mathbf{U}_i$ being of size $n_i \times n_i$ and time interval covariance matrix $\mathbf{V}$ being of size $t \times t$.

The **covariance of responses** for the split plot model has a simple form. Measurements within subjects are correlated but measurements between subjects are not. That is,

$$\begin{aligned} V(y_{ikm}) &= \sigma_P^2 + \sigma^2, \\ \operatorname{cov}(y_{i1m}, y_{i2m}) &= \sigma_P^2, \\ \operatorname{cov}(y_{ik3}, y_{ik4}) &= 0, \\ \operatorname{cov}(y_{i13}, y_{i24}) &= 0. \end{aligned}$$

For repeated measures, measurements within subject have a more complicated correlation structure, and measurements among subjects in the same treatment group may be correlated. Measurements between treatment groups are assumed to be uncorrelated. That is,

$$\begin{aligned} V(y_{ikm}) &= u_{imm} + v_{kk}, \\ \operatorname{cov}(y_{i1m}, y_{i2m}) &= u_{imm} + v_{12}, \\ \operatorname{cov}(y_{ik3}, y_{ik4}) &= u_{i34}, \\ \operatorname{cov}(y_{i13}, y_{i24}) &= u_{i34}. \end{aligned}$$

with u_{imm} the mth diagonal element of $\mathbf{U}_i$ and v_{kk} the the kth diagonal element of $\mathbf{V}$ and so on.

Thus for split plots, the time vector per subject, $\mathbf{y}_{im}^{\mathrm{T}} = (y_{i1m}, \cdots, y_{itm})$, has covariance

$$V(\mathbf{y}_{im}) = V(r_{im} + \epsilon_{im}) = \sigma_P^2 \mathbf{J} + \sigma^2 \mathbf{I},$$

with $\mathbf{J} = \mathbf{1}\mathbf{1}^{\mathrm{T}}$ being a matrix of all ones. For repeated measures, the covariances are

$$V(\mathbf{y}_{im}) = V(r_{im} + \epsilon_{im}) = u_{imm}\mathbf{J} + \mathbf{V}.$$

Subjects within treatment groups are often assumed to be uncorrelated ($\mathbf{V} = \sigma_P^2 \mathbf{I}$). In this case, there are a variety of ways to examine treatment differences by reducing data over time to a **single measurement per subject**. For instance, separate analyses by time interval can reveal consistent patterns or dramatic shifts in treatment effects over time.

Contrasts of treatment means over time, such as polynomial trends (particularly linear and quadratic) can reveal gradual changes. Determine separate slopes for each subject and use these as the measurements for a one-factor analysis. Other contrasts of interest, such as the quadratic coefficient or the change between the first and second halves of an experiment, could be handled in a similar fashion. Other summaries may not be linear in the response, but may capture important features of the time trend, such as the time of maximum response or the standard deviation per subject. Reducing measurements over time to subject summaries may in fact be the most appropriate way to address a scientist's key questions.

Analyzing single measurements per subject may be rather conservative, but it side-steps problems of correlated measurements over time. The distribution of contrasts depends on the correlation structure in bizarre ways that upset the approaches developed for contrasts with independent observations. Under certain conditions, it is necessary to examine contrasts with the aide of variance component estimates. However, the single-measurement approach is often much more workable.

A general multivariate approach for repeated measures, as examined briefly later in this chapter, is rather complicated. However, simplifying assumptions about $\mathbf{U}_i$ and $\mathbf{V}$, if tenable, lead to more tractable approaches using the split plot anova tables, as developed in the next two sections.

25.3 Split plot more or less

The previous sections noted the similarities and differences between the split plot and the repeated measures models. This section considers the situations where a split plot analysis is more or less appropriate for repeated measures designs. Certain conditions on the correlation structure of the random effects allow the use of overall tests developed for the split plot design. Essentially these conditions yield the same matching of terms in expected mean squares as found for nested designs. The analogy does not quite work when examining contrasts of means for treatments and time intervals.

Up to now, all models have assumed equal variance for random effects. If we keep this assumption among subjects and across time intervals, then the sufficient condition of **compound symmetry** validates the split plot anova and overall tests. This assumption is

$$\mathbf{V} = \sigma^2[\mathbf{I} + \mathbf{J}\rho/(1-\rho)],$$
$$\mathbf{U}_i = \sigma_P^2[\mathbf{I} + \mathbf{J}\rho_P/(1-\rho_P)],$$

with ρ_P and ρ appropriate correlation coefficients, between -1 and 1. In other words, the covariance matrices must both be proportional to a matrix with 1s down the diagonal and ρs off the diagonals,

$$\mathbf{V} = \frac{\sigma^2}{1-\rho} \begin{bmatrix} 1 & \rho & \cdots & \rho \\ \rho & 1 & \cdots & \rho \\ \vdots & \vdots & \vdots & \vdots \\ \rho & \rho & \cdots & 1 \end{bmatrix}.$$

Thus the variance and covariance over time within subjects and between subjects in a treatment group are, respectively,

$$\begin{aligned} V(y_{ikm}) &= \sigma_P^2/(1-\rho_P) + \sigma^2/(1-\rho), \\ \text{cov}(y_{i1m}, y_{i2m}) &= \sigma_P^2/(1-\rho_P) + \sigma^2 \rho/(1-\rho), \\ \text{cov}(y_{i13}, y_{i24}) &= \sigma_P^2 \rho_P/(1-\rho_P). \end{aligned}$$

An equivalent way to express this structure on the covariance is that the variance of the difference of two levels of either random effect is constant,

$$\begin{aligned} V(r_{i1} - r_{i2}) &= 2\sigma_P^2, \\ V(e_{i3m} - e_{i4m}) &= 2\sigma^2, \end{aligned}$$

for subject and time interval random effects, respectively. This **sphericity condition** is necessary and sufficient for the validity of overall tests based on a split plot anova. The condition is crucial for the development of inferential tools without recourse to multivariate methods. Note that all sums of squares in the split plot anova are really quadratic forms involving differences of ys, which simplify to expressions involving terms like $(r_{i1} - r_{i2})^2$ and $(e_{i3m} - e_{i4m})^2$. The above conditions ensure that the expected mean squares are the same as those for the split plot design. If the design is balanced ($n_i = n$) then all the usual sums of squares in the split plot anova table are proportional to independent (central or non-central) χ^2 variates. The next section shows that this correlation structure does not change the degrees of freedom, or the dimensionality, of the partitioning of the sample space.

Huynh and Feldt (1970) showed how the sphericity condition on the differences allows for unequal variances among the random effects. In terms of the covariances of the random effects, the sphericity condition imposes

$$\begin{aligned} V(e_{ikm}) = v_{kk} &= \sigma^2(1 + 2\lambda_k) \\ \text{cov}(e_{i3m}, e_{i4m}) = v_{34} &= \sigma^2(\lambda_3 + \lambda_4) \end{aligned}$$

for the time intervals with $\boldsymbol{\lambda}^T = (\lambda_1, \cdots, \lambda_t)$ arbitrary, and

$$\begin{aligned} V(r_{im}) = (\Lambda_i)_{mm} &= \sigma_P^2(1 + 2\omega_{im}) \\ \text{cov}(r_{i1}, r_{i2}) = (\Lambda_i)_{12} &= \sigma^2(\omega_{i1} + \omega_{i2}) \end{aligned}$$

for the subjects, with the a vectors $\omega_i^T = (\omega_{i1}, \cdots, \omega_{in_i})$ being arbitrary constants. In matrix form, write the time interval covariance as

$$\mathbf{V} = \sigma^2[\mathbf{I} + \boldsymbol{\lambda}\mathbf{1}^T + \mathbf{1}\boldsymbol{\lambda}] = \sigma^2 \begin{bmatrix} 1+2\lambda_1 & \lambda_1+\lambda_2 & \cdots & \lambda_1+\lambda_t \\ \lambda_1+\lambda_2 & 1+2\lambda_2 & \cdots & \lambda_2+\lambda_t \\ \vdots & \vdots & \vdots & \vdots \\ \lambda_1+\lambda_t & \lambda_2+\lambda_t & \cdots & 1+2\lambda_t \end{bmatrix}.$$

Under equal variance this reduces to compound symmetry with $\lambda_k = \rho/(2-2\rho)$.

The split plot analysis is still valid for overall tests if the design is unbalanced. However, analysis is more complicated. There is an important distinction between difficulties arising from imbalance among treatment groups (unequal number of subjects n_i) and imbalance among time intervals within subjects (fewer than t time measurements for some). The overall tests are analogous to those for split plot, but contrasts are considerably more involved. The remainder of this chapter considers only **balanced designs**.

The **variances of marginal means** of random effects for time intervals are

$$V(\bar{e}_{ik\cdot}) = \sigma^2(1+2\lambda_k)/n$$
$$V(\bar{e}_{i\cdot m}) = \sigma^2(1+2\lambda.)/t.$$

The marginal means of $\bar{e}_{i\cdot m}$ are averages of independent, identically distributed variates. Similarly, variances for the marginal means of the subject random effects can be derived:

$$V(\bar{r}_{i\cdot}) \doteq \sigma_P^2(1+2\omega_{i\cdot})/n$$
$$V(\bar{r}_{\cdot\cdot}) = \sigma_P^2(1+2\omega_{\cdot\cdot})/an.$$

Combining these variances for random effects leads to following expressions for the variances of marginal means of the responses:

$$nV(\bar{y}_{\cdots}) = \sigma_P^2(1+2\omega_{\cdot\cdot})/a + \sigma^2(1+2\lambda.)/at$$
$$nV(\bar{y}_{i\cdot\cdot}) = \sigma_P^2(1+2\omega_{i\cdot}) + \sigma^2(1+2\lambda.)/t$$
$$nV(\bar{y}_{\cdot k\cdot}) = [\sigma_P^2(1+2\omega_{\cdot\cdot}) + \sigma^2(1+2\lambda_k)]/a$$
$$nV(\bar{y}_{ik\cdot}) = \sigma_P^2(1+2\omega_{i\cdot}) + \sigma^2(1+2\lambda_k).$$

25.4 Expected mean squares under sphericity

While the sums of squares are the same as in the split plot, the expected mean squares are only slightly different. This only arises in the subject comparisons, when averaging over the time intervals. The $E(MS)$ entries

EXPECTED MEAN SQUARES UNDER SPHERICITY

source	df	$E(MS)$
trt	$a-1$	$(\sigma^2(1+2\lambda.)+t\sigma_P^2)(1+\delta_A^2/(a-1))$
subject error	$a(n-1)$	$\sigma^2(1+2\lambda.)+t\sigma_P^2$
time	$t-1$	$\sigma^2(1+\delta_T^2/(t-1))$
trt*time	$(a-1)(t-1)$	$\sigma^2(1+\delta_{AT}^2/(a-1)(t-1))$
time error	$a(t-1)(n-1)$	σ^2
total	$atn-1$	

Table 25.1. *Expected mean squares under sphericity*

in Table 25.1 follow from calculations only slightly different from those for the split plot.

The **expected mean squares** can be derived in a straightforward manner using the variances of marginals. The time interval or 'subplot' error can be partitioned as

$$\begin{aligned} SS_{SPE} &= \sum_{ikm}(y_{ikm}-\bar{y}_{ik\cdot}-\bar{y}_{i\cdot m}+\bar{y}_{i\cdot\cdot})^2 \\ &= \sum_{ikm}e_{ikm}^2 - n\sum_{ik}\bar{e}_{ik\cdot}^2 - t\sum_{im}\bar{e}_{i\cdot m}^2 + nt\sum_m\bar{e}_{i\cdot\cdot}^2 \end{aligned}$$

which leads to expected mean square

$$E(MS_{SPE}) = \sigma^2[\sum_k(1+2\lambda_k)-(1+2\lambda.)]/(t-1) = \sigma^2 .$$

Calculations in the whole plot are a bit more complicated as both random effects appear, and the parameters for the time interval do not disappear.

$$\begin{aligned} SS_{PE} &= a\sum_{im}(\bar{y}_{i\cdot m}-\bar{y}_{i\cdot\cdot})^2 \\ &= a\sum_{im}[(r_{im}-\bar{r}_{i\cdot})+(\bar{e}_{i\cdot m}-\bar{e}_{i\cdot\cdot})]^2 \\ &= a\sum_{im}r_{im}^2 - an\sum_i\bar{r}_{i\cdot}^2 + a\sum_{im}\bar{e}_{i\cdot m}^2 - an\sum_i\bar{e}_{i\cdot\cdot}^2 . \end{aligned}$$

This yields an expected mean square of

$$\begin{aligned} E(MS_{PE}) &= \sigma_P^2[(nt+2\omega..)-(t+2\omega..)]/(n-1) \\ &\quad + \sigma^2[n(1+2\lambda.)-(1+2\lambda.)]/(n-1) \\ &= t\sigma_P^2 + (1+2\lambda.)\sigma^2 . \end{aligned}$$

The other expected mean squares can be derived in a similar fashion. The reader may want to speculate about the effect of imbalance in the number of subjects or of time intervals per subject on these values.

Table 25.1 suggests that the **overall tests for main effects and interactions** developed for the split plot model would be appropriate under sphericity. The distribution of the factor main effects pivot statistic

$$F = MS_A/MS_{PE} \sim F_{a-1,a(n-1);\delta^2}$$

is modified slightly since the correlation over time alters the non-centrality parameter

$$\delta^2 = nt \sum_i (\alpha_i - \bar{\alpha}.)^2 / (\sigma^2(1 + 2\lambda.) + t\sigma_P^2) \ .$$

Note further that σ^2 and σ_P^2 are not strictly variances of the random effects, but they enter into computations in the same way as variances for the random effects did in the split plot tests.

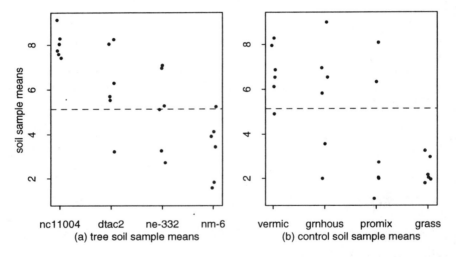

Figure 25.2. *Tree whole plot experimental units. Points are averages over the six* **dates** *for each soil sample.* **Soils** *are sorted by mean value within (a)* **tree** *or (b)* **control** **type**. *Dashed line is at the estimate of population grand mean.*

Example 25.2 Tree: The soil samples comprise the appropriate experimental units for comparing soil types. Figure 25.2 shows the average score over the six dates for the six soil samples from each soil type. That is, there is one measurement per experimental unit. The figures suggest that there is considerable variation with tree soils and within control soils, but only marginal, if any, evidence for differences between tree and control types. This is born out in Table 25.2. ◇

Example 25.3 Tree: Patterns over time among the soils can be examined after removing the whole plot average over dates (Figure 25.3). Notice

EXPECTED MEAN SQUARES UNDER SPHERICITY

source	df	SS	MS	F	p-value
type	1	76.8	76.8	4.08	0.050
soil(type)	6	833.0	138.8	7.37	<0.0001
error	40	753.1	18.8		

Table 25.2. *Tree whole plot anova*

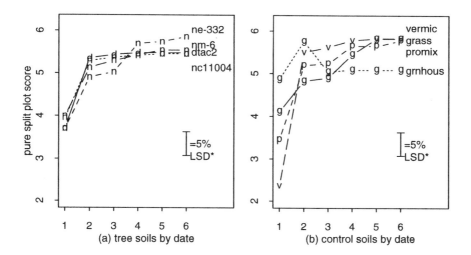

Figure 25.3. *Tree repeated measures within* soil *samples. Points are deviations of* date*soil *means from the* soil *means, with the estimated population grand mean added back.* LSD *values are based on subplot separate estimates of standard errors of differences for (a) tree (0.358) and (b) control (0.420)* soils, *appropriate under sphericity assumption.*

how the tree soils in Figure 25.3(a) have very similar patterns over dates, with the possible exception of ne-332. Control soils in Figure 25.3(b) show considerable variability in temporal pattern. All curves show an initial increase but level out to similar values, about 5–6 after removing the whole plot soil effect. This is partially reflected in the subplot analysis reported in Table 25.3, showing significant date*soil(type) interaction.

Table 25.3 is not completely satisfying. It does not specifically support the visual differences between the two plots in Figure 25.3. However, the significant interaction is evidence to support further investigation. Table 25.4 presents separate subplot analyses by type. Now the consistent pattern for tree soils is supported, and there is strong evidence that the controls are rather different from each other. ◊

source	df	SS	MS	F	p-value
date	5	116.3	23.25	101.68	<0.0001
date*type	5	0.6	0.13	0.56	0.73
date*soil(type)	30	34.2	1.14	4.99	<0.0001
error	200	45.7	0.23		

Table 25.3. *Tree subplot anova*

(a) tree soils

source	df	SS	MS	F	p-value
soil	3	415.0	138.331	10.23	0.0003
WP error	20	270.5	13.524	70.27	0.0001
time	5	52.7	10.539	54.76	0.0001
soil*time	15	3.0	0.199	1.03	0.43
SP error	100	19.2	0.192		

(a) control soils

source	df	SS	MS	F	p-value
soil	3	418.0	139.328	5.77	0.0052
WP error	20	482.6	24.130	91.09	0.0001
time	5	64.2	12.840	48.47	0.0001
soil*time	15	31.2	2.082	7.86	0.0001
SP error	100	26.5	0.265		

Table 25.4. *Tree separate split plot analyses by* **soil type**

25.5 Contrasts under sphericity

Contrasts of means raise the same issues that have appeared earlier. Usually it is advisable to perform an overall test, say for main effects, and then address the multiple comparison of marginal means. Some comparisons may not result in exact tests, requiring approximations using the method of moments or maximum likelihood and Satterthwaite approximations to degrees of freedom. The presence of two random errors has the same flavor as for split plots, with an added concern about the covariance structure. The sphericity conditions are quite natural for contrasts as the key concerns focus on differences of random errors.

Contrasts involving **time main effects** are fairly easy. Consider

$$\bar{y}_{..3} - \bar{y}_{..4} = \bar{\mu}_{.3} - \bar{\mu}_{.4} + \bar{e}_{..3} - \bar{e}_{..4} \ .$$

The variance of this difference is $2\sigma^2/an$ which has a natural estimate using

$E(MS_{SPE})$. Contrasts of **treatment main effects** require a bit more care. The contrast

$$\bar{y}_{1..} - \bar{y}_{2..} = \bar{\mu}_{1.} - \bar{\mu}_{2.} + \bar{r}_{1.} - \bar{r}_{2.} + \bar{e}_{1..} - \bar{e}_{2..}$$

has variance $2[\sigma^2(1+2\lambda.) + t\sigma_P^2(1+\omega_{1.}+\omega_{2.})]/nt$. This is not estimable unless $\omega_{i.} = 0$. While this condition may be assumed, it cannot be checked or tested. With this assumption, pivot statistics can be developed using $E(MS_{PE})$.

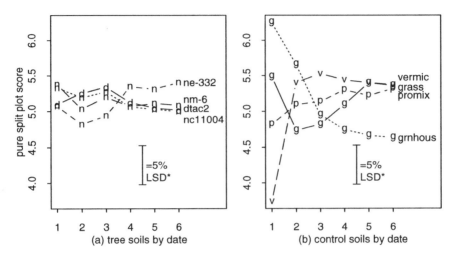

Figure 25.4. *Tree detrended repeated measures. Points are pure interaction* (date*soil) *after removing main effects of* soil *and* date, *with the estimated population grand mean added back.* LSD *values for (a)* tree *and (b)* control soils *are the same as in Figure 25.3.*

Example 25.4 Tree: It is somewhat difficult in Figure 25.3, to discern different patterns in the soils over dates because of the strong temporal trend. Figure 25.4 has this trend removed. Now the different pattern for ne-332 among tree soils (Figure 25.4(a)) is readily apparent. In addition, the grnhous control soil (Figure 25.4(b)) shows a distinct trend from the other three. This is supported by the analysis presented in tables discussed in Example 25.3. ◊

Interaction contrasts can be of several forms. Considering the change in treatment effect over time

$$\bar{y}_{1\cdot3} - \bar{y}_{1\cdot4} = \mu_{13} - \mu_{14} + \bar{e}_{1\cdot3} - \bar{e}_{1\cdot4}$$

the variance is straightforward, $2\sigma^2/n$, with natural estimate using MS_{SPE}.

However, the differential effect of two treatments for the same time interval
$$\bar{y}_{1\cdot 3} - \bar{y}_{2\cdot 3} = \mu_{13} - \mu_{23} + \bar{r}_{1\cdot} - \bar{r}_{2\cdot} + \bar{e}_{1\cdot 3} - \bar{e}_{2\cdot 3}$$
has a very complicated variance
$$2[\sigma^2 + \sigma_P^2(1 + \omega_{1\cdot} + \omega_{2\cdot})]/n$$
that is not estimable unless $\lambda_{\cdot} = \omega_{i\cdot} = 0$. Even with this, the variance must in general be approximated using REML or some other method for variance components. For balanced data, the estimate can be found by the method of moments
$$2[MS_{PE} + (t-1)MS_{SPE}]/nt \;.$$

25.6 Problems

25.1 Tree: Assess whether nm-6 has significantly lower score than the other tree soils for the autoclaved (acc = autocla) samples only. In the process, it is important to characterize the difference between the four tree soils and the four controls.
(a) Lay out the full model for repeated measures for this problem. Define terms. Briefly summarize the difference in assumptions between split plot and repeated measures under sphericity.
(b) Compare all eight soils for pathogen score using a split plot approach. Contrast the controls and the tree soils in some way that highlights the differences you observe in your interaction plot(s).
(c) Conduct formal contrasts over time. For instance, consider linear and quadratic and perhaps higher-order trends as if there were one observation per soil sample. Use contrasts other than polynomials if preferred.
(d) Now compare only the four soils from trees. You may do this either by selecting those four soils or with contrasts.
(e) Think simple. Summarize your findings in an elegant way which can inform the scientist about the main results.

25.2 Tree: Conduct a repeated measures analysis on the tree soils. Include a brief summary of separate analysis by time for the eight soils (can be done graphically).
(a) Focus on polynomial (or other) contrasts over time and the 'univariate tests' with and without sphericity correction.
(b) Look for complementary interpretations to the split plot method, and refer to interaction plots. Discuss any conflicting evidence.
(c) If you have not done so already, consider the tree soils separately for part of your analysis.
(d) Are there any shortcomings to the repeated measures approach for this problem? In particular, comment on the relevance of or problems with polynomial contrasts over time.

PROBLEMS

25.3 Berry: This problem comes from a horticulturist (Hagidimitriou and Roper 1994) investigating the relative energy balance between vegetative and flowering parts of cranberry uprights in the marshes of northern Wisconsin. It is very difficult to separate individual plants as they tend to overlap considerably. In fact, a large cranberry bed may be essentially one plant!

The researcher examined one cranberry bed which was divided into six `samples` for measurement purposes. Soluble sugar and starch (`conc`) were measured at various `periods` through the growing season for random samples of roots, stems and uprights. Later in the season, the upright portions were distinguished as either flowering or vegetative. For each measurement day, the scientist wished to compare photosynthesis in these plant parts. The scientist was interested in assessing the change in sugars and starches over the course of the summer. That is, was one consistently higher than the other, or does this depend on `Julian` date?

Your task is to address her question using a variety of repeated measures methods. Note that the `repeated` option does not work quite right for this problem in `proc glm`. That is, it considers the blocks (`sample`) as a factor and assumes you are interested in `period*sample` interactions. The second run of `proc glm` drops the `sample`, which is assumed to only affect the whole plot (as a blocking factor). The newer `proc mixed` seems to handle `sample` as a whole plot blocking factor appropriately.

(a) Analyze the data, date by date, for differences between flowering and vegetative parts. Discuss dates individually and give an overall assessment. You may find tables of means and/or interaction plots to be helpful. Include measures of variation as appropriate.

(b) Conduct an analysis as if this were a split plot design. [Use whatever segment of the output you think is appropriate, but clearly indicate your choice in terms of model and assumptions.]

(c) Examine the orthogonal polynomials individually. For instance, examine the linear slopes over times as a single measurement per experimental unit. As with separate analyses by dates, give some indication of where differences in `tissue` appear to be significant, and give an overall assessment.

(d) Based on this analyses, what are your conclusions? How would you share this with the scientist? Be brief, presenting relevant information in a short paragraph, perhaps with an adjoining table or figure.

CHAPTER 26

Adjustments for Correlation

Researchers for some time have been concerned about the effect of correlation on analysis. Short of a full-blown multivariate approach, it pays to recognize that highly correlated measurements have less 'information' than independent measurements. In the extreme, if two measurements are perfectly correlated, they together represent only one piece of information, or one degree of freedom. If there are b repeated measurements on a subject, these comprise somewhere between 1 and $b-1$ degrees of freedom. Knowing the amount of information can lead to some approximate adjustment of inferential methods.

The basic idea for analysis of repeated measures is to consider several complementary methods together. First, since treatments are applied to subjects, a conservative analysis with one measurement per subject can provide simple yet revealing insights. Split plot analysis is often justified if correlations have a certain form. Even if this is not strictly believable, there are some approximations which adjust the split plot analysis by accounting for the reduction in independent information due to correlation. A more 'sophisticated' approach is possible by using an array of multivariate tests which allow for arbitrary correlation structure among and within subjects.

Section 26.1 considers adjustments to the degrees of freedom to compensate for correlated measurements over time. Section 26.2 examines the effect of correlation on contrasts over time. It is assumed in these two sections, and most packages, that subjects within groups are independent. This is not often an issue, as many experiments by design have subjects that are independent with the main concern being about correlation of repeated measurements upon the same subject. The ideas of Section 26.1 could be developed further by assuming Huynh–Feldt dependence among subjects. Instead, Section 26.3 considers a full multivariate approach.

26.1 Adjustments to split plot

As noted in Chapter 25, if the sphericity conditions are met, then there are $t-1$ degrees of freedom per subject over time. If not, there should be somewhat less. The worst case scenario, or most conservative approach, would suppose that there is only one piece of information over time. On the other hand, minor violations of the sphericity conditions might encourage minor adjustments to the degrees of freedom. Further appeal to a Satterthwaite approximation leads to construction of tests and other tools of inference using these adjusted degrees of freedom.

Box (1954) first suggested the **conservative correction** for tests over time. Suppose there were $b = 2$ measurements over time. The experiment would involve before/after measurements, leading to one degree of freedom per subject over time. If there are more than two repeated measures, there must be at least one degree of freedom over time, but there is no guarantee that the third measurement adds much 'new' information. As a conservative approach, develop the same pivot statistics as for the split plot analysis, but treat them as if there were only one degree of freedom over time. That is, divide degrees of freedom in numerator and denominator by $t - 1$ for pivot statistics over time. Thus for time effects

$$F = MS_B/MS_{SPE} \approx F_{1,a(n-1)}$$

instead of using $df = t - 1, a(n - 1)(t - 1)$. For interaction of time and treatment, consider

$$F = MS_{AB}/MS_{SPE} \approx F_{a-1,a(n-1)} \, .$$

It is possible to show that more evidence is required to reject $H_0 : \beta_j = 0$ with this conservative approach. That is, for $t > 2$ the critical point for this distribution is larger,

$$F_{0.05;1,a(n-1)} > F_{0.05;t-1,a(n-1)(t-1)}$$

and hence the p-values are larger as well. Tests based on these distributions are conservative, in that any more 'liberal' test would reject H_0 more frequently. Rejecting the hypothesis of no difference in treatment over time with this conservative approach implies that any more liberal F test would also reject. However, failure to reject H_0 may not signify very much.

Box (1954) and Greenhouse and Geisser (1959) suggested adjustments to degrees of freedom which were later confirmed and extended by Huynh and Feldt (1970). This is sometimes known as the **Box correction**, summarizing the deviation of the correlation structure from the sphericity conditions.

The random vector of t measurements on a single subject $\mathbf{y}_{im}$ has distribution $N(\boldsymbol{\mu}, \mathbf{V})$. Suppose these are centered by their mean

$$\mathbf{y}_{im} - \bar{y}_{i \cdot m}\mathbf{1} = (\mathbf{I} - \mathbf{J}/t)\mathbf{y}_{im} = \mathbf{P}\mathbf{y}_{im}$$

using the projection matrix $\mathbf{P} = \mathbf{I} - \mathbf{J}/t$. The $t \times t$ variance-covariance matrix for the centered values $\mathbf{S} = (\mathbf{I}-\mathbf{J}/t)\mathbf{V}(\mathbf{I}-\mathbf{J}/t) = \mathbf{PVP}$ has diagonal elements $s_{kk} = \sigma_{kk}((t-1)/t)^2$ and off-diagonal elements $s_{kl} = -\sigma_{kl}/t^2$. The total variation for this subject $\sum_k(y_{ikm} - \bar{y}_{i \cdot m})^2$ has expectation $\sum_k s_{kk}$ and variance $2\sum_{kl} s_{kl}^2$. The Satterthwaite approximation matching these two moments to those of a χ^2 variate leads, under the sphericity conditions, to $df = t - 1$ degrees of freedom. In general

$$df = (t-1)\epsilon = \left(\sum_k s_{kk}\right)^2 / \sum_{kl} s_{kl}^2 .$$

Huynh and Feldt (1970) showed that $\epsilon = 1$ only if the sphericity condition holds, with $\epsilon < 1$ otherwise. Box (1954) suggested ϵ as a measure of the deviation of **V** from compound symmetry.

Greenhouse and Geisser (1959) set up approximate F tests by adjusting the numerator and denominator degrees of freedom by the estimator

$$\hat{\epsilon} = \frac{(\sum_j \hat{s}_{jj})^2}{(t-1)\sum_{ij} \hat{s}_{ij}^2}$$

based on the sample covariance matrix $\hat{\mathbf{S}}$. Unfortunately, simulations (see Crowder and Hand 1990, sec. 3.6) suggest this may have serious bias, being too small and leading to conservative tests, if the correction is moderate ($\epsilon > 0.75$) and there are few subjects relative to time periods ($an < 2t$). Huynh and Feldt (1976) suggested an alternative

$$\tilde{\epsilon} = \min\left(1, \frac{\hat{\epsilon}an - 2/(t-1)}{a(n-1) - \hat{\epsilon}(t-1)}\right)$$

which appears to overcome these problems. These two estimates, $\hat{\epsilon}$ and $\tilde{\epsilon}$, are known as the **G–G and H–F epsilon**, respectively, in some packages such as SAS. Adjustment using either estimated ϵ correction for examining time effects would use the usual pivot statistic but alter the distribution to

$$F = MS_B/MS_{SPE} \approx F_{\epsilon(t-1), \epsilon a(t-1)(n-1)} ,$$

and similarly for examining changes in shape of the time response as modified by treatment.

The worst case (Box's conservative test) arises with $\epsilon = 1/(t-1)$. The best case (sphericity) occurs with $\epsilon = 1$. In practice it is wise to check these extremes before appealing to approximate tests. If Box's conservative approach leads to significant differences, then any adjusted approach will also. If the liberal split plot approach shows no significance then neither will the adjusted approaches. If there are major discrepancies between liberal, conservative and ϵ-corrected tests then proceed cautiously to try to uncover the reasons behind this. Often a more simple approach such as analysis by time or contrasts over time can be very revealing.

Example 26.1 Tree: Examination of the repeated measures (subplot) error covariance matrix shows that the six scores over dates have partial correlations of 0.84 to 0.99 after removing effects of soil. Sphericity estimates across the eight soils are 0.41 (Greenhouse–Geisser) and

source	F	naive	G-G	H-F	Box
tree and control soil types					
date	101.68	.0001	.0001	.0001	.0001
date*type	0.56	.73	.58	.61	.46
date*soil(type)	4.99	.0001	.0001	.0001	.0007
adjusted DDF		200	81.4	100.8	40
tree soils					
date	54.76	.0001	.0001	.0001	.0001
date*soil	1.03	.43	.42	.42	.40
adjusted DDF		100	38.1	43.3	20
ϵ-adjustment				0.38	0.48
control soils					
date	48.47	.0001	.0001	.0001	.0001
date*soil	7.86	.0001	.0001	.0001	.0001
adjusted DDF		100	40.4	51.7	20

Table 26.1. *Tree sphericity-adjusted p-values*

0.50 (Huynh–Feldt), and slightly lower when estimated separately by **type**. Sphericity-adjusted tests give essentially the same results as uncorrected. Table 26.1 presents naïve p-values from earlier tables discussed in Example 25.3 along with the two estimates of sphericity-adjusted p-values and Box's conservative adjustment. Note that Box's adjustment actually gives smaller p-values when the F statistic is small; its p-values are larger for F values above 2. It is important to note here that there is evidence for significant deviation from sphericity for these data ($p < 0.0001$). ◇

26.2 Contrasts over time

It was earlier suggested that **contrasts over time** should be treated as if they were single measurements per subject and analyzed separately for each contrast. If the sphericity conditions hold, then contrasts over time behave as they would for split plot designs. Consider a contrast over time,

$$H_0 : \sum_k c_k \mu_{ik} = 0 \ ,$$

which is standardized, with $\sum_k c_k = 0$ and $\sum_k c_k^2 = 1$. The variance of the contrast for a single subject is

$$V(\mathbf{c}^T \mathbf{y}_{im}) = \mathbf{c} \mathbf{V} \mathbf{c}^T \ .$$

If the covariance $\mathbf{V}$ satisfies sphericity, then it can be readily seen that $\mathbf{cVc}^T = \sigma^2$. However, in general, the variance of a contrast over time depends in a complicated way on $\mathbf{V}$.

It is possible to construct an approximate test for how close a contrast is to satisfying this (even if $\mathbf{V}$ does not completely satisfy sphericity). However, in practice it is wise to perform separate contrasts. For instance, to investigate the effect of treatment on linear trends over time, compute the slope for each subject and consider these to be the response in one-factor analysis on subjects in treatment groups. That is, each contrast would have a separate mean square for error, unlike the liberal test which uses MS_{SPE} for all contrasts. The next section takes a multivariate approach to contrasts, showing how to develop theory for contrasts more formally.

When should the sphericity correction be used? Clearly, if the sphericity conditions seem reasonable, proceed with analysis appropriate to split plot designs as developed in the previous chapter. However, if sphericity is in doubt the above corrections or a multivariate approach may be appropriate. The **sphericity test** (see Milliken and Johnson 1992, sec. 27.1, or Crowder and Hand 1990, sec. 3.6) can be used to examine the hypothesis that the sphericity condition holds, or equivalently $H_0 : \epsilon = 1$. Unfortunately, the current test for sphericity developed by Mauchly (1940) is very sensitive to the assumption of normality, similar to tests of variance discussed in Chapter 14, Part E. Consider a $(t-1) \times t$ matrix of orthogonal contrasts over time, $\mathbf{C}$, that span the hypothesis space of time effects. The sphericity test statistic is

$$S = (n(a-1) - r) \, log \left(\frac{\det(\mathbf{C\hat{V}C})}{|\text{tr}(\mathbf{C\hat{V}C})/(t-1)|^{t-1}} \right) \sim \chi_r^2$$

with $r = (t-2)(t+1)/2$ degrees of freedom. In practice, this statistic may be considered advisory, given its sensitivity. That is, if there is any doubt about the correlation structure *before* doing the experiment, then keep an open mind. It is wise always to examine univariate split plot analysis, sphericity-corrected split plot analysis and multivariate analysis, noticing how they complement one another. In addition, always plot the data!

Example 26.2 Tree: The sphericity approach assume that the pattern of correlation depends only on the length of time between observations. This is assumed for many other methods, including autoregressive methods. However, Figure 26.1 shows evidence that the partial correlation is much stronger among the last two or three dates. Multivariate methods can implicitly account for this, but do not necessarily reveal the structure. They certainly do not enlighten without plots to augment formal statistical summaries. Figure 26.1 suggests that the last two dates contain little new information, and may in fact lead to a somewhat liberal interpretation. That is, mean square error estimates over dates are probably too small as

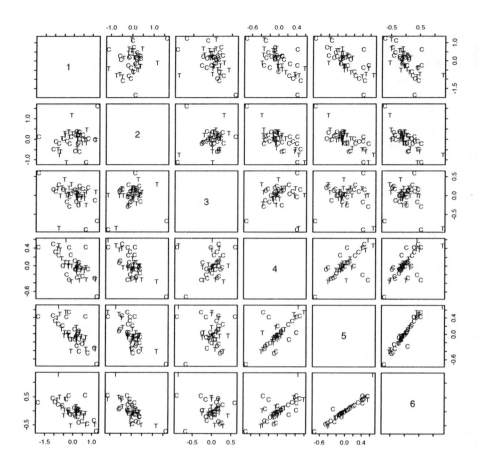

Figure 26.1. *Tree partial correlations over time. Pairs plot of repeated measures residuals over* dates *identified as coming from tree (T) or control (C)* soil *samples Fixed effects and the whole plot random effect have been removed. Notice the increasing correlation over time.*

a result. It may be worthwhile to reanalyze the first four dates. However, it would be important to state clearly that this was done, and remark explicitly on any discrepancies in results that might arise. ◊

26.3 Multivariate repeated measures

This section briefly considers multivariate analysis relevant to repeated measures. Multivariate analysis seems quite appealing as it makes few assumptions about the form of correlations over time (Krzanowski 1990, sec. 13.4; Morrison 1976, sec. 4.5–4.6). However, multivariate tests can be conservative if sphericity holds. Further, simulations do not provide a clear choice between the sphericity-corrected split plot analysis and multivariate analysis of variance (see Crowder and Hand 1990, sec. 3.6). That is, the power and robustness tradeoff depends on ϵ, subject sample size (an) and the number of time periods (b).

Multivariate analysis in repeated measures typically begins with the four overall statistics developed for multiple responses in Chapter 18, Part F. These multivariate tests complement plots over time and univariate statistics for each time period and for polynomial contrasts over time. Ideally, the multivariate results agree with the univariate results. If not, they may point to a more complicated interpretation. For instance, there may be a nonlinear trend over time that is not readily summarized with quadratic or cubic terms. This may suggest time periods for closer examination in future experiments.

The notation in this section is consistent with matrix notation in Chapter 12, Part D, but differs from the rest of the current chapter and part. The multivariate approach to multiple responses of Chapter 18, Part F, further sets the stage. There the reader can find general summary statistics for multivariate analysis. Let $m = 1, \cdots, n$, denote the subjects in the experiment, and assume they are independent. Suppose the factor level for subject i is encoded in the design matrix $\mathbf{X}$ defined below in the same fashion as presented for the linear model. Let $k = 1, \cdots, t$, denote the time intervals.

For fixed time k, the model for the mth subject is

$$y_{mk} = \sum_{j=1}^{p} X_{mj}\beta_{jk} + e_{mk}$$

with parameter vector $\boldsymbol{\beta}_m = (\beta_{1m}, \cdots, \beta_{pm})$, $n \times p$ design matrix $\mathbf{X}$ of rank $r \leq p$ and variance $V(y_{mk}) = \sigma_{kk}$. The errors may be correlated over time with some covariance $\mathrm{cov}(y_{mk}, y_{ml}) = \sigma_{kl}$. Let $\mathbf{V}$ be the $t \times t$ variance-covariance matrix over time. For convenience, normality is assumed. The model can be written as

$$\mathbf{y}_k = \mathbf{X}\boldsymbol{\beta}_k + \boldsymbol{\epsilon}_k \sim N(\mathbf{X}\boldsymbol{\beta}_k, \sigma_{kk}\mathbf{I}) \ .$$

A **least squares (LS) solution** of the normal equations is

$$\hat{\boldsymbol{\beta}}_k = (\mathbf{X}^{\mathrm{T}}\mathbf{X})^{-}\mathbf{X}^{\mathrm{T}}\mathbf{y}_k \ .$$

Estimable functions are linear combinations of $\mathbf{X}\boldsymbol{\beta}_k$ whose LS estimator is

$$\mathbf{X}\hat{\boldsymbol{\beta}}_k = \mathbf{X}(\mathbf{X}^T\mathbf{X})^-\mathbf{X}^T\mathbf{y}_k = (\mathbf{I}-\mathbf{P})\mathbf{y}_k \sim N(\mathbf{X}\boldsymbol{\beta}_k, \sigma_k^2(\mathbf{I}-\mathbf{P}))$$

with projection matrix $\mathbf{P} = \mathbf{I} - \mathbf{X}(\mathbf{X}^T\mathbf{X})^-\mathbf{X}^T$.

The multivariate model over time can be written in matrix form as

$$E(\mathbf{Y}) = \mathbf{XB}$$

with an $n \times t$ matrix of responses $\mathbf{Y} = (\mathbf{y}_1, \cdots, \mathbf{y}_t)$ and a $p \times t$ matrix of parameters $\mathbf{B} = (\boldsymbol{\beta}_1, \cdots, \boldsymbol{\beta}_t)$. Estimable functions over time have LS estimators which are linear combinations of

$$\mathbf{X}\hat{\mathbf{B}} = (\mathbf{I}-\mathbf{P})\mathbf{Y} .$$

The **error residuals** $\mathbf{Y} - \mathbf{X}\hat{\mathbf{B}} = \mathbf{PY}$ play a key role in the development of pivot statistics. The estimate of the time covariance matrix $\mathbf{V}$ is

$$\hat{\mathbf{V}} = (\mathbf{Y} - \mathbf{X}\hat{\mathbf{B}})^T(\mathbf{Y} - \mathbf{X}\hat{\mathbf{B}})/(n-r) = \mathbf{Y}^T\mathbf{PY}/(n-r)$$

with diagonal entries

$$\hat{\boldsymbol{\epsilon}}_k^T\hat{\boldsymbol{\epsilon}}_k = \mathbf{y}_k^T\mathbf{P}\mathbf{y}_k \sim \sigma_{kk}\chi_{n-r}^2$$

leading to natural estimates of error variances $\hat{\sigma}_{kk} = \mathbf{y}_k^T\mathbf{P}\mathbf{y}_k/(n-r)$.

The **general hypothesis of contrasts** is the starting point for useful pivot statistics. For a fixed time k, consider

$$H_0 : \mathbf{C}\boldsymbol{\beta}_k = \mathbf{0}$$

with the $g \times p$ matrix $\mathbf{C}$ concerning factor contrasts where subject is the experimental unit. The contrast $\mathbf{C}\boldsymbol{\beta}_k$ must be estimable. That is, $\mathbf{C} = \mathbf{AX}$ for some $g \times n$ matrix $\mathbf{A}$. The estimator of this contrast $\mathbf{C}\hat{\boldsymbol{\beta}}_k = \mathbf{A}(\mathbf{I}-\mathbf{P})\mathbf{y}_k$ has distribution $N(\mathbf{C}\boldsymbol{\beta}_k, \sigma_{kk}\mathbf{D})$ with $\mathbf{D} = \mathbf{A}(\mathbf{I}-\mathbf{P})\mathbf{A}^T$. Thus at time k the corresponding quadratic form

$$\begin{aligned}R(\mathbf{C}\boldsymbol{\beta}_k) &= (\mathbf{C}\hat{\boldsymbol{\beta}}_k)^T\mathbf{D}^{-1}\mathbf{C}\hat{\boldsymbol{\beta}}_k \\ &= \mathbf{y}_k^T(\mathbf{I}-\mathbf{P})\mathbf{A}^T[\mathbf{A}(\mathbf{I}-\mathbf{P})\mathbf{A}^T]^{-1}\mathbf{A}(\mathbf{I}-\mathbf{P})\mathbf{y}_k\end{aligned}$$

has distribution $\sigma_{kk}\chi_g^2$ under $H_0 : \mathbf{C}\boldsymbol{\beta}_k = \mathbf{0}$. This can be turned into an F statistic using the variance estimate above, yielding

$$F = \frac{R(\mathbf{C}\boldsymbol{\beta}_k)/g}{\hat{\sigma}_{kk}} \sim F_{g,n-r} \text{ under } H_0 .$$

A contrast at a fixed time k is an example of a more general hypothesis of contrasts of the form

$$H_0 : \mathbf{CBm} = \mathbf{0} .$$

The basic idea in the development of multivariate test statistics is to pick one-dimensional 'directions' $\mathbf{m}$ that convey important information. For a fixed time k, the vector $\mathbf{m}$ contains a 1 in the kth position and 0s in the other $p-1$ positions. Other contrasts of interest might include polynomial contrasts over time or the difference of the first half and last half corresponding to a change point. The statistic $\mathbf{C\hat{B}m}$ for this hypothesis is normally distributed with variance $(\mathbf{m^T V m})\mathbf{D}$. Therefore, the quadratic form

$$\begin{aligned} R(\mathbf{CBm}) &= (\mathbf{C\hat{B}m})^T \mathbf{D}^{-1} \mathbf{C\hat{B}m} \\ &= \mathbf{mY^T(I-P)A^T[A(I-P)A^T]^{-1}A(I-P)Ym} \end{aligned}$$

has distribution $(\mathbf{m^T V m})\chi_g^2$ under $H_0 : \mathbf{CBm} = \mathbf{0}$. The squared error residuals in direction $\mathbf{m}$ have distribution

$$\mathbf{m^T Y^T P Y m} = (n-r)\mathbf{m \hat{V} m} \sim \chi_{n-r}^2$$

yielding the F statistic

$$F = \frac{R(\mathbf{CBm})/g}{\mathbf{m^T \hat{V} m}} \sim F_{g,n-r} \text{ under } H_0 \ .$$

A multivariate hypothesis can now be formulated as

$$H_0 : \mathbf{CBM} = \mathbf{0}$$

in which the $t \times q$ matrix $\mathbf{M} = (\mathbf{m}_1, \cdots, \mathbf{m}_q)$ contains q contrasts over time. For simplicity of presentation, both $\mathbf{C}$ and $\mathbf{M}$ are assumed to be full rank g and q, respectively. With this knowledge, it is now possible to build a $q \times q$ multivariate hypothesis matrix

$$\begin{aligned} \mathbf{H} &= \mathbf{M^T \hat{B}^T C^T D^{-1} C \hat{B} M} \\ &= \mathbf{M^T Y^T(I-P)A^T[A(I-P)A^T]^{-1}A(I-P)YM} \end{aligned}$$

with elements having distribution $h_{ij} \sim (\mathbf{m}_i^T \mathbf{V} \mathbf{m}_j)\chi_g^2$ under $H_0 : \mathbf{CBM} = \mathbf{0}$. The $q \times q$ multivariate residual matrix for this hypothesis is

$$\mathbf{E} = \mathbf{M^T Y^T P Y M} = (n-r)\mathbf{M^T \hat{V} M}$$

with elements having distribution $e_{ij} \sim (\mathbf{m}_i^T \mathbf{V} \mathbf{m}_j)\chi_{n-r}^2$. Combining these element-wise leads to a matrix of pivot statistics, with elements

$$F = \frac{h_{ij}/g}{e_{ij}/(n-r)} \sim F_{g,n-r} \text{ under } H_0 \ .$$

Again, it is important to consider overall tests before delving into specific contrasts, even with a multivariate approach. The four common overall statistics presented in Chapter 15, Part E, are repeated below to remind

| | | | tree soils | | control soils | |
statistic	ndf	ddf	F	p	F	p
ROY	5	18	3.96	0.014	8.33	0.0003
WILKS	15	44.6	1.25	0.27	3.25	0.0011
HOTEL	15	44	1.33	0.23	3.31	0.0010
PILLAI	15	54	1.16	0.33	3.04	0.0014

Table 26.2. *Tree interaction multivariate tests by* soil *type*

the reader of typical summaries of multivariate information.

$$\text{ROY} = \lambda_1$$
$$\text{WILKS} = \det(\mathbf{HE}^{-1} + \mathbf{I})$$
$$\text{HOTEL} = \text{tr}(\mathbf{HE}^{-1})$$
$$\text{PILLAI} = \text{tr}(\mathbf{H}(\mathbf{H}+\mathbf{E})^{-1})$$

Example 26.3 Tree: Most of the time multivariate statistics reinforce patterns observed from univariate tests. However, the exceptions are worth noting. Table 26.2 summarizes the four multivariate statistics for date*soil interactions depicted in Figure 25.3. The formal results are generally similar to those found for split plot (Example 25.3) and for the sphericity-adjusted (Example 26.1) analyses.

However, notice that ROY is considerably more significant, especially for tree soils. Roy's greatest root picks out the strongest signal. It is likely that for tree soils, this is picking up the different pattern for ne-332 already noticed in Figure 25.3(a). This is the only interesting feature of differences among tree soils over dates.

Control soils show evidence of significant differences over dates with all four statistics, although ROY again picks up a major contrast, perhaps that grnhous soil has a fairly constant score over time relative to the other three. These could be investigated further by obtaining the linear combinations for the first eigenvectors for tree and control soils.   ◊

26.4 Problems

26.1 Nested: Verify that the variance for a repeated measures contrast is the same as that for split plot contrasts under sphericity.

26.2 Tree: There may be problems with assuming sphericity for this problem. Conduct an analysis allowing for more general correlation among dates in order better to assess whether nm-6 has significantly lower score than the other tree soils for either the non-autoclaved (acc = notauto) or autoclaved (acc = autocla) samples only. Characterize the difference between the four tree soils and the four controls. Address possible concerns from the scientist about how to communicate this more complicated analysis.

(a) Lay out the full model for repeated measures for this problem. Define terms. Briefly summarize the difference in assumptions between split plot and a multivariate repeated measures model.

(b) Test for deviation from sphericity. If this is significant, how does it affect interpretation of analysis in the previous chapter?

(c) Compare all eight soils for pathogen score using a multivariate repeated measures approach. Contrast the controls and the tree soils in some way that highlights the differences found in interaction plot(s).

(d) Compare the four standard multivariate test statistics. What are their relative merits? How do they perform in this problem? In particular, does ROY agree with the other three?

(e) How do the multivariate contrasts compare with polynomial (or other) contrasts over dates? Is the interpretation complementary?

(f) Indicate any shortcomings of the repeated measures approach for this problem.

(g) Think simple. Summarize your findings in an elegant way which can inform the scientist about the main results.

26.3 Berry: Return to Problem 25.3. Note that proc glm will not perform the multivariate tests if you include sample in the model. A message will appear in the log such as:

NOTE: No multivariate tests performed for JULIAN
      due to insufficient error degrees of freedom.

The newer proc mixed seems to handle sample as a whole plot blocking factor appropriately. However, proc mixed has somewhat different multivariate test options.

(a) Adjust the degrees of freedom for the split plot using the Greenhouse–Geiser ϵ. Perform relevant tests. How does this change the results?

(b) Adjust the degrees of freedom for split plot using the Huynh–Feldt ϵ (but use 1 if it is greater than 1). Comment on why this is so different from (a).

(c) Conduct the sphericity test using the printe option for the repeated statement. Interpret the results.

(d) As stated above, the multivariate analyses (and the sphericity test) had to be performed with sample removed from the model. Briefly consider the multivariate tests and compare these results with the earlier findings.

(e) Based on all analyses, what are your conclusions? Have they altered or strengthened from Problem 25.3? How would you modify your communication with the scientist?

26.4 Season: Dairy scientists were investigating the effect of a disinfectant on various diseases (Drendel *et al.* 1994). One response of general health is the somatic cell count (**scc**) as measured by the veterinarian. They examined cows in two **herds** of two **ages** (first or second lactation), assigning animals at random to treatment or control (**trt**). Cows might be measured over one to four **seasons** during the 49 week trial. Questions of interest concern whether treatment is effective throughout the year.

(a) Decide if any transformation of **scc** is advisable. You may want to fit a model and examine residuals first. That is, you might consider begining the following steps and returning to this issue later. Use plots to justify your choice.

(b) Carefully lay out an appropriate model for somatic cell count. It might be wise to begin by ignoring issues of repeated measures *per se.*

(c) Conduct a formal analysis. Report results to the scientists. Use plots to indicate any important relationships and raise any potential concerns about the experiment.

(d) If you have not done so already, address issues of repeated measures. What can you hope to do with these data? If you decide against formal inference, what plots could shed light on assumptions about correlations over **seasons**?

CHAPTER 27

Cross-over Design

Cross-over designs are used for experiments in which subjects are expensive and few in number, requiring a design that yields as much as possible from each. Subjects typically receive two or more treatments in succession, with possibly very little time between treatments. One concern in such a design is that the effect of the previous treatment (or treatments) may **carry over** and become confounded with the effect of the current treatment. Analysis of cross-over designs tries to sort out treatment effects from carry-over. While it is possible to do this with appropriate assumptions, it is best at the design phase to set up experiments to avoid carry-over effects.

Occasionally this approach is used for spatial rather than temporal designs, in which there is concern about carry-over from adjacent sites (e.g. from light shading or nutrient runoff). These designs are more common in animal studies than plant studies. They have certain aspects in common with split plot designs, repeated measures and Latin square designs.

The cross-over basically involves a sample of subjects chosen at random from some population and assigned at random to a sequence of treatments to be given over several time periods, one treatment per time period. Thus the **experimental unit** for a sequence of treatments is a subject, while the experimental unit for a treatment is a time period within a subject. Ideally (in a balanced design), every subject is assigned every treatment the same number of times. Further, it is best if every treatment follows every other treatment the same number of times. Two-period cross-over designs are somewhat simpler than multi-period designs.

Section 27.1 develops the basic idea of the cross-over model, showing how the concept of carry-over simplifies treatment interactions over time. Section 27.2 shows how sequence order, carry-over and treatment effects can be confounded in a two-period cross-over design. A partition of the total sum of squares in Section 27.3 leads to pivots for inference. Section 27.4 develops a three-period cross-over design as a type of replicated Latin square design. Section 27.4 makes some comments about more general cross-over designs.

27.1 Cross-over model

A cross-over design is similar to repeated measures except that the factor is assigned across the time periods. Usually, each subject receives all levels of the factor in some sequence order. Within each subject there may be a period effect as well as factor effects. Possible interactions involving period and the factor are commonly referred to as carry-over.

Another distinction from repeated measures is that any (random) correlation over time is usually incorporated into the (fixed) carry-over effect. While it is theoretically possible to allow some covariance among measurements over time within a subject, these covariances may not be estimable. In practice simplifying assumptions are made about the carry-over between time periods which hopefully capture any time dependence.

Consider the split plot model

$$y_{ijm} = \mu + (\alpha_i + r_{im}) + (\beta_j + \gamma_{ij} + e_{ijm})$$

with $i = 1, \cdots, a$ sequences of factor levels, $m = 1, \cdots, n_i$ subjects per sequence and $j = 1, \cdots, b$ time periods per subject. Thus α_i is the sequence effect, β_j is the time period effect, r_{im} the subject error and e_{ijm} the time period error. This form does not explicitly indicate the treatment assigned per time period. Consider a more complicated model which allows for treatment and carry-over

$$y_{ijkm} = \mu + (\alpha_i + r_{im}) + (\beta_j + \tau_k + \gamma_{ijk} + e_{ijkm})$$

with $k = 1, \cdots, t$ the treatment assigned in period j for subject assigned to sequence i. Note that we have added another subscript without changing the design since the treatment level $k = k(i, j)$ is determined by the sequence and period. The treatment effect of treatment k is τ_k, with γ_{ijk} the carry-over effect. Thus there is some redundancy in the triplet (i, j, k), much as there is in the Latin square design. In fact, the sequence of treatments over time periods is prescribed by the sequence i. If there are the same number of periods as treatments $(b = t)$, there are $t!$ $(= t \times (t-1) \times \cdots \times 2 \times 1)$ possible sequences, although only a subset might be used in some experiments.

Typically there are not enough periods to investigate two-factor interactions involving treatment and period. Instead, simplifying assumptions about the form of the **carry-over** γ_{ijk} are made, such as

$$\gamma_{ijk} = \sum_{l=0}^{j-1} \lambda_{j,k(i,l)}$$

the sum of carry-overs from all previous period, including an initial effect. Most applications simplify further

$$\gamma_{ijk} = \lambda_{j,k(i,j-1)} = \lambda_{k(i,j-1)}$$

to an effect associated with the treatment in the previous period. This leads to the **cross-over model**

$$y_{i1km} = \mu + (\alpha_i + r_{im}) + (\beta_1 + \tau_{k(i,1)} + e_{i1km})$$
$$y_{ijkm} = \mu + (\alpha_i + r_{im}) + (\beta_j + \tau_{k(i,j)} + \lambda_{k(i,j-1)} + e_{ijkm}) \, , \, j > 1 \, .$$

The relationship of treatment k to sequence i and period j for subject m is made explicit here to avoid ambiguity. The notation is identical to that used for nested effects.

27.2 Confounding in cross-over designs

Simplifying the treatment by period interaction to a carry-over is quite appealing. However, care in design is needed, since the carry-over can be confounded with sequence and treatment. This section simply suggests some cautions. In practice, it is best to study the carry-over explicitly during experimental design and later analysis.

Example 27.1 Carry: Consider a cross-over experiment with two periods per subject and two treatments. Each subject is assigned to either sequence 1 (A then B) or sequence 2 (B then A). The expected response for subjects is as follows:

sequence	period	mean
1	1	$\mu_{11A} = \mu + \alpha_1 + \beta_1 + \tau_A$
1	2	$\mu_{12B} = \mu + \alpha_1 + \beta_2 + \tau_B + \lambda_A$
2	1	$\mu_{21B} = \mu + \alpha_2 + \beta_1 + \tau_B$
2	2	$\mu_{22A} = \mu + \alpha_2 + \beta_2 + \tau_A + \lambda_B$

The difference in means between sequences is

$$\bar{\mu}_{1..} - \bar{\mu}_{2..} = (\alpha_1 - \alpha_2) + (\lambda_A - \lambda_B)/2 \, ,$$

which hopelessly confounds sequence effect with carry-over. In fact, the sequence effect is precisely the difference due to which treatment is given first, or the carry-over. Thus it is reasonable to think of carry-over as measuring the sequence effect, or rather to assume $\alpha_i = 0$. The difference of period means

$$\bar{\mu}_{.1.} - \bar{\mu}_{.2.} = (\beta_1 - \beta_2) - \bar{\lambda}.$$

is confounded with the average carry-over effect.

The difference in mean response between treatments is

$$\bar{\mu}_{..A} - \bar{\mu}_{..B} = (\tau_A - \tau_B) + (\lambda_B - \lambda_A)/2 \, ,$$

which confounds treatment and carry-over effects. Combining with the sequence mean differences, one can isolate the treatment effects as

$$\tau_A - \tau_B = (\bar{\mu}_{..A} - \bar{\mu}_{..B}) + (\bar{\mu}_{1..} - \bar{\mu}_{2..})$$
$$= \mu_{11A} - \mu_{21B},$$

which is the difference in mean responses for the first period only. ◇

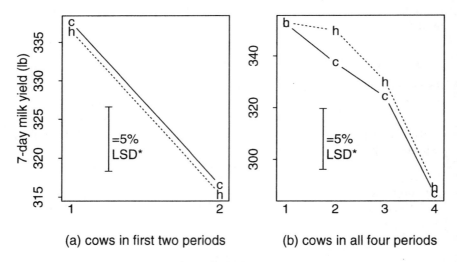

(a) cows in first two periods   (b) cows in all four periods

Figure 27.1. *Drink cross-over interaction plots: (a) data from the first two **periods** favor the cold **water** treatment slightly over the hot; (b) records for **cows** with four **periods** favor hot **water**. LSD bars appropriate for vertical comparisons within a **period**, suggesting hot/cold differences are negligible. Note differences in vertical scales.*

Example 27.2 Drink: A dairy scientist is interested in the effect on milk yield of feeding cows hot (lukewarm, actually) instead of cold water. This may have economic importance if the temperature of water can alter milk yield by even a pound per week. Animals were put on hot (or cold) water for three weeks, with measurements taken in the final week (as 7-day milk yield) of the period. Each cow was given both hot and cold water over a six week (two periods), with cows randomized as to whether they received hot or cold water first in each pair. Cows might be treated over several pairs of periods during the course of the study. Milk yield should gradually decrease over time, regardless of treatment. This decline is confounded with the hot/cold treatment for any given cow, but can be sorted out by comparing cows given hot or cold first.

The interaction plots in Figure 27.1 suggest a confusing pattern. When all cows are examined over the first two periods (Figure 27.1(a)), the hot water treatment seems to lower yield by a small amount. However, when cows with records for the first four periods are considered (Figure 27.1(b)), hot water appears to increase milk yield. The LSD bars caution that these differences are negligible, but it is still disconcerting. Further investigation points up that cows were recruited in different months of the year, which could in part explain the differences. In addition, cows could be removed from study for a variety of reasons possibly related to milk yield, including poor health. Note the difference in vertical scales for the two interaction plots. ◇

Thus the effect of sequence, the order of treatments, may be confounded with carry-over. With more than two periods, it is possible to separate sequence effect from carry-over. However, care is needed. The more periods there are, the more reliance is placed on assumptions about the form of carry-over (that is, the treatment by period interaction within sequence). Is the single-period carry-over model appropriate, or does it miss some longer-term effects? It may be possible to investigate this question using diagnostic plots of residuals against time identified by treatment.

Some care is needed in explicitly defining the carry-over. This may be done by defining a 'dummy' factor which is set to the treatment from the previous period, or to a unique, nonsense level for the first period. The carry-over factor has $c+1$ levels, the c treatment levels plus the first period nonsense level, but only $c-1$ degrees of freedom after adjusting for period.

27.3 Partition of sum of squares

The partition of variation and anova table pertinent to cross-over designs requires care once again. The carry-over induces an imbalance in the model, requiring the use of Type III sums of squares and corresponding adjusted inference. The partition of sum of squares can be summarized in terms of degrees of freedom and expected mean squares as shown in Table 27.1.

Inference for sequence, period, treatment and carry-over follows in a natural way. However, note that in particular situations (such as the two-period cross-over design), some terms may be redundant and some tests may not be possible. Further, sequence and carry-over both measure some form of period by treatment interaction which may signify some form of correlation over time.

Example 27.3 Carry: The anova for a two-period cross-over experiment with n subjects per sequence is shown in Table 27.2. Appropriate tests of hypotheses can be developed by paying attention to $E(MS)$. A significant sequence (=carry-over) effect would argue for caution in interpreting inference for treatments. In this case, the recommended approach is to examine only treatment differences in the first period using a t test. ◇

source	df	E(MS)
A=sequence	$a-1$	$(\sigma^2 + t\sigma_P^2)(1 + \delta_A^2/(a-1))$
subject error	$n.-a$	$\sigma^2 + t\sigma_P^2$
T=period	$t-1$	$\sigma^2(1 + \delta_T^2/(t-1))$
B=factor	$b-1$	$\sigma^2(1 + \delta_B^2/(b-1))$
C=carry-over	$b-1$	$\sigma^2(1 + \delta_C^2/(b-1))$
period error	rest	σ^2
total	$tn.-1$	

Table 27.1. *Cross-over expected mean squares*

source	df	E(MS)
A=sequence=carry-over	1	$\sigma^2 + 2\sigma_P^2 + n(\bar{\mu}_{1..} - \bar{\mu}_{2..})^2$
subject error	$2n-2$	$\sigma^2 + 2\sigma_P^2$
B=period	1	$\sigma^2 + n(\bar{\mu}_{.1.} - \bar{\mu}_{.2.})^2$
C=treatment	1	$\sigma^2 + n(\bar{\mu}_{..A} - \bar{\mu}_{..B})^2$
period error	$2n-2$	σ^2
total	$4n-1$	

Table 27.2. *Cross-over two-period anova*

Some investigators have recommended testing for the sequence effect to decide whether to analyze further the whole experiment or only the first period. This seems reasonable based on the argument in Example 27.1. However, recent literature cited in Crowder and Hand (1990, sec. 7.6) indicates that the size of the test may be much larger than the nominal significance level. That is, it is too easy to reject the hypothesis $H_0 : \tau_k = 0$ when it is true. This is similar to the dilemmas raised in deciding whether to pool interactions and in model selection.

27.4 Replicated Latin square design

Some cross-over designs are known as replicated Latin square designs when subjects are organized naturally into groups. With more than two periods, different variants can arise depending on how the possible sequences of

source	df	E(MS)
S=square	$s-1$	$(\sigma^2 + a\sigma_P^2)(1 + \delta_S^2/(s-1))$
A(S)=subject	s	$\sigma^2 + a\sigma_P^2$
B(S)=period	s	$\sigma^2(1 + \delta_{B(S)}^2/s)$
C=treatment	1	$\sigma^2(1 + \delta_C^2)$
period error	$s-1$	σ^2
total	$4s-1$	

Table 27.3. *Latin square with two treatments*

treatments are assigned to subjects across the groups.

Latin squares involving two periods follow the basic structure of Table 27.3. That is, each square consists of two subjects that are assigned the two sequences of treatments (1 then 2 or 2 followed by 1) at random. The random assignment of sequence is restricted to be within each pair, or 'square'. Each square is in fact a 2×2 Latin square with factors of sequence (= subject), period and treatment.

If there are more than two treatments and an equal number of periods per subject, it is possible to examine some interactions with replicated Latin squares. Table 27.4 displays the partition in which subjects are assigned into squares, perhaps on the basis of some factor, and randomly assigned unique sequences within the each square. Thus the square effect may correspond to some factor assigned to the subject of interest. There are now enough degrees of freedom to examine square by treatment interaction if desired. Note that there is no requirement *per se* that all squares be run together. That is, the actual time periods may differ among squares. If periods have a common interpretation across squares, it would be possible to separate this into period effect B and interaction with square $B*S$ or treatment $B*C$. However, there are usually a limited number of degrees of freedom leading to a compromise between examining interactions and increased degrees of freedom for error.

An alternative randomization arises in a design in which squares function as blocks of subjects. Subjects within a square may be randomly assigned to one of a subject groups and a sequence of treatments over time. Now the factor A has a common interpretation across squares. Various interactions are possible, as indicated in Table 27.5. Again, there is usually a premium on degrees of freedom in Latin squares, forcing the investigator to consider some interactions to be negligible at the outset.

source	df	$E(MS)$
S=square	$s-1$	$(\sigma^2 + a\sigma_P^2)(1 + \delta_S^2/(s-1))$
$A(S)$=subject	$(a-1)s$	$\sigma^2 + a\sigma_P^2$
$B(S)$=period	$(a-1)s$	$\sigma^2(1 + \delta_{B(S)}^2/s(a-1))$
C=treatment	$a-1$	$\sigma^2(1 + \delta_C^2/(a-1))$
$S*C$	$(a-1)(s-1)$	$\sigma^2(1 + \delta_{SC}^2/(a-1)(s-1))$
period error	rest	σ^2
total	$a^2 s - 1$	

Table 27.4. *Replicated Latin square with square effects*

source	df	$E(MS)$
S=square	$s-1$	$(\sigma^2 + a\sigma_P^2)(1 + \delta_S^2/(s-1))$
A=group	$a-1$	$(\sigma^2 + a\sigma_P^2)(1 + \delta_A^2/(a-1))$
subject error	$(a-1)(s-1)$	$\sigma^2 + a\sigma_P^2$
B=period	$a-1$	$\sigma^2(1 + \delta_B^2/(a-1)$
C=treatment	$a-1$	$\sigma^2(1 + \delta_C^2/(a-1)$
$A*C$	$(a-1)^2$	$\sigma^2(1 + \delta_{AC}^2/(a-1)^2)$
$B*C$	$(a-1)^2$	$\sigma^2(1 + \delta_{BC}^2/(a-1)^2)$
$S*C$	$(a-1)(s-1)$	$\sigma^2(1 + \delta_{SC}^2/(a-1)(s-1))$
period error	rest	σ^2
total	$a^2 s - 1$	

Table 27.5. *Replicated Latin square with interactions*

27.5 General cross-over designs

Some more general cross-over designs are briefly discussed here. For further detail and references, see Milliken and Johnson (1992, ch. 32) and Crowder and Hand (1990, ch. 7).

As the number of periods increases, there is an increased chance of missing data, leading to an unbalanced design. Here, many of the problems encountered with unbalanced nested or repeated measures designs are exacerbated. These are not pursued further in this text.

Balanced incomplete cross-over designs arise when the number of periods is not a multiple of the number of factor levels ($b \neq rc$ for some number of replicates r). In this case, analysis of the factor must involve the whole plot of subjects as well as the split plot over time. With two treatments and three periods some subjects must get sequences with two (or possibly three) of one treatment while others get sequences with one (or none) of the other. Thus sequence and treatment are partially confounded. See Milliken and Johnson (1992, sec. 32.3) or Crowder and Hand (1990, sec. 7.3) for further details.

Example 27.4 Rotate: A long-term crop rotation study in Lancaster, WI, examines, among other things, the effect of growing soybeans in one year on the nitrogen availability for corn and oat crops over the next two years (Vanotti and Bundy 1995). Crop rotations are in five-year cycles which have been repeated now for about 25 years. The rotations were CCCMM, CSCOM, CCOMM, with crops C = corn, M = alfalfa (*Medicago sativa* L.), S = soybean and O = oat. Thirty plots arranged in two blocks were randomly assigned to one of three rotations and to one of five phases in the cycle.

The previous year's crop has been shown to affect yield. Thus corn yield in the three phases of the 'control' rotation (CCCMM) should tend to decline due to depletion of the soil nutrients. Further, corn and oats following soybean in rotation (CSCOM) might do better than corn and oats following corn (CCCOM).

This design is not strictly a cross-over design, but it shares many features. For instance, there are repeated measures on the same plots, but treatments are changing over time. One- to two-year carry-over is of particular interest in this experiment. ◇

Example 27.5 Rotate: This experiment actually had each plot split into four units which received four levels of nitrogen fertilizer (control and three treatments in a doubling series) in spring tillage in every phase of corn each year. No nitrogen was added for other crops. ◇

27.6 Problems

27.1 Running: There was some concern about carry-over from one test session to the next.
(a) Examine this informally by plotting the response and/or residuals against session order. Be sure to identify the subject and the treatment combination. Make sure your plot(s) are not too complicated or messy. Is there any 'hint' of carry-over?
(b) Suppose you had funds to include more subjects. How would you modify the design to address carry-over concerns? Be sure to specify this carefully, either in words or using a mathematical model.

27.2 Drink: Consider the experiment introduced in Example 27.2.
(a) Write down the cross-over model for the first two `periods` of `milk` yield as influenced by hot/cold `water` treatment. Analyze for sequence (`first`), `period` and `water` effects. (Your interaction plot, if presented, should be similar to Figure 27.1(a).)
(b) Examining the first two `periods` of data, ascertain if there are differences in `milk` yield by `month`. Be sure to write out your model, with terms defined. Comment briefly on interpretation, noting in particular possible confounding.
(c) Four `periods` allow adjustment for confounding. Analyze all the data over four `periods`. Now carefully examine the data and analysis. Comment on the validity of interpretation in the presence of missing cells.
(d) Remove enough `cows` from the study to eliminate the missing cell problem found in (b). Redo the analysis. This corresponds to Figure 27.1(b).

27.3 Rotate: Consider one five-year cycle of crop rotation (Example 27.4). You may want to suppose that a whole cycle was completed before the data for the current cycle were gathered.
(a) Lay out a model similar to the that for the cross-over design with appropriate assumptions. Indicate clearly how a one-year carry-over could be properly defined.
(b) Construct an anova table for this experiment, indicating degrees of freedom and expected mean squares.
(c) The scientists were particularly interested in the possibility of two-year carry-over. Show how this can be handled.
(d) In a way, only a subset of the cross-over model is of direct interest. How would model and anova change if interest focused only on the oat `yields`?

27.4 Rotate: (a) Incorporate the `nitrogen` split plots mentioned in Example 27.5 into the model and anova for oat `yield`.
(b) Conduct appropriate analysis. Report results, with particular attention to the effect of `rotation` and `nitrogen` on oat `yield`. Include an interaction plot whether or not their interaction is significant.
(c) Perform a similar analysis for the corn that was grown in the phase immediately prior to oat.
(d) Interpret results in light of the scientists' questions articulated in Example 27.5.

27.5 Rotate: Can differences in oat `yield` be explained by corn `yield` in the previous phase? Conduct an ancova within the context of this cross-over model. That is, adjust for the regression of oat `yield` on corn `yield` when examining the the effect of `rotation` and `nitrogen` on oat `yield`.

(a) Plot oat `yield` against the corn `yield` from the previous phase for each plot. Identify points with plot symbols to indicate `rotation` or `nitrogen` level. Comment on any patterns.
(b) Average corn and oat `yield` over split plots and examine a scatter plot of oat against corn once again. Identify points by `rotation` and possibly by phase if relevant.
(c) Remove the whole plot effects from corn and oat `yield` to examine their relationship with respect to `nitrogen`. Are there still `rotation` effects to examine?
(d) Conduct a formal analysis of covariance. It might help to have separately identified covariates for the whole plot average corn `yield` and the split plot deviations of corn `yield`.
(e) Interpret your results in light of the above question.

References

Agresti A (1996) *An Introduction to Categorical Data Analysis*. Wiley, New York.
American Heritage Dictionary (1992) *American Heritage Dictionary of the English Language*, 3rd edn. Houghton Mifflin, Boston.
American Statistical Association (1995) 'Ethical Guidelines for Statistical Practice'. Office of Scientific & Public Affairs, Alexandria, VA.
Baskerville JC (1981) 'A systematic study of the consulting literature as an integral part of applied training in statistics', *The American Statistician* 35, 121–123.
Bates DM and Watts DG (1988) *Nonlinear Regression Analysis and Its Applications*. Wiley, New York.
Becker RA, Chambers JM, Cleveland WS and Wilks AR (1988) *The New S Language: A Programming Environment for Data Analysis and Graphics*. Wadsworth & Brooks/Cole, Pacific Grove, CA.
Belsley DA, Kuh E and Welsch RE (1980) *Regression Diagnostics*. Wiley, New York.
Bisgaard S (1991) 'Teaching statistics to engineers', *The American Statistician* 45, 274–283.
Boen JR and Zahn DA (1982) *The Human Side of Statistical Consulting*. Lifetime Books, Hollywood, FL.
Bollen KA (1989) *Structural Equations with Latent Variables*. Wiley, New York.
Box GEP (1954) 'Some theorems on quadratic forms applied in the study of analysis of variance problems, I. Effect of inequality of variance in the one-way classification', *Annals of Mathematical Statistics* 25, 290–302.
Box GEP, Bisgaard S and Fung C (1988) 'An explanation and critique of Taguchi's contributions to quality engineering', *Quality Reliability Engng. Intl.* 4, 123–131.
Box GEP and Draper NR (1987) *Empirical Model-building and Response Surfaces*. Wiley, New York.
Box GEP, Hunter S and Hunter WG (1978) *Statistics for Experimenters*. Wiley, New York.
Box GEP, Jenkins GM and Reinsel GC (1994) *Time Series Analysis: Forecasting and Control*, 3rd edn. Prentice Hall, Englewood Cliffs, NJ.
Box JF (1978) *R.A. Fisher, The Life of a Scientist*. Wiley, New York.
Bray JH and Maxwell SE (1985) *Multivariate Analysis of Variance*. Sage, Newbury Park, CA.

Broad W and Wade N (1982) *Betrayers of the Truth*. Simon & Schuster, New York.

Bronowski J (1965) *Science and Human Values*. Harper & Row, New York.

Bross IDJ (1974) 'The role of the statistician: scientist or shoe clerk', *The American Statistician* 28, 126–127.

Byrne BM (1989) *A Primer on LISREL: Basic Applications and Programming for Confirmatory Factor Analytic Models*. Springer-Verlag, New York.

Carroll RJ and Ruppert D (1988) *Transformation and Weighting in Regression*. Chapman & Hall, London.

Carroll RJ, Ruppert D and Stefanski LA (1995) *Measurement Error in Nonlinear Models*. Chapman & Hall, London.

Chambers JM and Hastie TJ (1992) *Statistical Models in S*. Wadsworth & Brooks/Cole, Pacific Grove, CA.

Chambers JM, Cleveland WS, Kleiner B and Tukey P (1983) *Graphical Methods for Data Analysis*. Wadsworth & Brooks/Cole, Pacific Grove, CA.

Chatfield C (1995) *Problem Solving: A Statistician's Guide*, 2nd edn. Chapman & Hall, London.

Chin SF, Storkson JM, Albright KJ, Cook ME and Pariza MW (1994) 'Conjugated linoleic acid is a growth factor for rats as shown by enhanced weight gain and imporved feed efficiency', *J. Nutrition* 124, 2344–2349.

Cleveland WS (1985) *The Elements of Graphing Data*. Wadsworth, Monterey, CA.

Cleveland WS (1993) *Visualizing Data*. Hobart, Summit, NJ.

Cochran WG (1977) *Sampling Techniques*, 3rd edn. Wiley, New York.

Cochran WG and Cox GM (1957) *Experimental Designs*, 2nd edn. Wiley, New York.

Conover WJ (1980) *Practical Nonparametric Statistics*, 2nd edn. Wiley, New York.

Cox DR (1958) *Planning of Experiments*. Wiley, New York.

Cox DR and Snell EJ (1981) *Applied Statistics, Principles and Examples*. Chapman & Hall, London.

Crowder MJ and Hand DJ (1990) *Analysis of Repeated Measures*. Chapman & Hall, London.

Daniel D (1969) 'Some general remarks on consulting in statistics', *Technometrics* 11, 241–245.

Deming WE (1965) 'Principles of professional statistical practice', *Annals of Mathematical Statistics* 36, 1883–1900.

Dhiman TR, Kleinmans J, Tessmann NJ, Radloff HD and LD Satter (1995) 'Digestion and energy balance in lactating dairy cows fed varying ratios of alfalfa silage and grain', *J. Dairy Science* 78, 330.

Dhiman TR and Satter LD (1996) 'Rumen degradable protein and its effect on microbial protein synthesis', *J. Dairy Science* 00, in review.

di Iorio FC and Hardy KA (1996) *Quick Start to Data Analysis with SAS*. Duxbury, Boston.

Diggle PJ (1983) *Statistical Analysis of Spatial Point Patterns*. Academic Press, New York.

Draper NR and Smith H (1981) *Applied Regression Analysis*, 2nd edn. Wiley,

New York.

Drendel TR, Hoffman PC, Bringe AN and Syverud TD (1994) 'The effect of a premilking teat disinfectant on somatic cell count and clinical mastitis', Technical Report R3598, Research Division, CALS, University of Wisconsin–Madison.

Efron B (1982) *The Jackknife, the Bootstrap and Other Resampling Plans.* SIAM, CBMS-NSF, Philadelphia.

Efron B and Tibshirani RJ (1994) *An Introduction to the Bootstrap.* Chapman & Hall, London.

Ehrenberg ASC (1982) *A Primer in Data Reduction: an Introductory Statistics Textbook.* Wiley, New York.

Eldridge MD, Wallman KK and Wulfsberg RM (1981) 'Preparing statisticians for careers in the Federal Government' (with discussion), *ASA Proc. Statist. Educ. Sect.*, 34–450.

Everitt B (1977) *The Analysis of Contingency Tables.* Chapman & Hall, London.

Feyerabend P (1988) *Against Method*, rev. edn. Verso, New York.

Fienberg SE (1980) *The Analysis of Cross-Classified Categorical Data*, 2nd edn. MIT Press, Cambridge, MA.

Fisher LD and van Belle G (1993) *Biostatistics: A Methodology for the Health Sciences.* Wiley, New York.

Fisher RA (1935) *The Design of Experiments.* Hafner/Macmillan, New York.

Fisher RA (1990) *Statistical Methods, Experimental Design, and Scientific Inference.* Oxford University Press, Oxford.

Freund RJ and Littell RC (1991) *SAS System for Regression*, 2nd edn. SAS Institute Inc., Cary, NC.

Fuller WA (1987) *Measurement Error Models.* Wiley, New York.

Gardner H (1985) *Frames of Mind: The Theory of Multiple Intelligences.* Basic Books / Harper Collins, New York.

Gardner H (1993) *Creating Minds: An Anatomy of Creativity Seen through the Lives of Freud, Einstein, Picasso, Stravinsky, Eliot, Graham and Gandhi.* Basic Books/Harper Collins, New York.

Gelman A, Carlin JB, Stern HS and Rubin DB (1995) *Bayesian Data Analysis.* Chapman & Hall, London.

Goldberg N (1986) *Writing Down the Bones: Freeing the Writer Within.* Shambhala, Boston.

Goldman IL, Paran I and Zamir D (1995) 'Quantitative trait locus anlaysis of a recombinant inbred line population derived from a *Lycopersicon esculentum* × *Lycopersicon cheesmanii* cross', *Theoretical & Applied Genetics* 90, 925–932.

Gonick L and Smith W (1993) *The Cartoon Guide to Statistics.* Harper Perrenial/Harper Collins.

Gowers E (1988) *The Complete Plain Words.* DR Godine, Boston.

Green PJ and Silverman BW (1994) *Nonparametric Regression and Generalized Linear Models: A Roughness Penalty Approach.* Chapman & Hall, London.

Greenhouse SW and Geisser S (1959) 'On the methods in the analysis of profile data', *Psychometrika* 24, 95–112.

Hagidimitriou M and Roper T (1994) 'Seasonal changes in nonstructural carbohydrates in cranberry', *J. American Society of Horticultural Science* 119, 1029–1033.

Hall ET (1966) *The Hidden Dimension.* Anchor/Doubleday, Garden City, NY.
Hall ET (1981) *Beyond Culture.* Anchor/Doubleday, Garden City, NY.
Haslett J, Bradley R, Craig R, Unwin A and Wills G (1991) 'Dynamic graphics for exploring spatial data with application to locating global and local anomalies', *The American Statistician 45,* 234–242.
Healy MJR (1973) 'The varieties of statistician', *J. Royal Statistical Society Ser. A 136,* 71–74.
Healy MJR (1984) 'Prospects for the future. Where has statistics failed?' (with discussion), *J. Royal Statistical Society Ser. A 147,* 368–374.
Healy MJR (1986) *Matrices for Statistics.* Clarendon Press, Oxford.
Herrmann N (1989) *The Creative Brain.* Ned Herrmann Group, Lake Lure, NC.
Higham NJ (1993) *Handbook of Writing in the Mathematical Sciences.* SIAM, Philadelphia.
Hoaglin DC, Mosteller F and Tukey JW, eds. (1991) *Fundamentals of Exploratory Analysis of Variance.* Wiley, New York.
Hoffman PC (1996) 'The effect of forage sample drying technique on ruminal disappearance of dry matter and crude protein', Extension Report, Marshfield Experiment Station, University of Wisconsin–Madison.
Hofstadter DR (1979) *Gödel, Escher, Bach: An Eternal Golden Braid.* Vintage / Random House, New York.
Hoyningen-Huene P (1993) *Reconstructing Scientific Revolutions: Thomas Kuhn's Philosophy of Science.* University of Chicago Press, Chicago.
Huff D (1954) *How to Lie with Statistics.* Norton, New York.
Hunter WG (1981) 'The practice of statistics: the real world is an idea whose time has come', *The American Statistician 35,* 72–76.
Hurlbert SH (1984) 'Pseudoreplication and the design of ecological field experiments', *Ecological Monographs 54,* 187–211.
Hurley C (1993) 'The plot-data interface in statistical graphics', *J. Comp. and Graphical Statist. 2,* 365–379.
Huynh H and Feldt LS (1970) 'Conditions under which mean square ratios in repeated measures designs have exact F-distributions', *J. American Statistical Association 65,* 1582–1589.
Huynh H and Feldt LS (1976) 'Estimation of the Box correction for degrees of freedom for sample data in randomised block and split-plot designs', *J. Educational Statistics 1,* 69–82.
Jeanne RL, Graf CA and Yandell BS (1995) 'Non-size-based morphological castes in a social insect', *Naturwissenschaften 82,* 296–298.
Jiang J (1996a) 'REML estimation: Asymptotic behavior and related topics', *Annals of Statistics 24,* 255–286.
Jiang J (1996b) 'Asymptotic properties of the empirical BLUP and BLUE in mixed linear models', *Annals of Statistics 24,* 000–000.
Johnson JL and Kotz S (1972) *Distributions in Statistics: Continuous Multivariate Distributions.* Wiley Interscience, New York.
Joiner BL (1982) 'Consulting, Statistical' *Encyclopedia of Statistical Sciences,* Kotz S and Johnson JL, eds. Wiley, New York.
Jöreskog KG and Sörbom D (1988) *LISREL 7: A Guide to the Program and Applications.* SPSS Inc., Chicago.

REFERENCES

Katsiotis A (1993) 'The determination and use of pollen grain size in four ploidy levels and the cytogenetics of tetraploid-octoploid hybrids in search of $2n$ gametes of *Avena*', PhD Dissertation, Department of Agronomy, University of Wisconsin–Madison.

Kennedy BW (1991) 'C. R. Henderson: The unfinished legacy', *J. Dairy Science* **74**, 4067–4081.

Kirk RE (1991) 'Statistical consulting in a university: dealing with people and other challenges', *The American Statistician* **45**, 28–34.

Krol E (1992) *The Whole Internet*. O'Reilly & Associates Inc., Sebastapol, CA.

Krzanowski WJ (1990) *Principles of Multivariate Analysis: a User's Perspective*. Oxford University Press, Oxford.

Kuhn TS (1962) *The Structure of Scientific Revolutions*. University of Chicago Press, Chicago.

Lehmann EL (1975) *Nonparametrics*. Holden-Day, San Francisco.

Lerner G (1993) *The Creation of Feminist Consciousness from the Middle Ages to Eighteen-Seventy*. Oxford University Press, Oxford.

Littell RC, Freund RJ and Spector PC (1991) *SAS System for Linear Models*, 3rd edn. SAS Institute Inc., Cary, NC.

Littell RC, Milliken GA, Stroup WW and Wolfinger RD (1996) *SAS System for Mixed Models*. SAS Institute Inc., Cary, NC.

Lurie W (1958) 'The impertinent questioner: The scientist's guide to the statistican's mind', *American Scientist* **46**, 57–61.

Mallow C (1973) 'Some comments on C_p', *Technometrics* **15**, 661–675.

Markova D (1991) *The Art of the Possible: A Compassionate Approach to Understanding the Way People Think, Learn and Communicate*. Conari Press, Berkeley.

Marquardt DW (1979) 'Statistical consulting in industry', *The American Statistician* **33**, 102–107.

Mauchly JW (1940) 'Signicance test for sphericity of a normal n-variate distribution', *Annals of Mathematical Statistics* **29**, 204–209.

Maxwell D and Stanosz G (1996) 'Assay to determine if weed suppressiveness is an inherent property of soil in which NM-6 is growing', Department of Plant Pathology, University of Wisconsin–Madison.

McCullagh P and Nelder JA (1989) *Generalized Linear Models*, 2nd edn. Chapman & Hall, London.

McGrayne SB (1993) *Nobel Prize Women in Science: Their Lives, Struggles, and Momentous Discoveries*. Carol Publishing Group, Secaucus, NJ.

Miller A (1993) 'Pyridate tolerance in *Brassicaceae* crops', PhD Dissertation, Department of Horticulture, University of Wisconsin–Madison.

Miller RG Jr (1981) *Simultaneous Statistical Inference*, 2nd edn. Springer-Verlag, New York.

Miller RG Jr (1997) *Beyond ANOVA, Basics of Applied Statistics*, 2nd edn. Chapman & Hall, London.

Milliken GA and Johnson DE (1992) *Analysis of Messy Data vol. 1: Designed Experiments*. Chapman & Hall, London.

Milliken GA and Johnson DE (1989) *Analysis of Messy Data vol. 2: Non-replicated Experiments*. Chapman & Hall, London.

Morrison DF (1976) *Multivariate Statistical Methods*. McGraw-Hill, New York.
Mosteller F and Tukey JW (1977) *Data Analysis and Regression*. Addison-Wesley, Reading, MA.
Myers MJ, Steudel K and White SC (1993) 'Uncoupling the correlates of locomotor costs: a factorial approach', *J. Experimental Zoology* 265, 211–223.
Novy RG (1992) 'Characterization of somatic hybrids between *Solanum etuberosum* and diploid, tuber-bearing *Solanums*', PhD Dissertation, Department of Plant Pathology, University of Wisconsin–Madison.
Palta JP and Weiss LS (1993) 'Ice formation and freezing injury: an overview on the survival mechanisms and molecular aspects of injury and cold acclimation in herbaceous plants', in *Advances in Plant Cold Hardiness*, ed. by PH Li and L Christersson. CRC Press, Boca Raton, LA.
Penslar RL (1995) *Research Ethics: Cases & Materials*. Indiana University Press, Bloomington.
Rao CR (1965) *Linear Statistical Inference and its Applications*. Wiley, New York.
Reid C (1982) *Neyman – from Life*. Springer-Verlag, New York.
Rohlf FJ and Sokal RR (1995) *Statistical Tables*, 3rd edn. Freeman, New York.
Rose-Hellekant TA and Bavister BD (1996) 'Roles of protein kinase A and C in spontaneous maturation and in forskolin or 3-isobutyl-1-methylxanthine maintained meiotic arrest of bovine oocytes', *Molecular Reproduction & Development* 44, 241–249.
Ryan BF, Joiner BL and Ryan TA Jr (1985) *Minitab Handbook*, 2nd edn. Duxbury, Boston.
Samuel MD, Goldberg DR, Thomas CB and Sharp P (1995) 'Effects of *Mycoplasma anatis* and cold stress on hatching success and growth of mallard ducklings', *J. Wildlife Diseases* 31, 172–178.
Samuels ML, Casella G and McCabe GP (1991) 'Interpreting blocks and random factors' (with discussion), *J. American Statistical Association* 86, 798–821.
SAS Institute (1992) *SAS/STAT User's Guide*, version 6. SAS Institute Inc., Cary, NC.
SAS Institute (1994) *JMP User's Guide*. SAS Institute Inc., Cary, NC.
Savage LJ (1976) 'On rereading R.A. Fisher (with discussion)', *Annals of Statistics* 4, 441–500.
Scheffé H (1959) *The Analysis of Variance*. Wiley, New York.
Schilling EG (1987) 'One approach to statistical training in quality', *ASA Proc. Statist. Educ. Sect.*, 21–23.
Searle SR (1971) *Linear Models*. Wiley, New York.
Searle SR (1987) *Linear Models for Unbalanced Data*. Wiley, New York.
Searle SR, Casella G, McCulloch CE (1992) *Variance Components*. Wiley, New York.
Seber GAF (1977) *Linear Regression Analysis*, Wiley, New York.
Senn S (1993) *Cross-over Trials in Clinical Research*. Wiley, New York.
Sigma Xi Society (1991) *Honor in Science*. Sigma Xi Scientific Research Society, Research Triangle Park, NC.
Snedecor GW and Cochran WG (1989) *Statistical Methods*, 8th edn. Iowa State University Press, Ames.
Snee RD (1993) 'What's missing in statistical education?', *The American Stat-*

REFERENCES

istician 47, 149–154.

Song K, Slocum MK and Osborn TC (1995) 'Molecular marker analysis of genes controlling morphological variation in *Brassica rapa* (syn. *campestris*)', *Theoretical & Applied Genetics 90*, 1–10.

Sprent P (1970) 'Some problems of statistical consultancy' (with discussion), *J. Royal Statistical Society Ser. A 133*, 139–164.

Stieve SM, Stimart DP and Yandell BS (1992) 'Heritable tissue culture induced variation in *Zinnia Marylandica*', *Euphytica 64*, 81–89.

Stigler SM (1986) *The History of Statistics: the Measurement of Uncertainty before 1900*. Belknap / Harvard University Press, Cambridge, MA.

Strunk W Jr and White EB (1979) *The Elements of Style*. Macmillan, New York.

Tierney L (1990) *Lisp-Stat: An Object-Oriented Environment for Statistical Computing and Dynamic Graphics*. Wiley Interscience, New York.

Tukey JW (1977) *Exploratory Data Analysis*. Addison-Wesley, Reading, MA.

Tufte ER (1983) *The Visual Display of Quantitative Information*. Graphics Press, Chesire, CT.

Tufte ER (1990) *Envisioning Information*. Graphics Press, Chesire, CT.

Turabian KL (1973) *A Manual for Writers of Term Papers, Theses and Dissertations*, 4th edn. University of Chicago Press, Chicago.

Vanotti MB and Bundy LG (1995) 'Soybean effects on soil nitrogen availability in crop rotations', *Agronomy J. 87*, 676–680.

Velleman PF and Velleman AY (1988) *Data Desk Professional*. Odesta Corp., Northbrook, IL.

Venables WN and Ripley BD (1994) *Modern Applied Statistics with S-Plus*. Springer-Verlag, New York.

Wattiaux MA, Combs DK and Shaver RD (1994) 'Lactational responses to ruminally undegradable protein by dairy cows fed diets based on alfalfa silage', *J. Dairy Science 77*, 1604–1617.

Weerahandi S (1995) *Exact Statistical Methods for Data Analysis*. Springer-Verlag, New York.

Woodward WA and Schucany WR (1977) 'Bibliography for statistical consulting', *Biometrics 33*, 564–565.

Wright S (1934) 'The method of path coefficients', *Annals of Mathematical Statistics 5*, 161–215.

Wright SP (1992) 'Adjusted p-values for simultaneous inference', *Biometrics 48*, 1005–1013.

Yandell BS (1991) 'Quantitative trait loci in *Brassica rapa*', *Proc. 8th Symp. Interface Comp. Sci. Statist. 12*, 258–261.

Zelen M (1983) 'Biostatistical science as a discipline: a look into the future' (with discussion), *Biometrics 39*, 827–837.

Index

additive model, 182, 189
additive models, 125–129, 183
   three-factor, 141–142
   two-factor, 113–114
   unbalanced design, *164–167*
Agresti A, 7
Ahn HS, xiv
Albright KJ, 256
American Heritage Dictionary, 6
American Statistical Association (ASA), 27, 28
American Statistician, 13, 27
analysis of covariance, 202, 204, *255–273*, 288
analysis of variance (anova), 4
ancova, *see* analysis of covariance
anova table
   balanced designs, *see* balanced designs, anova table
   three factors, 140
assumptions, *58–59*, 209–219
attenuation, *see* error in variables, classical error

backward elimination, *see* model selection, stepwise procedures
balanced designs, 125
   anova table, *130–132*
   effects model, *129*
   incomplete block (BIBD), 42, *344–345*
   nesting, *337–354*
   one observation per cell, 125–127, 129–130, 142, *151–157*, 341, 344
   three factors, *139–142*, 148
balanced experiments, 108

balanced incomplete block designs (BIBD), *see* blocking *and* balanced designs
bar graph, 10
Bartlett MS, *see* variance, unequal
Baskerville JC, 26
Bates DM, 7, 243
Bavister BD, 163
Becker RA, 13
Behrens–Fisher problem, 221, 225
Belsley DA, 210
best linear unbiased
   estimator (BLUE), 304, *309–310*, 310, 311, 332
   predictor (BLUP), *310*, 318, 319
Bisgaard S, 27, 191
blocking, 12, 198, 235, *340–345*, 357
Boen JR, 26
Bollen KA, 245, 293
Bonferroni, *see* multiple comparisons
Box correction to df, *see* repeated measures
Box GEP, 42, 44, 139, 189–191, 215, 227, 400, 401
Box JF, 27, 235
Box–Cox transform, *230–231*, 234
Bradley R, 14
Bray JH, 275
Bringe AN, 410
Broad W, 28
Bronowski J, 27
Bross IDJ, 26
Bundy, 419
Byrne BM, 293

Carlin JB, 236

INDEX

Carroll RJ, 229, 231, 233, 248, 252
Casella G, 79, 304, 307, 308, 310, 313, 327, 332, 344
causal models, *see* multivariate
cell means model, *see* factorial designs
central limit theorem, 216
Chambers JM, 11, 13, 14
Chang SC, xiv
Chatfield C, 3, 26
Chin SF, 256
Clayton MK, xiv
Cleveland WS, 13, 14, 49, 210
Cochran WG, 42, 44, 75, *see* variance, unequal, 231, 252
completely randomized designs (CRD), 8, *40*, 40–41, 43, 53, 68, 86, 105, 125, 130, 143, 158, 193, 196, 199, 200, 275, 298, 341, 347, 357, 366, 369
compound symmetry, *see* repeated measures, sphericity
confidence interval, *60–63*
connected cells, *180–184*
  balanced subsets, 186
Conover WJ, 7
constraints, *see* linear constraints
consulting, *see* statistical consulting
contrasts, *see* linear contrasts
Cook ME, 256
Cox DR, 7, 235
Cox GM, 42, 44
Craig R, 14
critical value, *60*, 61
cross-over designs, *411–419*
  replicated Latin square, *416–417*
Crowder MJ, 401, 403, 405, 416, 418, 419
cumulative distribution, 11

Daniel D, 26
Data Desk, 14
degrees of freedom, 56
Deming WE, 26
design structure, *38–42*
deviance, *see* maximum likelihood
Dhiman TR, 67, 359
di Iorio FC, 13, 18

Diggle PJ, 236
discriminant analysis, *see* multivariate
Draper NR, 7, 44, 75, 145, 190
Drendel TR, 410
Dunnett, *see* multiple comparisons
dynamic graphics, 14

effects model, *see* factorial designs
Efron B, 236
Ehrenberg ASC, 8
Eldridge MD, 27
error in variables, 241
  classical error, *250–252*
  regression calibration, *247–250*
error rate, *89–91*, 92
  comparison-wise, *90*, 104
  experiment-wise, *89–90*, 90, 104
estimable functions, 105, 107–108, *110–114*, 114, 116, 117, 127, 168, 175, 178, 180, 182, 183, 189, *196–197*, 197, 202, 204, 332, 347, 349, 395, 396, 406, 412
  general form, *117–124*, 180–182
ethics, *26–28*
Everitt B, 7
Ewing B, xiv
examples, xiii, *15*
  bacteria, 128–129, 132–137, 142, 173–175, 342–344, 354, 378
  berry, 397, 409
  biotron, 345
  brassica, 367, 370, 378–379
  budding, 119–120
  carry, 413–415
  cloning, 54–56, 58, 62–63, 74, 76–77, 80, 87, 95–97, 99–101, 103, 210–211, 214, 217–218
  company, 157–158
  design, 18, 37, 39–41, 44–45, 108, 111–112, 114–123, 179–184, 188–191, 253
  diet, 68, 103, 260, 274, 378
  drink, 414–415, 420
  drying, 379–380
  feed, 256, 262, 264–265, 269, 271–272, 274

forage, 67–68, 87, 253–254, 359–364, 367
growth, 177–179, 184–186, 193–194, 212, 219–220, 227, 237
hardy, 175–176
infer, 61, 65–66, 68–69, 75–76, 86, 90–92, 124, 130, 147–148, 157, 176, 205, 217, 219, 230–233, 332–333
interact, 18–19, 32–33
iron, 252–253
nested, 341, 346–347, 349, 351–353, 357–359, 363, 369, 372, 408
nutrient, 242–244, 247, 249–251
oocyte, 163–166, 169, 171–173, 175, 355
path, 277, 280–281, 289–293
permute, 236
power, 59, 66–67, 83–85, 222, 339–340, 354
product, 191–192, 194
random, 298–299, 310–311, 313, 317–318, 324–326
rotate, 419–421
running, 142–143, 237, 356, 419
season, 410
size, 258, 277, 288
tomato, 50–51, 53, 58, 62, 71–72, 74, 79–80, 354–355, 368
tree, 86–87, 103–104, 124, 384–385, 392–393, 395–396, 401–404, 408–409
tukey, 151, 153–157
variance, 228, 237
wasp, 274, 278–279, 285–287, 293–294
wheat, 353–356
expected mean squares, *81–82*, 130
three factors, 140
experimental design, 35–44
explained variation, see R^2 and model selection

F test, *77–78*
factor, *4–5*
factorial designs, *107–125*
balance, see balanced designs
cell means model, *53–55, 107–108*
central replicate, 190–191
fixed effects, *108–114*
fractional, 43, 139, *189–192*
Latin square, 42, 43, *187–189*
missing cells, *177–192*
random effects, *298–301*
three factors, see balanced designs
unbalanced, see unbalanced designs
Feldt LS, 389, 400, 401
Feyerabend P, 27
Fienberg SE, 7
Fisher LD, 36
Fisher RA, 27, 39, 64, 235
Fisher's LSD, see multiple comparisons
forward selection, see model selection, stepwise procedures
fractional factorial designs, see factorial designs
Freund RJ, 13, 117, 121, 123, 283
Fuller WA, 248, 252
Fung C, 191

Gardner H, 28, 29
Geisser S, 400, 401
Gelman A, 236
general F test, 145, 149, 172–175, see sums of squares, Type III, 186, 276
general F-test, *80–81*
general guidelines, *31*
generalized inverse, *195*
Goldberg DR, 128
Goldberg N, 30
Goldman IL, 50
Gonick L, 10
Gosset W, 98
Gowers E, 30
grand mean
sample, *73*
Green PJ, 243
Greenhouse SW, 400, 401
Greenhouse-Geisser (G-G), see repeated measures, sphericity

Hagidimitriou M, 397
Hall ET, 29

INDEX

Hand DJ, 401, 403, 405, 416, 418, 419
Hartley HO, see variance, unequal
Haslett J, 14
Hastie TJ, 11, 13
Healy MJR, 26, 112, 196
Henderson H, xiv
Herrmann N, 28
Higham NJ, 15, 30
histogram, see plot
Hoaglin DC, 72, 137, 149, 150, 210, 231
Hoffman PC, 379, 410
Hofstadter DR, 27
Hotelling–Lawley trace, see multivariate
Hoyningen-Huene P, 27
HSD, see multiple comparisons
Hsiao CF, xiv
Huff D, 9, 10
Hunter S, 42, 44, 139, 189
Hunter WG, 26, 42, 44, 139, 189
Hurlbert SH, 39
Hurley C, 14
Huynh H, 389, 400, 401
Huynh-Feldt (H-F), see repeated measures, sphericity
hypothesis
  compound, *73*
  general linear, 201
  test, *63–65*
  three-sided, 64
  Types I–IV, see sums of squares

idempotent, *196*
incomplete designs, see nested models or factorial designs, missing cells
interaction, 109, 113, 116, 182
  missing cells, 186
  pooling, *145–147*, 416
  Tukey's one-degree-of-freedom test, *155–157*
interaction plot, see plot
Internet, xiii, *15*

Jenkins GM, 215
Jiang J, 308
jitter, 55, *210*, 211, 212, 219

JMP, see SAS, JMP
Johnson DE, xiii, 97, 134, 155, 157, 189, 227, 228, 308, 322, 373, 385, 403, 418, 419
Johnson JL, 306
Joiner BL, 14, 26
Jöreskog KG, 293

Katsiotis A, 353
key questions, 3, *22*
Kimmel J, xiv
Kirk RE, 26
Kleiner B, 14
Kleinmans J, 359
Kotz S, 306
Krol E, 15
Krzanowski WJ, 275, 283, 405
Kuh E, 210
Kuhn TS, 27
Kutner MH, xiii

Latin square, 412
  replicated, 189
Latin square designs, see factorial designs
LD Satter, 359
least squares
  estimates, 56
  ordinary, *196–197*
  weighted, *197–198*
Lehmann EL, 7, 234
Lerner G, 27
Levene H, see variance, unequal
linear combinations, *73*
linear constraints, 112, *114–117*, 165
  missing cells, 179
  set-to-zero, 115–117, 119, 120, 205
  sum-to-zero, 114–117, 119, 123, 137, 140, 154, 205
linear contrasts, *73–77*
  orthogonal, *76–77*
  orthogonal polynomials, *75*
linear models, 110, *195–204*, see regression or analysis of covariance or factorial designs
Lisp-Stat, 14
LISREL, 293

Littell RC, 13, 117, 121, 123, 283, 308
longitudinal studies, *see* repeated measures
LSD, *see* multiple comparisons
Lurie W, 26

main effects, 113
Mallow's $C(p)$, *see* model selection
Mandel interaction, *see* balanced designs, one observation per cell
manova, *see* multivariate
many-to-one comparisons, *see* multiple comparisons, Dunnett
margin plot, *see* plot
Markova D, 28
Marquardt DW, 26
MathSoft, 13
matrix algebra, *195–196*
Mauchly JW, 403
maximum likelihood, 195, *198–199*, 306, 307
   deviance, *203–204*, 204
   restricted (REML), 195, *199*, 306, 308–310, 317, 318, 321, 330, 332, 354, 366, 396
Maxwell D, 384
Maxwell SE, 275
McCabe GP, 332, 344
McCullagh P, 233
McCulloch CE, 79, 304, 307, 308, 310, 313, 327, 332
mean
   population grand, 73, 110
   population marginal, 109–110
   sample grand, 73
Miller A, 370
Miller RG Jr, 89, 97–99, 209, 221–222, 227, 252
Milliken GA, xiii, 97, 134, 155, 157, 189, 227, 228, 308, 322, 373, 385, 403, 418, 419
Minitab, 14
MINQUE, *308*, 317, 332
missing cells, 108, 117, *see also* factorial designs *and* sums of squares, Type IV
MIVQUE, *see* MINQUE

model selection, 145–157, 171, 416
   hierarchy, 145
   Mallow's $C(p)$, 150, *150–151*
   parsimony, 145, 215
   R^2, *150*
   rule of 2, *149*
   stepwise procedures, *148–149*
Morrison DF, 275, 283, 284, 405
Mosteller F, 7, 72, 137, 149, 150, 210, 231
multiple comparisons, 394
   based on F tests, *91–97*, 100–103, 162
   based on range of means, *97–103*
   Bonferroni, 91, *93–94*, 95, 100–102
   Duncan, *98–100*, 102
   Dunnett, *98*, 104
   Fisher's LSD, 8, 91, *92–93*, 94, 95, 98, 100–103
   more than one factor, 125
   REGWF, *94–95*, 96, 97, 100, 102
   REGWQ, 102
   Scheffé, 91, *94*, 97, 100–103
   SNK, *98*, 100, 102
   studentized range, *97*
   tradeoffs, *100–103*
   Tukey's HSD, *97–98*, 100, 102
   Waller–Duncan, *95*, 97, 99, 102
multiple responses, 198, 202, *see also* multivariate *and* regression, 405
multiple-stage tests, 91
multivariate
   causal models, 288–293
   discriminant analysis, *284–288*
   Hotelling–Lawley trace, 282–284, 408
   latent variables, 292–293
   manova, *276–284*
   mixed effects, *330–332*
   Pillai–Bartlet trace, 282–284, 408
   random effects, *322–324*
   repeated measures, *see* repeated measures, multivariate
   Roy's greatest root, 282–284, 408
   Wilks's likelihood ratio criterion, 282–284, 408
Myers MJ, 142

INDEX

Nelder JA, 233
nested designs, *337–378*
  split plot, 41, *357–367*, 369–371
  strip plot, 41, *371–372*
nested models, *145*
Neter J, xiii
Neyman J, 27
non-central χ^2, *81*, 82
non-centrality parameter δ^2, *81–82*,
  82–83, 126, 131–132, 200–202, 268,
  269, 303, 329, 330, 339, 349, 351,
  357, 362–363, 392
Nordheim EV, xiv
normal equations, 56, 111, 114, 119
normal scores, *234–235*
Novy RG, 54

observational study, 22, *35*, 36, 255
Omori Y, xiv
orthogonal polynomials, *see* linear
  contrasts
Osborn TC, 119
overall test, *77–78*

Palta JP, 175
Paran I, 50
Pariza MW, 256
parsimony, *see* model selection
partition of sum of squares, *78–81*
path diagram, 147
Penslar RL, 28
permutation test, *see* randomization
pie chart, 10
Pillai–Bartlett trace, *see* multivariate
pivot statistic, *59*, 71, 315
plot, 9–13, 22–24, 150
  box, 11, 49, *52–53*, 211, 216, 217,
    364
  dot, *52*
  effect, *137*, 364
  half-normal Q-Q, *137*, 190
  histogram, 10, 49, *51–52*, 59, 211
  interaction, 11, *133–135*, 139, 213
  margin, *134*, *152–153*, 157, *214*
  normal Q-Q, 190, 211, *216–217*,
    217–218, 234
  one-factor, 49–53

residual, 11–12, *209–213*, 216, 226,
  227, 364
scatter, 11, 209, *211–212*
stem-and-leaf, 10, 49, *49–51*, 52,
  150, 211, 217
population grand mean, 109, 116, 120,
  123
population marginal mean, 123
power, *82–85*
precision, *see* confidence interval
profile analysis, *see* repeated measures
pseudo-replications, *see* sub-sampling
p-value, *65*

Q-Q plot, *see* plot, normal Q-Q
quadratic form, 196

R^2, 150, *150*
  adjusted, *150*
Radloff HD, 359
random effects, 149, 198
randomization, 39, *235–237*
randomized complete block designs
  (RCBD), *40*, 42, 43, 341–345, 355,
  356, 369, 370, 378, *see also* nested
  designs
Rao CR, 155
Regal R, xiv
Regard, 14
regression, 7, 12, 14, 76, 149, 151, 171,
  202, 204, 209–212, 239, 241, 242,
  244, 248, 255, 257, 266, 287, 288
  multiple responses, *275–293*
  ordered groups, *241–252*
Reid C, 27
Reinsel WG, 215
REML, *see* maximum likelihood
repeated measures, 12, 198, 202,
  383–408
  adjustment to split plot, *399–404*
  Box conservative test, *400*, 401
  compound symmetry, *see* sphericity
  multivariate, *405–408*
  sphericity, *388–390*, 393, 399–404,
    408
replicated Latin square, *see* cross-over
  designs *and* factorial designs

Ripley BD, 14, 18, 124
Rohlf FJ, 79
Roper T, 397
Rose-Hellekant TA, 163
Roy's greatest root, *see* multivariate
Rubin DB, 236
rule of 2, *see* model selection
Ruppert D, 229, 231, 234, 248, 252
Ryan BF, 14
Ryan TA Jr, 14

S, *see* S-Plus
S-Plus, 13–15, 18, 114, 123–124, 386
sample report outline, *31*
sample size, *82–85*
Samuel MD, 128
Samuels ML, 332, 344
SAS, 5, 13, 15, 18, 75, 76, 91, 98, 104, 109, 114, 115, 117, 120–122, 184–186, 194, 226, 253, 283, 285, 287, 294, 308, 321, 324, 332, 347, 363, 373, 375, 376, 386, 401
  JMP, 14–15
SAS Institute, 13–15, 89, 115
Satter LD, 67
Satterthwaite approximation, 219, 221, *224–225*, 304, 306, *307*, 308, 309, 317, 318, 321–322, 329, 330, 332, 366, 394, 400
saturated model, 137, 151
Scheffé H, xiii, 97, 152, 155, 209, 219, 222, 231, 328, *see also* multiple comparisons
Schilling EG, 27
Searle SR, xiii, 65, 79, 165, 195–196, 258, 304, 307, 308, 310, 313, 322, 327, 332
Seber GAF, 7, 75, 195–196, 252
Seiferheld W, xiv
Sharp P, 128
Sigma Xi, 27
significance level, *64*, *82*
significant digits
  class variance, *308*
  error variance, *307*
  mean, *61*
  standard deviation, *67*

Silverman BW, 243
Slocum MK, 119
Smith H, 7, 75, 145
Smith W, 10
Snedecor GW, 75, 231, 252
Snee RD, 28
Snell EJ, 7
Sokal RR, 79
Song K, 119
Sörbom D, 293
Spector PC, 13, 117, 121, 123, 283
sphericity, *see* repeated measures
split plot, *see* nested designs
Sprent P, 26
SPSS, 14
standard deviation (SD), 8, *55*, 61
standard error (SE), 8, *60–61*
Stanosz G, 384
statistical consulting, *21–31*
Statistical Science, 13, 27
StatLib, 14, 15, *15*
Stefanski LA, 248, 252
stem-and-leaf, *see* plot
stepwise procedures, *see* model selection
Stern HS, 236
Steudel K, 142
Stieve SM, 177
Stigler SM, 27
Stimart DP, 177
Storkson JM, 256
strength of evidence, *see* significance level
strip plot, *see* nested designs
Stroup WW, xiv, 308
Strunk W Jr, 30
Student, *see* Gosset W *or* t test *or* multiple comparisons
Student–Newman–Keuls (SNK), *see* multiple comparisons
studentized range, *see* multiple comparisons
sub-sampling, 198, 227, 235, *338–340*, 351, 357
summary, *30–31*
sums of squares
  distribution, *65–66*

INDEX

missing cells, *see* Type IV
Type I, 76, *168–170*, 184, 186
Type II, *170–172*, 184, 186–187
Type III, 167–168, *172–175*, 184, 186–187, 201, 321
Type IV, 175, *184–187*
survival curve, 11
Systat, 14
Syverud TD, 410

t test, *59–60*
Tessmann NJ, 359
test of hypothesis, *see* hypothesis
Thomas CB, 128
Tibshirani RJ, 236
Tierney L, 14
time series, *see* repeated measures
transformations, 209, *229–234*
   Box–Cox, *230–231*
   for interactions, *231*
   for normality, *233–234*
   proportions and counts, *232–233*
   variance stabilizing, 216, *231–233*
treatment structure, 38, *42–43*
Tufte ER, 9, 10
Tukey interaction, *see* balanced designs, one observation per cell
Tukey JW, 3, 7, 68, 72, 137, 149, 150, 152, 155, 210, 231
Tukey P, 14
Tukey's HSD, *see* multiple comparisons
Turabian KL, 30
Type II interaction plot, *see* plot, margin

unbalanced designs, *161–192*
   additive models, *164–167*
   grand mean, *162–163*
   marginal means, *162–163*
   nesting, 354
   sums of squares, *see* sums of squares
unequal
   variance, *see* variance
unequal sample sizes, *see* unbalanced designs
Unwin A, 14

van Belle G, 36
Vanotti MB, 419
variance
   estimate, 56–58
   formal inference, *65–67*
   stabilizing transformations, *see* transformations
   unequal, *221–227*
Velleman AY, 14
Velleman PF, 14
Venables WN, 14, 18, 124

Wade N, 28
Wallman KK, 27
Wardrop R, xiv
Wasserman W, xiii
Watts DG, 7, 243
Weerahandi S, 221, 225–226, 228
Weiss LS, 175
Welsch RE, 210
White EB, 30
White SC, 142
Wilks AR, 13
Wilks's likelihood ratio criterion, *see* multivariate
Wills G, 14
Wolfinger RD, 308
Woodward WA, 26
World Wide Web, *see* Internet
Wright S, 245, 246
Wright SP, 103
Wulfsberg RM, 27

XPro, 226

Yandell BS, 119, 177
Yang Y, xiv

Zahn DA, 26
Zamir D, 50
Zelen M, 27